Stormwater

Stormwater

A RESOURCE FOR SCIENTISTS, ENGINEERS,
AND POLICY MAKERS

William G. Wilson

THE UNIVERSITY OF CHICAGO PRESS • CHICAGO AND LONDON

William G. Wilson is an associate professor in the Department of Biology at Duke University. He is the author, most recently, of *Constructed Climates: A Primer on Urban Environments,* also published by the University of Chicago Press.

The University of Chicago Press, Chicago 60637
The University of Chicago Press, Ltd., London
© 2016 by William G. Wilson
All rights reserved. Published 2016.
Printed in the United States of America

25 24 23 22 21 20 19 18 17 16   1 2 3 4 5

ISBN-13: 978-0-226-36495-7   (cloth)
ISBN-13: 978-0-226-36500-8   (paper)
ISBN-13: 978-0-226-36514-5   (e-book)
DOI: 10.7208/chicago/9780226365145.001.0001

Library of Congress Cataloging-in-Publication Data

Names: Wilson, Will, 1960– author.
Title: Stormwater : a resource for scientists, engineers, and policy makers / William G. Wilson.
Description: Chicago ; London : the University of Chicago Press, 2016. | Includes bibliographical references and indexes.
Identifiers: LCCN 2016015886 | ISBN 9780226364957 (cloth : alk. paper) | ISBN 9780226365008 (pbk. : alk. paper) | ISBN 9780226365145 (e-book)
Subjects: LCSH: Urban runoff. | Urban runoff—Environmental aspects. | Urban runoff—Management.
Classification: LCC TD657 .W55 2016 | DDC 628/.21—dc23 LC record available at https://lccn.loc.gov/2016015886

♾ This paper meets the requirements of ANSI/NISO Z39.48-1992 (Permanence of Paper).

To Frances

Contents

6 Nutrients 109

Preface

A majority of people live in cities. That fact makes an understanding of the context, mechanisms, and consequences of urban environments important for making sound decisions for human health and welfare, as well as the protection of the broader environment.

Stormwater is a large part of that urban environment, and this book covers its background science and applied concerns. In cities, rain falls on impervious surfaces, dislodges and suspends pollutants, and then flows downhill. This stormwater runoff is the equivalent of a liquid avalanche and enters stormwater systems and streams where organisms try to survive and reproduce. In my home of Durham, North Carolina, these streams lead to reservoirs that serve people with drinking water, meaning there's a direct benefit in promoting cleaner runoff, and one approach is the wise use of urban open space.

This book follows the structure of a seminar I've offered for several years at Duke University. A list of primary papers given at the end of each chapter is a sampling of the weekly readings covering the various topics. I summarize these and many other primary publications within the chapters. Although I present the book's material in a quantitative way, I've strived to make these basic concepts of stormwater science accessible to a broad range of interested students, citizens, policy makers, and environmentalists.

I divide the book into three parts: urban conditions, environmental harms, and solutions. The first part, urban conditions, involves the ultimate reason for our stormwater problems—the greatly expanded human population—as well as natural and human-altered aspects of precipitation, emissions and deposition of pollutants, and the imperviousness of urban land use. Regarding emissions, I've focused on nonpoint sources while mostly ignoring point sources. A leaking underground fuel tank, a chemical spill at a factory, and a coal ash dump site are all examples of point sources. Yes, all of these problems can reduce water quality, but the sources of the problems aren't very complicated, and they're not really stormwater-related issues. Nonpoint sources include cars and coal-fired power plants. The use of cars is distributed across the landscape in proportion to human density, and the point source atmospheric emissions of power plants become nonpoint depositions wherever the wind blows. Particulates from both sources land on impervious surfaces to be washed away in the next storm.

These urban conditions lead to the environmental harms presented in the second part of the book. The most direct harm comes from the hydrological changes and environmental damage caused by flashy urban runoff. Since rain can't infiltrate parking lots and roofs, large volumes of water run off these surfaces in one big rush, flushing streams of organisms and material. The runoff also carries a variety of pollutants, namely, nutrients, metals, and what I've called "emerging" pollutants, a collection that includes both new chemicals humans have introduced to the environment and well-known pollutants that have a lower profile, such as PAHs, that may be less familiar to the average reader. Thermal effects are another problem that can either directly or indirectly harm ecosystems. The net effects of these many harms are summarized in chapters covering stream and groundwater responses and the changes to organisms and ecosystems.

The third part of the book covers solutions to stormwater problems separated into two chapters. The first chapter examines the benefits of riparian ecosystems, covering some of the harms stormwater brings to them, but primarily outlines the many ecosystem services provided by enhancing riparian buffers. The second chapter discusses stormwater control measures with a focus on approaches that fall under the umbrella of low-impact development or water-sensitive urban design.

I've come to the topic of stormwater by a complicated path. As a child growing up on a farm in central Minnesota, I wasn't exposed to a hotbed of pro-environmental thinking, but that rural environment was the next best thing to a natural one, certainly compared with urban environments. Though people drastically altered the landscape for farming, large areas of forest and wetlands remained relatively unbothered. Rural Minnesota folks also care deeply about preserving natural resources for fishing, hunting, and farming. Cities, on the other hand, are the antithesis of a natural environment.

My academic training was in theoretical physics, which unknowingly prepared me for a transition to ecological and evolutionary theory. Earning tenure within that discipline then permitted me more time to engage within the Durham community, and my service on Durham's City and County Open Space and Trails Commission was my first introduction to the human-related issues behind open space. Open space means exactly what it sounds like: forests, pastures, and fields in rural areas, and anything ranging from small gardens to large parks, or even single trees, in urban environments. Despite large-scale development, many parts of Durham County remain quite rural, with plenty of privately owned open spaces, and dedicated county and city staff seek long-term preservation of key resources. Local conservation groups also protect natural areas quite forcefully.

An early 2000s collapse in science funding and higher education's enhanced focus toward revenue generation and away from basic science closed some academic doors for me and theoretical evolutionary ecology, but the situation also provided an opportunity to examine urban open space issues in academic detail. That pursuit led to my urban environments book with the University of Chicago Press, *Constructed Climates* (2011). Writing that book introduced me to a few of the problems with urban stormwater. Given the recognized water quality needs throughout this region, along with state-level protections that are becoming steadily weaker, the topic of stormwater seemed worthy of detailed study, hence this book.

I'm visual, and I believe that graphs and plots provide the best summaries of scientific results. I use them liberally, without hesitation. Each of the plots I show comes from one or more peer-reviewed or agency-published publications and, on occasion, new and unpublished data. Each of the chapters, save the introductory one, lists several primary publications that serve as discussion papers for my seminar class. An appendix provides the barest outline of the progression of U.S. laws pertaining to stormwater. I've also produced three indexes, one providing the pages where various locations are mentioned, a standard topic index, and, lastly, an index of acronyms. The acronym index appears last for easy access, with a page number that provides the best explanation of the acronym. Additional comments, clarifications, corrections, downloads, and updates will be accessible from my academic web page, biology.duke.edu/wilson.

I thank many people for their help in different ways over the several-year span that I wrote this book. Many folks kindly provided papers and shared unpublished or extended data sets, and others reviewed various sections of early drafts. Miguel Alvarez-Cobelas and two anonymous reviewers read the entire book and gave many helpful suggestions and edits. I gratefully acknowledge the contributions of Paula Bailey, Roger Bannerman, Kathi Beratan, Revital Bookman, Derek Booth, Whitney Broussard, Stanley Buol, Allen Burton, Steve Corsi, John Cox, Charlie Driscoll Jr., Mike Dupree, Tim Fletcher, Bill Herb, Alex Huryn, Lars Hylander, Taylor Jarnagin, Krista Jones, Chris Konrad, Barbara Mahler, Pierre Marmonier, Gradie McCallie, Don Monteith, K aren Nelson, Jake Peters, Geoff Poole, Frank Princiotta, Alison Purcell O'Dowd, Ken Reckow, David Sample, Bridget Scanlon, Jon Schoonover, Shirlee Tan, Christian Torgersen, John Vucetich, Chris Walsh, and Bruce Webb. No doubt

I have left someone off of this list—my apologies. I also benefited from the fortitude of my stormwater science seminar students. Despite all of this help, I am certain that errors will be found, I accept complete ownership, and I will greatly appreciate feedback and clarifications.

William G. Wilson
Durham, NC
August 20, 2015

Part I
Urban Conditions

Chapter 1

More people, more pavement

Out in the Arizona desert in the midst of a storm, there's stormwater. So-called soils in that area feel and act like concrete surfaces, and it's a wonder that plants can even find a toehold. But when it comes to rain, the drops just splat on the surface, with the first few drops wetting the dry surface, perhaps with some evaporation, all the remaining rain runs downhill, collects in channels, and comes rushing through long-scoured stream beds. Seemingly very little water infiltrates into the "soil" despite the welcome of the wonderful cactus forests in the region.

In this desert backdrop, building a house or road with surfaces that don't allow rainwater infiltration isn't much of a change for stormwater: rain doesn't infiltrate those surfaces either, and the only change might be some vegetation loss.

Contrast this with nearly any other area of the country besides these deserts. Here in Durham, North Carolina, I've seen two inches of rain disappear into the ground during a dry spell in spring. Vegetation grabs a bit on leaf surfaces and bark, some evaporates away to cool surfaces, but much of it flows through the pores and channels in dry soils emptied of any water. Flowing quickly downward, this precipitation makes up for water lost through the transpiration of abundant vegetation that pulls water through plant tissues and out, as vapor, through open stomata on leaves, enabling the absorption of carbon dioxide and formation of more plant tissue.

Build houses or roads in this area, and the entire soil surface changes to one that's impervious to the infiltration of water. Development seals off soil channels to the downward flow of water, causing rains to wash surfaces clean of pollutants and to flow and collect downhill into large quantities, creating flashy flows like those in the desert. Anything that intercepts raindrops before they hit the ground counts as an impervious surface: parking lots, roofs, roads, sidewalks, even umbrellas count against our landscape alterations.

Though true in a sense, that definition would mean that trees and their leaves could count as impervious surfaces, but that water is needed for growth, one of nature's benefits welcomed like the water collected by people using roofs to fill cisterns. Rain falling on those structures counts as a resource, by humans or nature, something desired and valued, to be concentrated for its beneficial uses like drinking, washing, and watering crops and animals.

It's counterintuitive to think of water as a pollutant. We drink it, we bathe in it, we water our gardens and crops with it. But a flood of it in our homes and cities leads to problems with terrible consequences.

Water collected in the depressions of Durham's Carolina clay, for example, turns into a forbidding mess for any and all footed creatures or wheeled vehicles, leaving puddles where mosquitoes hatch in about three minutes. In excess, water becomes an undesirable pollutant, something to be hurried downhill and out of the way. The great flush of high water volume from impervious urban surfaces then scours and erodes streams, washing away the organisms and, after the water's gone by, drying out streamside habitats. As with our structures, too much water harms natural systems as well, or at least disturbs these systems in new ways.

This book broadly covers the science behind stormwater. Part I examines the background conditions of the water cycle, emissions and depositions of pollutants, and the basics of land use patterns that create impervious surfaces. Part II covers the harms brought on by stormwater, beginning with the runoff itself, covering the various associated harms, including contaminants and thermal pollution, and finishing with an overview of the organismal and ecosystem responses. Part III considers a small sampling of solutions for the problems, with an emphasis on the ecosystem services provided by streams and riparian systems, as well as coverage of constructed stormwater control measures. I provide a particular emphasis on landscape-distributed measures that fall under the guise of low-impact development.

Facets of stormwater.

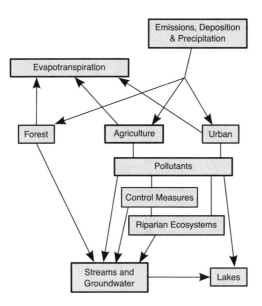

Figure 1.1: Summary of stormwater-related aspects of the water cycle. Part I examines rainfall and atmospheric deposition onto various land use types, setting the background conditions for the various environmental harms, indicated by "pollutants," to surface waters discussed in Part II. Part III covers solutions to these harms through the use of riparian ecosystems and stormwater control measures.

For hundreds of millions of years, rain fell and nobody cared, mainly because there were no people. Then, for a couple million more years, rain fell, but people still didn't care, except when the rain didn't fall. Today, our dramatically larger human population causes all of the environmental problems that we talk about, and the recent population increase has indeed been invasive. When my oldest grandparent was born in 1888, there were just one to two billion people in the world. Now there are more than 7.2 billion people.[1] Our global human population increased sixfold in 130 years, just a few generations, and proportional with that increase grew energy use, chemical emissions, land use changes, impervious surfaces, water use, and crop production.

---

[1] The U.S. Census Bureau summarizes historical population numbers from several sources at www.census.gov/popclock/.

All these people, mostly concentrated into cities, put tremendous stresses on our natural ecosystems. In particular, our large population affects stormwater because modern human land use changes the flow of water, and the really damaging pollutants come about from people and their activities—the more people, the more pollutants. When precipitation falls on the impervious surfaces that characterize urbanization, the water makes its way through constructed stormwater systems to urban streams.[1] The impervious surfaces of cities create stormwater, which produces high peak flows and low base flows and, along with the pollutants carried with those flows, leads to bioaccumulation, toxicity to organisms, and poor habitat quality. These effects of urbanization also kill sensitive organisms living in urban streams. Sensitive species represent a set of organisms clearly separate from tolerant species when examined along an axis representing urbanization.[2]

Gaining a better understanding of stormwater science benefits from a basic discussion of the water cycle, parts of which are characterized in Figure 1.1, with more details covered in Chapter 2. It's a cycle, and since I often wish it would rain, let's start there. Rain falls on the ground and then has several fates. First, either it can go back up into the air by evaporation—turning into a vapor and becoming part of the atmosphere once again—or it can be drawn into a plant's roots to become part of its sap, destined to evaporate out of one of the plant's many stomata, a process called transpiration. Evaporation and transpiration, taken together, are called evapotranspiration. A little bit of the water in plants gets split apart and transformed into plant tissue, becoming a part of the biosphere. Rain can also soak into the ground, immediately becoming groundwater, or it could flow along the surface, collecting as it moves downhill, thereby being labeled stormwater, eventually merging with a nearby stream or water body. As we'll discuss later, streams and groundwater interact strongly. Water can also evaporate from streams, lakes, and oceans.

Keep in mind the many different types of "used" water: stormwater, wastewater (or sewage), groundwater, drinking water, and reservoirs. Rain falling on permeable surfaces like forests and prairies soaks in and can recharge groundwater sources, though heavy rainfalls can cause water to run off these surfaces. In contrast, rain that falls on impervious surfaces automatically becomes stormwater because it has no allowance for infiltration. Water running off of surfaces—called overland flows—becomes stormwater, often draining to streams that connect with rivers to refill lakes, reservoirs, or oceans. Lakes and reservoirs recharge from water flowing through streams, including stormwater flushed into urban streams. Wastewater originates from our houses, businesses, and hospitals, flushed down toilets, showers, and sinks, partly cleaned while passing through wastewater treatment plants before entering streams. Yet another term is gray water, which means water coming from sinks, washing

---

[1] NRC (2009). [2] Purcell et al. (2009).

Water and warmth make plants grow.

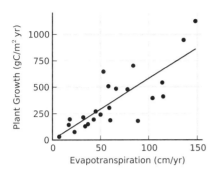

Figure 1.2: Plant growth, more formally called net primary productivity (NPP), increases across 23 ecological areas across the globe with increasing levels of evapotranspiration, which combines water availability and warmth (after Cleveland et al. 1999).

machines, and, perhaps, also showers and bathtubs becomes a resource used for underground irrigation of shrubs and trees, an option allowed in some jurisdictions. Groundwater, which we're most familiar with from wells drilled into aquifers, includes the water within aquifers and the moisture in the unsaturated vadose zone. Groundwater recharges by rainwater filtering through the ground throughout watersheds. Drinking water can come from both reservoirs and groundwater sources.

At various places in the water cycle, water can be stored. Lakes, ponds, and aquifers are obvious places, but so are puddles, leaf surfaces, and cisterns. Each storage pool affects how water flows, as well as ground and air temperatures because of the cooling brought about by evapotranspiration.

Our topic here is stormwater, but our large human population needing food leads to agricultural practices deeply dependent on water, fertilizers, and technology. Regarding the first of these factors, Figure 1.2 uses data from 23 vastly different ecological areas ranging from deserts to rainforests[1] to show how plant growth increases in response to increases in evapotranspiration. Water is important to plant growth, and that's why we irrigate crops.

Critically important, too, are the technological advances that include the electricity that runs wells and pumps, the changes that mechanized agriculture, and the biological sciences that revolutionized the domestication and alteration of crops and animals we eat.

In the early 1900s electrical lines grew in cities: more people per mile of transmission line meant higher profits for private providers. To solve the problem, the 1936 Norris–Rayburn Rural Electrification Act formed electrical cooperatives that subsidized building rural electrical lines and provided 100% loans for electrifying and plumbing our nation's farms. An important

---

[1] Cleveland et al. (1999).

aspect of rural electrification was the conversion of wells driven by wind-mills to electric pumps, greatly increasing the amount of water available for crop production. Increasing the water available to crops directly invokes the biology contained in Figure 1.2.

Technological developments also increased yields. Speaking directly to past technology, a historical connection comes from the very definition of an acre, centered on the amount of land workable in a day by a team of horses and a farmer. First divide a square mile (1,760 yards by 1,760 yards) into 16 squares, each 440 yards on a side. Each of these squares constitutes 40 acres, roughly the amount of land that could be worked in a spring season. To get to a single acre, a 40-acre parcel is divided into two rows of 20, one-acre strips, each measuring 220 yards (one furlong) by 22 yards (one chain). The area of an acre is then 4,840 square yards, or 43,560 square feet. Practically speaking, a furlong was how far a team of horses could pull a plow in one pass without resting. Today a typical farm has thousands of acres, worked by satellite-guided tractors costing hundreds of thousands of dollars.

As for fertilizers, amino acids, the building block of all proteins, have an NH_2 group attached to a carbon atom attached to everything else, meaning that without nitrogen, growing plants have a problem. We owe increased yields, in large part, to reactive nitrogen, and we owe reactive nitrogen to fossil fuels. The atmosphere consists mostly of very stable N_2 that makes up some 78% of our atmosphere. That form of molecular nitrogen has a strong triple bond joining the two atoms together, and that triple bond makes the N_2 molecule nearly irrelevant to biological processes.

Lightning makes important contributions of nitrogen into the biosphere as it splits the powerful triple bond of N_2 with even more powerful electrical currents, producing the reactive nitrogen compounds, NO and NO_2 (nitric oxide and nitrogen dioxide, respectively). Similarly, as air and vaporized gasoline are burned in the high-temperature and high-pressure environments of an engine's cylinders, that process also generates reactive nitrogen compounds. So do all processes involving the burning of fossil fuels, like coal-fired power plants, since nitrogen comes along with the oxygen needed for burning. Burning fossil fuels represents about half of atmospheric deposition of nitrogen onto the ground, all of which could be removed through emissions controls.[1] These very reactive forms of nitrogen, together called NO_x, combine with volatile organic compounds to cause problems in our urban air and our health.[2]

Nitrogen also enters the biosphere naturally when fixation takes place in some free-living bacteria, as well as in the symbiotic root nodules of some plants, importantly the legumes like beans and clover, and rice production creates excellent anaerobic conditions that help fix nitrogen. Nitrogen fixation

---

[1] Galloway et al. (2003). [2] Olszyna et al. (1997) provides a detailed discussion of nitrogen emissions and ozone production in the southeastern U.S. Also see Wilson (2011).

Nitrogen fertilizer produces high yields.

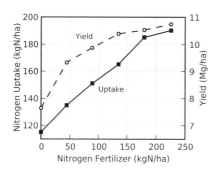

Figure 1.3: Nitrogen fertilizers make corn grow, with yield saturating at high applications (after Cassman et al. 2003). Even without any applied fertilizer, corn still takes up nearly 120 kgN/ha, providing a yield of nearly 8 Mg/ha, about 0.8 kg/m^2. Applying 200 kgN/ha increases yields by about 3,000 kg/ha.

also takes place in root nodules of riparian vegetation, for example, trees of the *Alnus* genus (birch trees), where the nitrogen-fixing bacteria have a mutualistic interaction.

Farmers have known for millennia that increasing nitrogen availability, or at least some proxy for nitrogen like beans, increases the productivity of their crops.[1] However, just as nitrogen helps your garden, it fosters the growth of algae in lakes and streams, discussed later in Chapter 6.

About a century ago came the discovery of an artificial means of converting nitrogen in the air into a form that plants could use, called the Haber–Bosch process, but the cost comes in terms of the energy of a high-pressure, high-temperature, fossil-fuel-burning furnace. Half of human food now comes from synthesized nitrogen, providing the roughly 2 kgN/year each person needs.[2] Figure 1.3 shows that corn takes up lots of nitrogen and gives a saturating yield as nitrogen becomes less of a limiting factor with increasing applied fertilizers.[3] Water, weeds, and pests make up other concerns for suboptimal yields in cereal grains.

We're rather sloppy with all of this nitrogen. A careful accounting shows that only about 12% of synthesized reactive nitrogen for agricultural production makes it all the way to people's mouths, while some of the lost nitrogen enters the environment.[4]

Figure 1.4 shows how total nitrogen use since 1850 by humans increased along with the sixfold increase in human population: since just 1950 total nitrogen increased sixfold. Some nitrogen comes from fossil fuel use—think

---

[1] Citations to ancient Greece and Roman agricultural practices are found in White (1970). [2] Galloway and Cowling (2002). [3] Cassman et al. (2003) discuss protecting ecosystem services while meeting human needs for cereal crops. [4] Galloway and Cowling (2002) describe the flow of N through the food production chain.

Synthetic nitrogen swamps the biosphere.

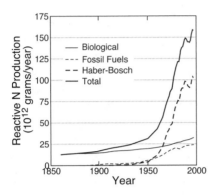

Figure 1.4: Recent reactive nitrogen contributions from burning fossil fuels and synthetic production via the Haber–Bosch process have greatly exceeded fixation through biological means over the last century (after Galloway et al. 2003), enabling the increased crop yields of Figure 1.3.

emissions—and some from biological routes like planting legumes, but most comes from commercial fertilizers produced via the energy-intensive Haber–Bosch process.[1] A plowed-under crop of alfalfa can add up to 300 KgN/ha, compared with 200 kgN/ha for high application levels of synthetically produced nitrogen. This biological source, of course, comes at the cost of one year's crop, although the income loss can be mitigated to an extent via crop rotation with soybeans, which helps soil quality while also producing a marketable crop.

Nitrogen can fall from the sky when nitrogen gas gets converted to a useful form via lightning or emissions, but the amount is minimal compared with fertilizer applications, at around 7 kg of available nitrogen per hectare per year (0.7 gm/m^2), or about 6 pounds per acre. In urban areas these deposition levels are higher, and the pollutants include heavy metals and flame retardants. All of the contaminants that settle onto impervious surfaces are transported downstream with stormwater runoff.

Figure 1.5 shows a small section of Durham's stormwater pipe system[2] that flows toward the top of the image and drains into the Ellerbe Creek,[3] looking very similar to other urbanized regions with poor water quality (see, for example, Figures 11.12 and 11.13). Small circles identify stormwater drains, thin white lines connecting them label the pipes, and I've drawn thick white lines separating basins of different stormwater pipelines. Think of these thick white

---

[1] Galloway et al. (2003) document the growth of nitrogen use over the last century. [2] This stormwater pipe system map was kindly provided by Durham's GIS department. Walsh et al. (2005a) display a section of Melbourne, Australia's stormwater system, and similar GIS maps are available online for other cities. [3] Protected by the venerable Ellerbe Creek Watershed Association, www.ellerbecreek.org.

Constructed creeks drain urban watersheds.

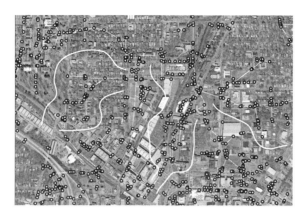

Figure 1.5: A small section of the stormwater pipeline in downtown Durham, NC, imaged in 2007 (courtesy of the Durham GIS department). Pipes (thin white lines) connect drains (small circles) and terminate as stormwater outfalls draining into an urban stream. The thick white lines separate sub-basins, much like ridges between valleys.

---

lines as ridgetops separating small creeks. Other dark lines denote streams or stream remnants. We see lots of impervious surface collecting and piping pollutant-laden stormwater into small creeks, ending up in the nearby Falls Lake reservoir.

The Neuse River basin includes about 40% of the city of Durham and includes the upper right portion in Figure 1.5, while the Cape Fear River basin takes a small portion in the bottom left. Much of Durham's Neuse River basin stormwater flows into the Falls Lake reservoir, the drinking water source for several cities in Wake County. Estimates report that up to 43% of rainwater falling in a city exits through its stormwater pipes,[1] which implies, perhaps, that eventually 43% of all the dust and petroleum products that fall onto the roads and parking lots exit with that rainfall right into the streams. And then our neighbors downstream in Raleigh, North Carolina, drink that water. The stormwater also seems to be a wasted resource: that runoff represents water diverted from the area's actual evapotranspiration, acting like irrigation in reverse. That lost water results in a warmer, drier city.[2]

In summary, we have a large human population that generates many contaminants that settle on impervious surfaces that wash off in heavy rains that

---

[1] Fractions of impervious surfaces are discussed by Pickett et al. (2001). [2] See Durham's urban heat island in Wilson (2011).

Problems with U.S. streams and lakes.

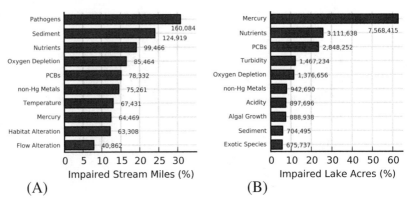

Figure 1.6: Top 10 problems that impair U.S. streams and lakes using data provided in 2014 that summarizes state data current from 2006–2012 (USEPA). (A) Of 3.53 million miles of U.S. streams, 519,000 miles are impaired; (B) of 41.6 million acres of lakes, reservoirs, and ponds, 12 million acres are impaired. Oxygen depletion includes organic enrichment.

flush down urban streams. This situation causes problems for our streams and lakes, which Figure 1.6 outlines for the U.S., according to 2014 estimates provided by the states to the U.S. Environmental Protection Agency (USEPA).[1]

The USEPA estimates that the U.S. has a total of 3.53 million miles of rivers and streams and 41.6 million acres of lake, reservoir, and pond surface area. Of these amounts, states assess 27.8% of the stream miles and 42.9% of lake surface for impairment status. Of these assessed waters, 52.8% of stream miles and 67.3% of the lake surface are deemed impaired. These are not random samples of lakes and streams because the states monitor these waters, primarily focusing on those bodies that they suspect to be impaired. Figure 1.6A,B shows the percentages of these impaired streams and lakes affected by the top 10 problems for each surface water type. All of these problems are quite self-explanatory, although oxygen depletion includes the precursory problem of organic enrichment.

A quick note on a few of these problem sources in Figures 1.6 and 1.7: Hydromodification refers to the changes people make in our waterways by building ponds and dams, creating channels, dredging bottoms, and diversions. Habitat alterations include removal of riparian vegetation, creating stream

---

[1] The USEPA's National Water Quality data are online at iaspub.epa.gov/waters10/attains_nation_cy.control.

Sources of problems with U.S. streams and lakes.

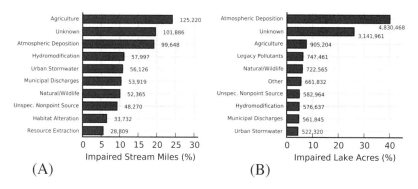

Figure 1.7: Sources of the problems shown in Figure 1.6 that impair U.S. (A) streams and (B) lakes (USEPA).

bed changes (including processes like the urban stream syndrome), wetlands drainage, and stream bank erosion. Clearly these two features go together when things like the drainage of wetlands alters stream hydrology. Examples of natural and wildlife problems include those due to flooding, drought, and waterfowl. Point sources include factories and sewage treatment plants, things that stay fixed on the ground, generating effluent or emissions from a single location, and, perhaps for regulatory reasons, above some emission level. In contrast, nonpoint sources are distributed broadly across the landscape where people live and drive. These sources move around on the ground and are not just cars and trucks but also trains, planes, and tractors—off-road vehicles that add up to a lot of emissions. Nonpoint sources also include products made in a factory, like household cleaners and paints, but transported to where people live and the emissions take place. Finally, once point source emissions enter the atmosphere, the deposition might become a nonpoint source distribution.

Of course, if the streams leading to lakes have problems, it's not a surprise that the lakes have problems, too, since they're the recipients of pollutants flowing down the rivers. Shared problems include nutrients, metals, and oxygen depletion, but streams and lakes suffer from unique problems, too. For example, pathogens cause problems in streams, coming from sewage and livestock, but the problem goes away in large water bodies. Multiple causes can impair the same stream mile or lake acre, so the percentages don't add up to 100.

At first glance of Figure 1.7, agriculture looks bad, but we need to consider the sources on a per-acre land use basis. Urban land use accounts for 60 million acres in the 48 U.S. states, but agricultural croplands hold more than 400 million acres, with another 600 million acres in pasture and rangeland.[1] These numbers come from a total of 1.9 billion acres, meaning that urban land use accounts for just 3.2% of land use compared with agriculture's half, yet urban stormwater accounts for a vastly disproportionate share of the stream mile impairments compared with agriculture. Adding in the many other specific sources attributable to urban living clarifies urban land use's harms to U.S. surface waters.

In the following chapters, much of the discussion reflects back to these problems and sources.

---

[1] Land use data are online at www.ers.usda.gov/data-products/major-land-uses.aspx.

Chapter 2

Precipitation and Evapotranspiration

Stormwater, of course, begins as rain, but rain begins as surface water that was either evaporated or transpired, which then became rain. These repeating steps in the water cycle can pick up pollutants: raindrops pick up dust and gases, stormwater rinses surfaces clean, contaminated groundwater seeps into streams. This chapter lightly covers the water cycle, with its pools and fluxes, from surface water to precipitation.

Outside the urban world, forests transpire and waters evaporate, and the combined evapotranspiration recycles roughly 60% of precipitation back into the atmosphere. In part, removing trees from urban environments eliminates half of this water removal pathway, resulting in greater runoff.

An important factor for stormwater consequences is the great variability in droughts and wet periods in freshwater availability, affecting everything from trees to people. Though limited in short-term extent, stormwater issues have some climate change-induced consequences, too, with climate change having already influenced the water cycle enough to increase precipitation levels by 6–8% over the last century.

The variability we see in rainfall represents just one of the challenging aspects that lead to difficulties in solving our stormwater problems, with structures tolerating 1-inch rains having great difficulties with the rare 5- or 10-inch storms. Goals for controlling, or at least dealing with, stormwater flows must reflect the wide range of potential water volumes.

Durham rainfall varies.

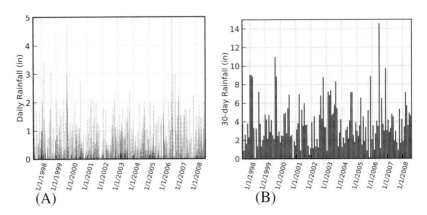

(A)   (B)

Figure 2.1: (A) Nearly 4,000 24-hour rainfall measurements from Durham County, and (B) 30-day periods summed throughout A. Daily rainfall rarely exceeds 4 inches but often exceeds 1 inch. Tremendous variability on all time scales leads to droughts and floods.

Precipitation, ultimately, is local. Durham's yearly rainfall reaches nearly 48 inches, almost 4 inches per month, 1 inch per week, or 1/7 inch per day. I found these 1997–2008 rainfall data from the National Climatic Data Center[1] for Durham's National Weather Service Cooperative Observer Station no. 312515. Data for this station go back to 1899, but only these more recent data were freely available. The data aren't perfect—some dates lacked entries, and "comfortable" values like 1/4 inch, 1/2 inch, and 1 inch have higher than expected frequencies of occurrence due to observer biases.

Durham's daily rainfall rarely exceeds 4 inches, as shown in Figure 2.1A, but often exceeds 1 inch (these likelihoods are quantified in Figure 2.8). The plot includes nearly 4,000 24-hour rainfall measurements during the 11-year span. These daily values are added up to give rainfall during 30-day periods throughout the data set in Figure 2.1B. Despite the 4-inch average per month, Durham has many months with much less rain, and many months with much, much more rain. It's during these very wet periods, after Durham's clay soils have absorbed all the water they can possibly hold, that additional rains flood rural and urban streams. And, as we'll read in later chapters, land use changes also affect the flows of water, from the urban heat island affecting storm for-

---

[1] See ncdc.noaa.gov.

Los Angeles rainfall varies.

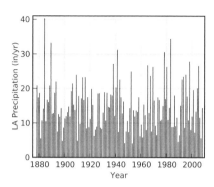

Figure 2.2: Annual precipitation falling in Los Angeles over the last century generalizes the daily and monthly rainfall variation seen in Figure 2.1 for Durham, NC (National Weather Service via *Los Angeles Times*, 2/2/2009).

mation to the impervious surfaces in the city that cause less water to infiltrate soils and more water to run off into stormwater systems.[1]

Variability characterizes rainfall. On the other side of the North American continent, Los Angeles, California, has a very different climate and rainfall pattern, (Figure 2.2) but shares the feature of tremendous variability.[2]

Let's take a step way back and first ask, where *is* water?[3] The oceans hold 1.35 billion km^3, ice throughout the world holds 33 million km^3, and groundwater (water-saturated soil) holds the next largest pool but an uncertain amount somewhere between 4.2 and 15.3 million km^3. The soils above groundwater contain another 122,000 km^3, and the air holds a piddling 13,000 km^3. Indeed, the atmosphere holds so very little that if all the water in the atmosphere fell to the ground at once, it would make a very small 3-cm (a little more than an inch) rainfall.

Water moves from these pools as either vapor, liquid, or solid. Perhaps the best visualization of the scale of this flux is that 1 m of water evaporates from the oceans each year, or nearly 3 mm per day. In response, each year there's movement of 111,000 km^3 as precipitation over land and 385,000 km^3 over the oceans. Of this annual precipitation, 71,000 km^3 entered the atmosphere due to evapotranspiration from land, and 425,000 km^3 due to evaporation from oceans. Oceans return the 40,000 km^3 of water flowing out of streams and rivers through precipitation brought back over the land. These two numbers sum to about 500,000 km^3 pouring into an atmosphere that holds only 13,000 km^3—it's like filling a teacup with a garden hose! This result means

---

[1] Hollis (1988) provides a great discussion of urban hydrology. [2] The data for Figure 2.2 came from an article in the *Los Angeles Times* but, as cited within, originated with the National Weather Service: articles.latimes.com/2009/feb/02/local/ me-annual-rain-graphic2. [3] Schlesinger (1997) provides numbers for pools and fluxes in the water cycle.

More rain with global warming.

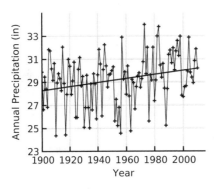

Figure 2.3: Precipitation trend across the U.S. over the last century (after Kunkel et al. 1999; data from NOAA). Global warming increases evaporation rates, leading to more precipitation. The increase amounts to about 2 inches in 30, or roughly a 7–8% over the last century.

a molecule of water spends only 10 days or so in the air after it enters the atmosphere until it falls back to land or ocean.

Humans affect the water cycle as they change the climate, speeding up the water cycling ever so slightly as the climate gets warmer (Figure 2.3). Essentially, global warming increases evaporation rates, and what goes up must come down. As the atmosphere warms up it can also hold more water vapor. The consequences of these two changes include more intense storms, more intense droughts, wet places getting wetter, and dry places getting drier.[1] As Figure 2.3 shows, the continental U.S. is getting more precipitation, an increase that amounts to about 2 inches in 30, or roughly a 7–8% increase over the last century.[2]

Data for Figure 2.3 came from a NOAA website and includes around 1,300 weather stations.[3] The fit to the data shows an increase in precipitation of about 1.3% per decade. The frequency of extreme precipitation events, those lasting 7 days with more-than-expected rainfall for a year's duration, increases even faster at around 3% per decade, with a particularly strong increase over the Southwest and central Great Plains and toward the northeastern U.S. Despite the national trend, much variation exists across the nation and over the decades. Authors mention that standards from the early 1960s govern designing structures dealing with runoff, and that reliance on these standards will eventually lead to underperforming stormwater control systems. Another study demonstrates that heavy precipitation has increased, with more precipitation on rainy days, increasing more than 2 mm/day since 1950.[4]

---

[1] Trenberth (2011). [2] Kunkel et al. (1999) examined 1931-1996 precipitation changes in the U.S. [3] NOAA, www.ncdc.noaa.gov/oa/climate/research/cag3/na.html. [4] Peterson et al. (2008) examined precipitation extremes across North America since 1950. Kunkel et al. (1999) examined extreme precipitation events associated with tropical cyclones.

From the 1950s on, the U.S. overall has seen a decrease in extremely high snowfall events (the highest 10% since the 1930s), while various areas had increasing trends in extreme low-snowfall events (the lowest 10%).[1] Average temperature between November and March appears to play the main role, a value that's increasing with climate change, but a connection to El Niño years is also important. Extreme cold temperatures are warming faster (about 3.5°C since 1960s) than extreme warm temperatures (about 1°C since 1960s).[2]

Important mechanisms driving these extreme events were the Arctic Oscillation (opposite pressure over the North Pole compared with the mid-latitudes) and the El Niño-Southern Oscillation (unusually high sea surface temperatures in the central Pacific Ocean).[3] On top of those changes, there's a decrease in "extreme temperature range" by about 0.7°C per decade, mainly due to increasing nighttime temperatures. For example, one study examining 40 stations in the northeastern U.S. during 1870–2005 found more frequent warm nights and summer days and fewer cold days and cold spells.[4] Another aspect of climate change includes an increase in growing season length, 3.8 days per decade during 1961–2000.

Concerning even more aspects of climate change, another study of the northeastern U.S. examined temperature and precipitation variables.[5] After removing weather stations with incomplete data, 38 stations provided temperature data, and 34 provided precipitation data. Researchers found highly significant trends during 1961–2000 among the following climatic measures:

Climatic Measure	Station Observations
Frost days	16% decreasing, 3% increasing
Extreme temperature range[6]	40% decreasing, none increasing
Growing season length	17% increasing, none decreasing
Days >1 cm precipitation	29% increasing, none decreasing
Consecutive dry days	12% decreasing, none increasing
Maximum 5-day precipitation	21% increasing, none decreasing

Of these variables, growing season length increased by 3.8 days per decade, and consecutive dry days decreased by 0.5 days per decade. Overall, we are seeing a wetter and warmer Northeast.

---

[1] Kunkel et al. (2009) studied extreme snowfall rates in the U.S. [2] Peterson et al. (2008) also examined temperature and snowfall extremes across North America since 1950. [3] Griffiths and Bradley (2007) studied indicators for extreme temperature and precipitation events for the northeastern U.S. between 1926 and 2000. [4] Brown et al. (2010) studied climate change in the northeastern U.S. [5] Griffiths and Bradley (2007) studied climate trends in the northeastern U.S. [6] The extreme temperature range measures the difference between the highest summer temperature and the coldest winter temperature.

A study of 50 years of data taken during Indiana summers also showed trends of increasing precipitation.[1] At these 31 stream-measuring stations, both low and medium streamflows increased as well, but many important variables played important roles. Despite the concurrent increases in precipitation and streamflow, in these drained agricultural fields streamflow seemed to be only weakly affected by precipitation trends, and high flows not at all. In part the decoupling of the two phenomena makes sense: drainage between precipitation events helps buffer precipitation by providing storage space. This mechanism should help increase base flows and decrease peak flows, but increasing high flow trends are seen in some regions. That means that, at least in part, drainage is responsible for streamflows, not just precipitation. It's not a simple and direct connection between precipitation and streamflow: watershed properties are important.

An extensive study examining precipitation and various streamflow statistics from 36 stations covered several Minnesota watersheds[2] In addition to precipitation in the state increasing around 10 cm to about 70 cm over the last century, there has been a general increase in peak flows and the frequency of low flows. All of the following streamflow measures increased by 1% or more per year over the last 90 years of the 20th century (1913–2002): mean annual streamflow, maximum daily streamflow during snowmelt (March–May), maximum daily streamflow during the rainfall season (June–November), the 7-day low flow during the summer period (May–October) and during the winter period (November–April), the high flow days, and the extreme flow days. Global warming causes much of this increase by making winters warmer, causing more abrupt snow melts. It's a messy picture, however, and although there are clearly increasing averages over the 90 years, all of the measures (including precipitation) show a periodicity of about 13–15 years, with the strongest changes taking place since the 1980s. Only one of the 36 stations had an urban watershed, and the study makes no conclusions regarding land use effects.

As these studies demonstrate, complicated effects may occur with climate change since the atmosphere is the source of precipitation, and precipitation is the source of groundwater.[3] Changes in precipitation patterns will alter groundwater availability and lead to potential changes in streams.

An important part of the watershed is land cover, so let's move on to trees, interception, and evapotranspiration. Regarding interception, think about what you do when you're walking on the sidewalk and a light rain starts falling. If there's a large tree nearby, ducking under its canopy can keep you dry because it acts a bit like an umbrella. Or in a light rain on a small, tree-lined

---

[1] Kumar et al. (2009) studied Indiana streamflows and precipitation trends. [2] Novotny and Stefan (2007) show increasing precipitation over Minnesota and increasing stream flow measures. [3] Green et al. (2011) link groundwater changes with climate change.

Rain falls freely on the trees.

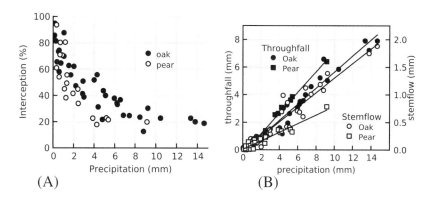

Figure 2.4: (A) Trees intercept a small amount of rain before the rain hits the ground, and (B) some of that intercepted rain travels along the tree stems and trunk on its way to the ground (Xiao et al. 2000).

road, there are dry spots under trees while the unshaded road spans get wet. These phenomena demonstrate the interception of rain by trees. The leaves get wet and retain a bit of precipitation.

Davis, California, was the location of the study in Figure 2.4, which used a 9-year-old, 22-cm-DBH (diameter at breast height), 4.8-m-crown-diameter Bradford flowering pear, *Pyrus calleryana*, and an eight-year-old, 12.5-cm-DBH, 3.2-m-crown-diameter cork oak, *Quercus suber,* during the 1996–1998 rainy seasons.[1] Study authors built a huge inverted roof-like structure around each tree to measure these components of tree-intercepted rainfall.

As measured, interception accounts for 15% and 27% of the total rainfall for the pear and oak trees, respectively (Figure 2.4A). All of the remaining rain ended up on the ground, and that includes 8% and 15% of the total rain flowing down the stems and trunks of the pear and oak trees, respectively (Figure 2.4B). The remaining rain falls through the tree canopies, though perhaps after bouncing off a leaf or two.

Effects of individual trees quickly add up to cause local influences on the water cycle. A wonderful demonstration comes from a study done in the 1940s in the North Carolina Appalachian range, comparing water fluxes before and after a logging operation.[2] In this study the scientists cleared a very small watershed, measuring the seasonal water flows before and after the vegetation removal. Yearly rainfall in the area roughly equaled 150 cm. When the forest

---

[1] Xiao et al. (2000).   [2] Hoover (1944).

Forests transpire, evaporate, store, and intercept water.

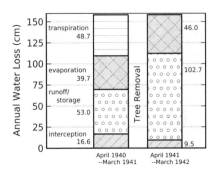

Figure 2.5: Partitioning of annual rainfall before and after the logging of a small watershed in western North Carolina (Hoover 1944). After logging, transpiration ceases while runoff and storage increase greatly.

was intact, evapotranspiration totaled nearly 90 cm, or 60% of the rainfall (Figure 2.5). As a comparison to Figure 2.4A, another 16.6 cm (11%) went to interception—the rain that hits and sticks to leaves and branches. Just 53 cm of rain actually hit the ground *and* was subsequently either infiltrated into the ground or ran off into a stream.

After removing the vegetation, transpiration was gone, evaporation increased a bit, and some interception by woody debris remained, while runoff and storage accounted for the greatest amount of rainfall fates. For example, in the year after removal, annual runoff alone increased by about 17 inches, or 43 cm. All in all, the results showed that the annual transpiration by the intact forest totaled 48.7 cm over an entire year—dividing by 365 days gives an average of about 1.3 mm per day.

This discussion leads us to the source of precipitation—evapotranspiration. Ecologists sweep all the biologically relevant ways of turning liquid water into vapor under the term "evapotranspiration," but only two really important ways catch my interest for the questions at hand: simple evaporation of liquid water, which accelerates when water's heated, and transpiration, which takes place when plants use and lose water while growing.

Two main ingredients limit evapotranspiration: heat and water. A location's potential evapotranspiration measures how much liquid *could* change into vapor if an unlimited water supply existed. Assuming infinite water means that heat ultimately limits potential evapotranspiration: how much water can a location's heat evaporate and transpire? In the Arctic, for example, there's plenty of (frozen) water, and if there were more heat there'd be more evapotranspiration. Actual evapotranspiration, in contrast, tells how much evapotranspiration *actually* takes place in a spot given both its heat and water.

Precipitation limits evapotranspiration.

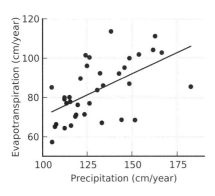

Figure 2.6: About 60% of the precipitation falling on watersheds in the southeastern U.S. reenters the atmosphere through evaporation and transpiration (after Lu et al. 2003). The remainder either infiltrates soils or runs off into streams.

The large difference between potential evapotranspiration and actual evapotranspiration of the desert, for example, motivates a farmer to irrigate the fields.

Evapotranspiration at the level of an individual tree is just transpiration, and there's a sap flow of roughly 30 kg per day from a 25-cm-DBH tree. A tree that size has a natural crown width of about 6 m, meaning a crown projection area of about 30 m^2.[1] Dividing that sap flow by the crown projection area gives roughly 1 kg/m^2/day, working out to about 1 mm of water moving from that area to the tree's roots, up its trunk, and out into the air through the pores in its leaves. This equivalence between 1 kg/day/m^2 and 1 mm/day is best understood by noting that 1 L of water $(10 \text{ cm})^3$ has a mass of 1 kg. Imagine a square box 10 cm on a side, cut into 100 slices, each 1 mm thick (10 cm = 100 mm). Those 100 slices cover 1 m^2.

Figure 2.6 arises from measures of the precipitation falling into watersheds scattered across the southeastern U.S. and the lesser amount of water flowing out of the streams draining the watersheds. Evapotranspiration makes up the lost water, and as the plot shows, we see that nearly two-thirds of the rain evapotranspires away.

Things change in cities because of the reduced tree cover, increased imperviousness, and diverse human activities. One study of a "prosperous" suburb (with 40% imperviousness) of Vancouver, British Columbia, performed in the early 1980s, indicated that urban irrigation was responsible for high levels of urban evapotranspiration.[2] During the winter months with plenty of rain, irrigation averaged 0.76 mm/day of water over the land area, but the highest

---

[1] Wilson (2011).   [2] Grimmond and Oke (1986) studied urban water balance in Vancouver, British Columbia.

The fates of rain.

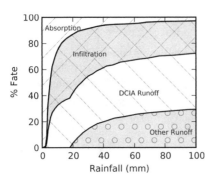

Figure 2.7: Depending on the amount of precipitation, different fractions have different fates in Miami, FL (Lee and Heaney 2003). Small amounts are absorbed, and some infiltrates soils; then with increasing rainfall greater amounts run off. DCIA, directly connected impervious area.

irrigation levels in the driest month of June reached 7 mm/day after a prolonged lack of rain. Overall, irrigation accounted for 52% of annual water use of daily water use, levels that were just under 400 L per resident.

The various fates of rain under conditions of different rainfall intensities in a highly urbanized landscape are summarized in Figure 2.7.[1] Initial dribbles of rain go toward the interception by trees discussed earlier, and wets surfaces, a portion called the "initial abstraction" by these researchers. Imagine that the surface was dry but the soil was so saturated that no more water could infiltrate. A little bit more rain could fall to fill up all the little puddles and holes, and that amount is the initial abstraction. As an example, no runoff is observed for urban Los Angeles rainfalls less than 1.5 mm.[2]

The next rainfall portion gets split between infiltration and runoff from directly connected impervious surfaces. We speak of two types of impervious surface measurements. The first type measures total impervious surface covering land in a drainage basin, regardless of where rainwater goes after falling, be it a lawn, ditch, or stormwater drain. The second type, effective or connected impervious surface, is more difficult to measure[3] but also more relevant to water quality because it harms streams the most of all impervious surface factors. This impervious surface fraction drains directly into stormwater systems and thus directly into streams. With these connected surfaces, stormwater doesn't touch terrestrial soils at all, implying very little chance if any to filter out any impurities. For example, a road's storm drain captures rain that falls and washes particulates and trash and oil from the road, flushing it right into the stream wherever the pipe's outfall sits. Certainly, some filtration takes place in the stream bed, as discussed later, but often the drainage takes place

---

[1] Lee and Heaney (2003).   [2] Sabin et al. (2005).   [3] Lee and Heaney (2003).

Durham rainfall often exceeds stormwater ponds.

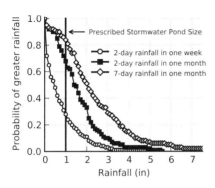

Figure 2.8: Wet periods affect soil conditions that influence successful stormwater treatment. These curves show the probability of exceeding a given amount of rainfall in any several-day period within either a week or a month compared with stormwater pond volume requirements (Wilson 2011).

under high rainfall and high streamflow, minimizing even this small opportunity. As we'll see later, even very low amounts of effective impervious surface seriously degrade streams—when it covers a little less than just 10% of the watershed's area.

Infiltration depends on soil moisture levels, and we all know that precipitation comes in wet periods and dry periods of varying length. Plots like that of Durham's daily rainfall (Figure 2.1) or yearly rainfall for Los Angeles (Figure 2.2) provide some limited information on this variation. Figure 2.8 examines the Durham rainfall data in more detail. A bit complicated, the plot shows the likelihood of expecting a certain amount of rain over a few days in a longer period of time. Specifically, the curves give the likelihood that total rainfall over a 2- or 7-day stretch of time within any given 7-day or 1-month period exceeds a given amount.[1]

For example, the curve marked by diamonds shows that there's a 50-50 chance that any given month will have a 7-day rainfall that exceeds 2 inches. The same curve shows that any given month has a 25% chance that a 7-day rainfall will exceed 3.3 inches. Likewise, during any given week there's a 25% chance of a 2-day rainfall exceeding 1 inch.

Given the runoff that takes place, stormwater control measures (SCMs) try to buffer the flows into streams. Detention ponds serve, in part, to prevent runoff from immediately entering streams by capturing it and then slowly releasing the stormwater over a longer time period. During that time chemical, biological, and physical processes degrade pollutants (see Figure 13.13). Of course, emptying the pond is critical to making space available to capture water from the next rainfall, as well as providing base flow for creeks below the pond.

---

[1] Wilson (2011).

Different rainfalls, different flows, different problems.

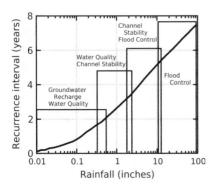

Figure 2.9: Dealing with light rainfalls has a different objective than dealing with heavy rainfalls (after Claytor and Schueler 1996). Approaches to promote water quality and groundwater recharge give way to flood control at high rainfall amounts.

If it sat there filled with water from a previous storm, and didn't empty, then the water from the next storm would simply displace the same volume of stale pond water, with no amelioration of high peak flows downstream of the pond.

Development rules and ordinances must reflect the objectives of Figure 2.9, and balance, for example, detention pond size, runoff volumes, water quality needs, and costs and benefits. Ponds can't be too small to provide any benefit (a teacup-sized pond to hold 100 acres' runoff) or too big to inhibit development (a 10-acre pond for an 11-acre watershed). They need to be sized just right and account for drainage area, land uses, rainfall patterns, runoff volumes and flow rates, soil types, slopes, geology/topography, land availability, future land use, depth to groundwater, supplemental water sources, freezing patterns, safety and community acceptance, and maintenance considerations—it's a long and likely incomplete list of factors.

Are Durham's stormwater detention ponds (see, e.g., Figure 13.16), sized properly for Durham's rainfall? Durham's development rule sizes ponds to hold a 1-inch rainfall, which means 3,600 ft^3 of pond volume per acre. Historical precipitation levels exceed that volume roughly 25% of the time by 2-day rainfalls in any week, and about 84% of the time by 7-day rainfalls during any given month. In contrast, a 2-inch rainfall requirement would mean 7,250 ft^3/acre, but within any week the pond would be exceeded only 10% of the time during a 2-day period, and 50% of the time during a 7-day period in a month. That's a balance.

Given these widely varying rainfalls, the goals of SCMs, according to an excellent USEPA publication, includes controlling the drastic change in flows from urban areas, removing pollution, and reducing pollution from the various sources.[1] On the other hand, the conceptual plot in Figure 2.9 identifies the realistic goals for SCMs given a certain level of precipitation and local rainfall frequency.[2] Of course, these goals overlap, and the details depend on local soils and land use, so the various numbers might change, yet the qualitative guidance remains. If your city gets a 10-inch rainfall, then flooding is the biggest concern bar none, whereas with a 0.5-inch rainfall, pollutants in the first flush could reasonably be dealt with—but if there's an ocean of runoff to come after that, then much hope is lost.

The four objectives of Figure 2.9 meet clear desires and different expectations. Groundwater recharge seeks to infiltrate precipitation. Water that runs off the watershed's surfaces should be clean, a reasonable objective when flows are low. When flows increase from either high precipitation or high imperviousness, stable stream channels minimize the erosion and sedimentation problems that lead to stream and riparian ecosystem degradation. Sediments smother mussels and macroinvertebrates, along with problems discussed elsewhere, and we seek to prevent that harm.

Likewise, high flows that rise above stream banks push water into places water shouldn't be, causing damage to built structures and bringing even more pollutants into the waters from sources that shouldn't have gotten wet. Regulations preventing development of floodplains solve a different but related problem than SCMs that seek to infiltrate the initial small fraction of a rainfall, yet a comprehensive stormwater plan incorporates both objectives.

Chapter Readings

Burian, S.J., and J.M. Shepherd. 2005. Effect of urbanization on the diurnal rainfall pattern in Houston. *Hydrol. Proc.* 19: 1089–1103.

Hoover, M.D. 1944. Effect of removal of forest vegetation upon water-yields. *Trans. Am. Geophys. Union* 25: 969–977.

Kunkel, K.E., D. Easterling, D.A.R. Kristovich, B. Gleason, L. Stoecker, and R. Smith. 2012. Meteorological causes of the secular variations in observed extreme precipitation events for the conterminous United States. J. Hydrometeorol. 13: 1131–1141.

Trenberth, K.E. 2011. Changes in precipitation with climate change. *Clim. Res.* 47: 123–138.

Xiao, Q., E.G. McPherson, S.L. Ustin, M.E. Grismer, and J.R. Simpson. 2000. Winter rainfall interception by two mature open-grown trees in Davis, California. *Hydrol. Proc.* 14: 763–784.

---

[1] USEPA (1999). [2] Claytor and Schueler (1996) is a huge publication with extensive discussions of many SCMs, the basic concepts behind them, and design specifications, as well as the basic issues of stormwater.

Chapter 3

Emissions, Deposition, and Accumulation

Humans certainly dominate emissions in urban areas, but the persistent presence of many human-created chemicals arises from a cycle of deposition and resuspension. These disturbances range from trucks stirring up roadside dust to forest fires releasing heavy metals. As a result, even the most remote areas of Earth receive depositions of human-created chemicals that continues for decades after terminating their production and use. On the positive side, however, the U.S. undertook serious efforts to improve air quality, with good reductions measured and reported by the USEPA.

Discussion in this chapter includes mercury, pesticides, PCBs, and others, which have multiple sources that contribute chemicals to the air. Atmospheric chemistry drives the formation of other compounds, while physical processes combine these chemicals with dust and water, which then drop out of the atmosphere onto the surfaces below.

The deposition process involves a concept called "deposition velocity," which measures how quickly or efficiently a surface strips particulates out of the air. Generally, obstructions reduce air speeds and allow particulates to settle, but some pollutants get pulled into the leaves of trees. As a result, urban vegetation has measurable levels of air pollutants.

Mercury mining peaked in the 1970s.

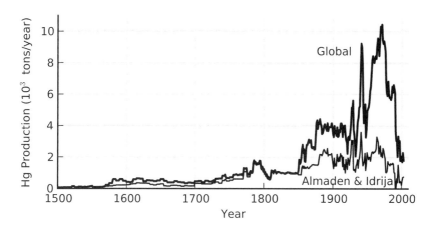

Figure 3.1: Mercury production through mining operations in Spain and worldwide (Hylander and Meili 2003). These production numbers exclude emissions from coal-fired power plants (about 1,000 tons/year). The drop-off occurred because new processes in a variety of areas sidestep mercury.

Many of these particulates and pollutants collect on impervious surfaces like roads, which transport not just cars but pollutants with rainwater runoff into streams. Indeed, automobiles and the shedding of material from their sundry parts and fluids represent an incredibly large source of heavy metals and hydrocarbons that make their way into our streams.

Mercury (Hg) pollutes the environment, and after its deposition onto impervious surfaces, stormwater provides one route. (Chapter 7 covers more issues surrounding mercury and other metals.) The demand for electricity means that near many of our urban areas, coal-fired power plants emit large amounts mercury present in coal, though much of the historical mercury sources come from silver and gold mining.

Figure 3.1 compiles accounts from a large number of global sources, summarizing 500 years of Hg production.[1] The plot includes only new sources and by-products and does not double count recycled mercury, such as previously deposited mercury released from forest fires. Total production throughout human history sits around 1 million tons, and the cinnabar (mercury ore) mines in Almaden, Spain, account for one-third of that total. One-quarter of mercury originates from North and South America.

---

[1] Hylander and Meili (2003) discuss the history of Hg production, use, and transport.

Throughout the 20th century, nearly half of the produced mercury went toward the production of chlorine and sodium hydroxide (NaOH), while the other half was used for silver and gold mining. Fortunately, new mercury-free techniques for both uses make mercury obsolete, and this technological change caused the very large decrease in mercury production seen in Figure 3.1 from the mid-1900s on. Mercury continues to be used for pigments in many products; for example, cosmetics allow 65 ppm mercury, even though it causes cognitive deficits, heart attacks, and general premature death.

The real concern, at this point in time, becomes one of reducing emissions from coal and other sources. Regardless of the origins, we'll deal with lingering ecosystem effects of our past mercury use for some time to come, as we'll see in later figures.

I find it fascinating that for thousands of years people knew that mercury dissolved gold flecks. This knowledge drove them to separate mercury from ore by heating cinnabar and condensing the mercury vapor, and then to mix the liquid mercury with gold ore to make a mercury–gold amalgam. To get the gold, a person takes this mix and burns off the mercury. That person also inhales the mercury vapor: one's body retains about 80% of inhaled mercury.

Liquid mercury even evaporates at room temperature, and that vapor gets scrubbed from the air and deposited on Earth's surface. Certain bacteria convert this elemental mercury into methylmercury—think methane, CH_4, but with one hydrogen atom replaced with a mercury atom—which then becomes readily available to the food chain. This methylation is a highly variable process dependent on the chemistry in soils and water. Methylation of mercury occurs in anoxic sediments and involves sulfate-reducing bacteria,[1] and the addition of sulfates to wetlands can increase mercury methylation two- to threefold over control levels.[2] Methylmercury is taken up by phytoplankton, which moves the mercury up the food chain as successive predator species consume their prey (see Figure 7.5). In 2008, 43% of U.S. lake acres had fish with mercury levels right at a critical level for human ingestion.[3] Indeed, median mercury concentration in predatory fish in U.S. lakes is just under 300 ppb, a level considered unsafe for human consumption. A rough estimate gives 25 tonnes of Hg cycling through fish each year.[4]

Thermostats that contained mercury always fascinated me as a child. Those thermostats were such clever devices: a coil of metal rotated as its temperature changed, and that rotation changed the balance of a small puddle of mercury inside a glass tube. When the room temperature got too cold, the mercury rolled down the tube and established contact between two wires, energizing the switch on a furnace. As the room heated up, the coil reversed its rotation, and the mercury moved in the other direction, breaking electrical contact and shutting off the furnace. What a dangerous device: even a gram of spilled

---

[1] Lovett et al. (2009). [2] Jeremiason et al. (2006). [3] USEPA (2009). [4] Swain et al. (2007).

Sources of anthropogenic mercury emissions.

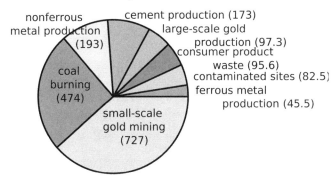

Figure 3.2: Human-related mercury emission sources (tonnes/year) for 2010 (UNEP 2013). Only sources exceeding 1% of global emissions are included, and total 1,888 tonnes annually, while another 72 tonnes comes from smaller sources. Large uncertainties surround some of these estimates.

---

mercury can elevate indoor concentrations to dangerous levels, making these household items quite risky. (Indeed, I recall an elementary school classmate bringing a small vial of mercury for everyone to play with during recess.) Technology has eliminated such events, at least in developed nations.

Addressing the problems of environmental mercury by eliminating its use requires understanding the sources of mercury deposition. Figure 3.2 shows the major sources of human-caused mercury emissions for 2010, estimates put together by the United Nations Environment Programme and published in an excellent discussion of mercury, *Global Mercury Assessment 2013: Sources, Emissions, Releases and Environmental Transport.*[1] For legibility purposes, I only included sources that exceeded 1% of global emissions, which excluded oil and gas refining and burning, mercury mining, and an obsolete process winding down in the chlor-alkali industry. Perhaps a bit surprising, human cremation contributes 3.6 tonnes each year, a result of the 340 tonnes of mercury used for dental fillings. About 1,000 tonnes of mercury is emitted as an unintentional by-product of some processes involving mercury, such as burning coal, but emissions from small-scale "artisanal" gold mining operations are considered intentional.

---

[1] UNEP (2013).

Here's a quick look at the global mercury picture, which has been rapidly changing over the last couple of decades. The atmosphere contains an estimated 5,000 tonnes of mercury, while the residence time in the atmosphere is just 1 year.[1] How much is 5,000 tonnes spread out over Earth's surface? Since Earth has a surface area of 510 million km^2, or 510 x 10^{12} m^2, and 5,000 tonnes of mercury equals 5 x 10^9 g, dividing the mass by the area gives about 0.01 mg/m^2, or 10 μg/m^2. This estimate compares well with lake sediment measurements, such as those in Figure 7.2.

The 1-year residence time represents a really fast cycling of mercury.[2] In the 1990s the greatest emissions, by far, came from the burning of coal in Asia, 5–10 times more than any other such large region, but those emissions are slowly coming under control with enhanced emissions regulations. These emissions roughly equal the recycling of mercury from past use—the mercury that's already present on land and water and contained within the organisms of the biosphere. Around 2,000 tonnes each comes from emissions of the biosphere and emissions from burning fossil fuels. These numbers are also a roughly threefold increase over "natural" mercury cycling.[3] Natural cycling depends on the number of volcanic eruptions: average annual geological emissions sit around a few hundred metric tonnes, which roughly equals the reemissions from burning biomass. Ocean-to-atmosphere "evasion" (emissions from "natural" processes on either land or sea) represents about one-third the current total emissions, about the same as the anthropogenic sources. Wildfires are thought to be the major source of emissions in the southern hemisphere.

On a global scale, 6,600 tonnes/year are emitted, with one-third direct anthropogenic (an estimate matching that in Figure 3.2) and two-thirds from natural or historical anthropogenic sources. Coal plants represent a global emissions source of about 1,450 tonnes/year, a number that dates to the early 2000s.[4] Reemissions from the biosphere—through forest fires and decomposition—returns mercury to the atmosphere in an amount roughly equal to current anthropogenic emissions.

Many emissions of mercury came from the U.S. gold rush days in the 1800s.[5] Gold and silver mining between 1800 and 1920 released about 60,000 tons, compared with all of the industrial sources that have released about 10,000 tons summed up to 1998.

Overall, the emissions and precipitation concentration were reduced by a third or more from the early 1990s to the early 2000s. Across the U.S., emissions declined across 8 years in the Midwest (28%) and Northeast (14%), but with no trend in the Southeast.[6] The origins of the mercury emissions include the biggest fraction coming from coal boilers, and lesser contributions from

---

[1] Swain et al. (2007). [2] Hylander and Meili (2003). [3] Lindberg et al. (2007). [4] Driscoll et al. (2007). [5] Pirrone et al. (1998) discuss mercury emissions. [6] Butler et al. (2008).

waste burning and fuel combustion. These latter two routes have been shut down, but in 2002 they represented about 100 tonnes/year. Emissions from burning coal did not change over that decade. Within the U.S., mercury emissions in 1990 totaled 222 tonnes/year, decreasing to just 103 tonnes/year by 2002. Much of the reduction came through the control of medical and municipal waste-burning routes. A study from Seattle demonstrated a nearly 60% decrease in wet deposition coincident with the closure of waste incineration sources, a decrease not reflected elsewhere: local emissions are important sources for deposition. On a longer-term scale, however, present deposition estimates indicate levels four to six times higher than in 1900.[1]

In terms of relating emissions to deposition, long-range transport comes to the rescue: deposition from U.S. sources onto U.S. soils was estimated at 52 tonnes/year in mid-1990s, much less than the 100–200 tonnes of emissions. Of course, the rescue from the emission sources comes with the harm of deposition somewhere else.

Coal accounts for about 60% of human mercury emissions, about 1,500 tonnes per year, and mining accounts for 20–30%. Emissions controls on coal-fired power plants, or "scrubbing" as it's sometimes called, captures 30–85% of the would-be total emissions through a variety of complicated processes— both dry and wet—that deal with a host of additional pollutants from those emission sources.[2] Coal has a mercury content up to 0.1 mg/kg, dependent on the type of coal. The ash that remains behind after burning has a similar concentration, but adding the residue from scrubbing the emissions brings the resulting fly ash mercury concentrations up to around 1 mg/kg. An analysis of the leaching potential of mercury and other pollutants from fly ash appears limited to the range of 0.1% to less than 5% over a century.[3]

Mercury's certainly bad, but it isn't our only problem, and we now move on to other pollutants. Figure 3.3 shows high PCB (polychlorinated biphenyl) concentrations in the air around industrialized areas that have high PCB use.[4] PCB emissions are measured here in a 1° quadrat. Given a sphere, visualize a cone emanating from its center. That cone subtends a certain amount of solid angle, which is the three-dimensional counterpart of the two-dimensional angle. Given an entire sphere, there's a total solid angle of $(2\pi)^2$ steradians (the unit of solid angle), which could also be stated as $(360°)^2$. The actual surface area of a 1° quadrat depends on the sphere's radius, say, an orange versus Earth, but for Earth that quadrat holds roughly 12.4 km^2. Hence, 100 tonnes/1° quadrat equals 100 million g spread over 12.4 million m^2, or just under 10 g/m^2 at the highest emission levels in Figure 3.3.

Simply put, high local emissions mean high local air concentrations. During 1930–1993 more than 1.3 million tons of PCBs was produced, mostly

[1] Evers et al. (2007) cite Perry et al. (2005). [2] USEPA (2005a). [3] USEPA (2005a, 2006). [4] Pozo et al. (2009); this figure comes from the supplementary information.

Emissions pollute the air.

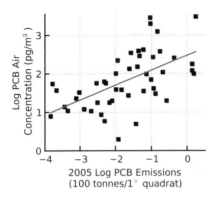

Figure 3.3: Atmospheric concentration of polychlorinated biphenyls (PCBs) versus local emissions (Pozo et al. 2009). A $1°$ quadrat corresponds to about 12.4 km^2, meaning 100 tonnes/$1°$ quadrat is just under 10 g/m^2.

involving electrical equipment.[1] PCB emissions peaked in 1970 along with production, but all uses are anticipated to be terminated by 2050. Depending on assumptions, emissions represent between 1.3% and 11.8% of total PCB production.[2]

Often PCBs are called "persistent organic pollutants," a class that includes pesticides like DDT (dichlorodiphenyltrichloroethane, or a couple of benzene rings with five chlorine atoms), herbicides, and flame retardants. And they are persistent. For example, decades after the ban on their use, the concentration of DDT in Los Angeles in stormwater suspended solids ranged up to 70 ng/g, and PCBs up to 119 ng/g.[3] Particulate matter at the outfall in Boston Harbor of Boston's combined sewer and stormwater system had a PCB concentration averaging around 1μg/g.[4] A Switzerland study revealed stormwater concentrations of up to 400 ng/L, which implied deposition rates of 2.9 ng/m^2/day dry, and 35.3 ng/L wet.[5] These numbers implied a total PCB flux of roughly 116 kg for 2008 in Switzerland, about 70% of that flowing with stormwater. PCBs that go through wastewater treatment plants mostly end up in the sludge, and the subsequent processing of the sludge determines where the PCBs go next.

Another pollutant, atrazine, is one of the most commonly found agricultural herbicides or pesticides, for which concentrations of 0.1–1.0 ppb (μg/L) caused *Daphnia* population declines in German lakes and various fish health problems.[6] At just 0.04 ppb, atrazine causes observable effects on salmon sperm production, and 0.12 ppb reduces marine algal photosynthesis.[7]

[1] Breivik et al. (2007). [2] Breivik et al. (2007). [3] Curren et al. (2011). [4] Eganhouse and Sherblom (2001). [5] Rossi et al. (2004). [6] Blann et al. (2009) provide a comprehensive table with toxicity levels for various contaminants. [7] Cox (2001).

Regulations decrease the emissions of many pollutants.

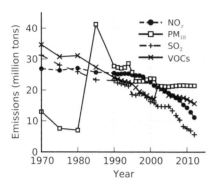

Figure 3.4: Annual pollution emissions from human sources show gradual decreases over the last several decades (data from the USEPA). Pollutants include volatile organic compounds (VOCs), reactive nitrogen (NO_x = NO + NO_2), particulate matter (with condensables) <10 μm in size (PM_{10}), and sulfur dioxide (SO_2).

Concentrations in drinking water sources across Iowa are in the range from below 0.6 to 2.2 ppb.

Studies like that shown in Figure 3.3 measuring pollutants at such very low concentrations add a "blank," which involves inserting and removing a collection disk but not letting it collect any air sample. These blanks, or controls, assure that the downstream analyses beyond the sample collection point are working correctly and not adding contaminants to the samples. In other words, what if the post office the samples shipped through on their way to the lab was inundated with fumes of some sort? Blanks define the "level of detection," or LOD—essentially, the lower bound of the detection limit. Values from real samples that fall below the LOD have to be taken as zero because blanks have just as much of the substance in question. Yet another type of control places known levels of the substance into the set of samples, to determine the "recovery" accuracy of the process for a particular substance.

Regulations are reducing emissions. The underlying USEPA emissions data for Figure 3.4 separate these plotted numbers into many source categories, including fuel, chemical production, metals processing, transportation, and solvent use, among others. In fact, the USEPA labels more than 85% of PM_{10} (particulate matter < 10 μm in size) emissions as "miscellaneous."[1] The USEPA compiled these emissions *estimates* from many information sources; the sharp changes arise from new methodologies, measurements, and regulations, not necessarily from actual increases or reductions.[2] The numbers I plot

---

[1] Emissions data from the USEPA are available at www.epa.gov/ttn/chief/trends.   [2] Likens et al. (2005) discuss some of the mysterious, sharp changes in the emissions plot.

Forests scrub the air clean.

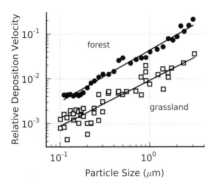

Figure 3.5: The efficiency of land surfaces to strip pollutants from the air, called the "deposition velocity," depends on particle size and differs for forests and grasslands (McDonald et al. 2007). Forests efficiently remove large particles.

for particulate matter include "condensables," which means vapor emissions, not really solid or liquid particles at the moment of emission. This vapor may later condense into very small particles. USEPA data also include numbers for particulate matter excluding condensables, showing that about 40% of particulate matter constitutes condensable emissions.

What do these emissions numbers mean on a personal level? Consider, for example, emissions of VOCs, presently 15 million tons, or 30 billion pounds, per year. That's a few thousand times more mass than the 2,000 tonnes (4.4 million pounds) of mercury, but those numbers compare apples and oranges. These VOC emissions run the gamut of solvent use, highway vehicles, other fuel combustion, and so on. Dividing this huge number by the U.S. population of 300 million gives 100 pounds per person per year, or VOC emissions of about one-third pound per American per day.

Notable successes lead to the decreasing trend, giving hope for further air quality improvements. Implementation of the 1970 Clean Air Act[1] reduced the various pollutants over this period, showing the power of effective legislative regulation. Most impressive, perhaps, legislation reduced VOC emissions from highway vehicles from 17 million tons in 1970 to 1.9 million tons in 2012. Fewer emissions means less deposition, and that means cleaner water.

We might think of pollution dropping onto impervious surfaces as a more or less passive process, but the results in Figure 3.5 demonstrate that the situation depends on the surface and the particle size of the pollutants. In this example contrasting forests and grasslands (also called "moorland" in this study from the U.K.), these two environments filter the pollutants out

---

[1] The USEPA website, www.epa.gov/air/caa, provides an overview of the 1990 Clean Air Act.

Rough biomes scrub the air clean.

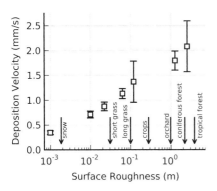

Figure 3.6: Deposition velocities increase with increasing surface roughness across Scotland, Germany, and The Netherlands (Gallagher et al. 2002). Arrows indicate deposition velocity estimates for various landscape covers.

directly.[1] "Filter" isn't really the right word, but we'll come to that in a bit. Crops were intermediate between forests and grasslands (see Figure 3.6).

What's a "deposition velocity"? It's a neat concept, and I think of an eraser analogy. Back in the Dark Ages when people used pencils that left lead or graphite streaks on paper, often in the shape of letters and numbers, occasional mistakes were made and minds were changed. Rubber erasers (or some sort of rubbery material) brushed across the paper, magically removed the marks. Good erasure depended on the quality of the eraser, how dried out it was, how hard one rubbed, and the quality of the paper and pencil. In this deposition velocity analogy, removal of the eraser material represents pollution removal from the air. Erasers can have the consistency of whipped cream or wood, and the paper might be rough sandpaper or greased glass. Whipped cream on sandpaper wears down very fast, but wood on greased glass has no wear at all.

Measurements of deposition velocity rely on a phenomenon called "eddy correlation." Imagine some measurement location up in the air above the tree canopy, grasslands, or other environment of interest. Generally, the wind blows in one horizontal direction, but we're interested in the vertical fluctuations: sometimes little eddies cause upward air flow, and sometimes they cause downward air flow. Over long periods of time over level ground, though, these vertical air flows have to average to zero because air isn't, for example, bunching up on the ground. Particles are also being carried by the wind, and at that height the particle density takes on some mean value, but just like the vertical component of the wind, there are density deviations that must average

---

[1] McDonald et al. (2007) examined deposition velocities in forests and grasslands.

to zero. Taken separately, these means and deviations don't provide us any information regarding the deposition of the particulates.

However, if we expect that particulates are sticking to the ground, then particulate density below our measurement height is less than the density above our location. The opposite would be true for a quantity like water vapor being released by the vegetation below our location and being diluted in the drier air above. What this expectation means is that the upward-moving air currents would be carrying lower densities of particulates, while downward moving air currents would be carrying higher particulate densities. And this idea means that the average of the product of vertical wind deviations and particulate density deviations yields an estimate of the overall downward flow of particulates at our measurement location. The estimate of the correlation between these two deviations directly leads to the deposition velocity.[1]

The deposition velocity measures how quickly the eraser wears down and the rate at which it leaves shavings (or particulates or mercury) behind. A low deposition velocity means few shavings, like wood on greased glass, and a high deposition velocity means lots of shavings, like whipped cream on sandpaper. A high deposition velocity means pollutants drop out of the air very efficiently.

Air is pretty much the same all over the world, though certainly temperature, humidity, and acidity vary dramatically. Figure 3.5 shows that different land use types in the same location have different deposition velocities for different sizes of particles, in this case by a factor of nearly 10. The values are given relative to another important factor for life as a particle, which is the wind condition when a particle would find itself lifting off of a surface, specifically, the wind shear velocity at which particles would start blowing around via saltation processes (described Chapter 5), like when sand dunes start forming because sand grains can lift off the soil surface. We feel this condition during those days at the beach when the wind is just strong enough that sand grains start pelting our ankles.

Another interesting study, shown in Figure 3.6, zeroed in on small particles just 0.1–0.2 μm in diameter but considered their deposition velocity in different environmental areas.[2] In this plot, surface roughness increases, as one can imagine redundantly, as a surface gets bumpier. The formal term is the "roughness length" that sets the scale at which the wind stops "feeling" the influence of the local land effects.[3] Put together, Figures 3.5 and 3.6 give a good sense of how deposition takes place.

A clever study used these different values to determine the value of trees with respect to their utility in reducing air pollution. The study considered

[1] Swinbank (1951) called it simply "vertical flux."   [2] Gallagher et al. (2002) studied the deposition velocity of small particles in different environments.   [3] Grimmond and Oke (1991) provide a bit of an insight into the physics behind surface roughness calculations.

two urban centers in the U.K., West Midlands and Glasgow, and then computationally added trees to those cities. In summary, adding trees to the city increases the deposition velocity and cleans the air.[1] However, more than doubling Glasgow's tree coverage to 8% decreases PM_{10} by only 2%, though planting trees to the maximum extent reduces PM_{10} by as much as 30% in some areas.

These results for urban areas mesh with data from studies involving forest edges[2] showing increased deposition rates ranging from 17% to 56% in low-elevation forests in the northeastern U.S. dominated by deciduous trees. Essentially, speeding winds over open plains come to a sudden stop at forest edges and deposit their wind-blown particles on the leaves and ground of the bordering trees. Urban trees often epitomize an edge, sitting alone or in small collections, which likely enhances their pollution-trapping capabilities. Adding together nitrate and ammonium, forest edge deposits were 43% greater than that within the forest, and 38% greater than deposition in the open plains. For whatever reason, the effects are strongest for nitrate and calcium (compared with sulfate and ammonium). Details of the configuration are important, though, and removing the vegetation near the ground at the edges removed the enhanced deposition at that spot but increased deposition deeper into the forest. This thinning essentially cut the brake lines that slowed down the wind, letting it slip in under the tree canopy. The implications for urban tree deposition processes don't seem directly relevant, though, since wind will likely just sidestep a solitary tree.

However, it seems relatively straightforward that wide stream buffers on the windward sides of urban streams could trap airborne nutrients and pollutants and subsequently sequester and denitrify the deposits.

Urban areas have high levels of the class of pollutants called polycyclic aromatic hydrocarbons (PAHs), and it shows up in trees.[3] Figure 3.7A summarizes measurements taken near the Sam Houston Parkway in Houston, Texas, using an aggregated measure of total PAHs. Statistical measures show levels in the inner city of $1{,}220 \pm 571$ ng/g dry weight of the pine needles sampled; in the outer city, 760 ± 282 ng/g; and in the rural and nonresidential areas, 393 ± 35 ng/g.[4] Interestingly, pine needles have 30% higher PAH concentrations in winter than in summer, presumably because PAHs evaporate away when it's hot.

These values are in line with those in other areas, including South Korea and Mexico, where concentrations ranged from 31 to 563 ng/g.[5] Sixteen PAHs measured in samples of pine needles collected from 29 rural, industrial, and

---

[1] McDonald et al. (2007).   [2] Weathers et al. (2001) measured increased deposition in forest edges. Citations within describe enhanced deposition for additional forest edge types.   [3] Hwang and Wade (2008) measured PAHs in Houston, Texas.   [4] Significant digits listed here as reported.
[5] Hwang et al. (2003) compared PAHs in pine needles from the U.S., South Korea, and Mexico.

Pollutants in leaves and needles.

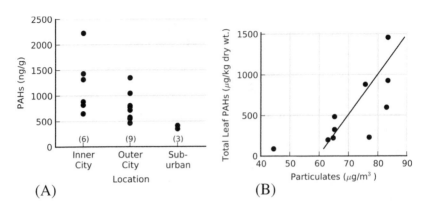

(A)   (B)

Figure 3.7: (A) Polycyclic aromatic hydrocarbon (PAH) concentrations in pine needles across three areas of Houston, TX (Hwang and Wade 2008). (B) PAHs measured in *Quercus ilex* leaves in Palermo, Italy, versus the concentration of particulates in the surrounding urban air (Orecchio 2007).

---

urban sites across Portugal ranged from 114 to 1,944 ng/g, compared with remote sites measuring around 77 ng/g.[1] Shocking news: urban air is dirty!

Pine needles constitute a "passive sampler" of local air quality, in part because pine trees collect particulates better than other tree species due to their great amount of surface area and dense structure.[2] Careful analysis can use the distribution of two- to seven-ring PAHs found in the air to fingerprint the different sources of PAHs.[3] Not all come from human emissions. Trees themselves have PAH emissions, especially in the southeastern U.S., and the emissions can be quite extensive.[4] Imagine a leaf-chomping larva and the sap oozing from the wounds of its bite: while those drops of sap dry out, PAHs evaporate into the air.[5]

Another study, shown in Figure 3.7B, was done in Palermo, Italy, during the summer of 2003, performed on evergreen trees called holm oak, or holly oak.[6] Leaves were collected from trees very near traffic, about 2–3 m off the ground, and then analyzed for PAH content, similar to measures of pine needles in Figure 3.7A. At the same time, air was sampled at those sites for particulate matter from the dust deposited on a plate in a device that samples a specific volume of air. The data point at the far left, with a particulates value around 45 μg/m^3, represents a control site far from human influences. As the

---

[1] Ratola et al. (2010). [2] Beckett et al. (2000). [3] Hwang et al. (2003). [4] See Wilson (2011).
[5] Copolovici et al. (2011). [6] Orecchio (2007) examined PAH uptake by roadside vegetation.

Pollution moves between air and soils.

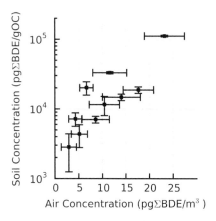

Figure 3.8: PBDE (polybrominated diphenyl ether) concentrations in the air and soil across a rural–urban–rural gradient through Birmingham, U.K. (Harrad and Hunter 2006). PBDEs are chemicals used as flame retardants in consumer products. Each point is a different site.

intercept at about 60 $\mu g/m^3$ indicates, particulates appear to be a fact of life, independent of the PAHs generated by urbanization, but PAHs and additional particulates increase together rather dramatically. The data were originally plotted with leaf PAH content as the independent variable, with the intent to use leaves as predictors of long-term air quality (estimated by measures of air particulates), despite the mechanistic reality that the air quality determines, in part, the leaf PAH content.

In summary, tree leaves serve as useful air quality monitors.

Figure 3.8 displays information on a class of flame retardants that goes by the initials BDEs and PBDEs (brominated and polybrominated diphenyl ethers).[1] These chemicals were used as a flame retardant in many products, such as carpet underlay and seat cushions, and as these indoor foams (including car interiors) break down, PBDEs evaporate from their many sources and then move outdoors.

What are the parts of flame retardants? A phenyl is just a benzene ring—six carbon atoms connected in a ring, with a hydrogen atom attached to each—except that something other than a hydrogen atom is attached to one carbon. Diphenyl is just two phenyls connected by a common attachment, like a rod connecting two rings. Diphenyl ether means the two phenyls are connected by an oxygen atom instead of something more arbitrary. Finally, brominated means that bromine atoms attach to various other carbon atoms of the phenyls (in place of the hydrogens), and having more than one bromine means it's

---

[1] Law et al. (2006) provides a comprehensive overview of flame retardants.

polybrominated. Each ring can have several bromine atoms—up to a total of 10—hanging off of the carbons at various places.

Dropping the "P" gives an acronym "BDE," usually also indicating the "poly" with a preface "penta," "octa," or "deca" (e.g., deca-BDE), to denote the number of bromine atoms in the molecule (respectively, 5, 8, or 10). A variant can be further specified by a number like 47 or 99 that refers to different ways these bromine atoms can be arranged in the molecule.

The specific "congeners" BDE-47, -99, -100, -153, and -154 are found most frequently in the environment and roughly match the emissions from foam production plants.[1] Low-bromine versions reside in polyurethane foams, and high-bromine versions reside in the hard plastic cases of consumer electronics. When you think about the polyurethane foams that make your couch and car seat comfortable, you should think about flame retardants, as these foams are made up of about one-third flame retardants. Of the total distribution of PBDEs, 30% reside in cars and 10% in furniture, and the rest is distributed widely in the things we really don't want to have burn up: textiles, building materials, packaging, and so forth. Production numbers cited include an estimate that in 2001 global PBDE use was 7,500 tonnes, with 7,100 t used in North America and just 150 t in Europe; deca-BDE-209 was 56,000 t (tons or tonnes unclear).[2]

So, what does this flame retardant do? A fire propagates when chemically bonded molecules split, releasing energy as heat and light, as well as "free radicals," or molecular shrapnel, that seek to balance out their electron shells by reacting with other molecules, thereby creating a chain reaction. As we'll see later with nitrification, in which microbes transform ammonium to nitrate (via nitrite) and use the chemical potential energy for growth, the process requires oxygen to sweep up electrons. Likewise, respiration of hydrocarbons by microbes, fungi, and animals for their energy also requires oxygen to take up electrons. With fire, for the chain reaction to continue, fire needs oxygen to take up the electrons while turning, say, a wood pile's hydrocarbons into carbon dioxide. Flame retardants interrupt the chain reaction by spewing counteracting bromine "free radicals" as they heat up in a fire, which combine with the flame's free radicals to stop the chain reaction.[3]

The curve in Figure 3.8 arises from 11 sites running through Birmingham, U.K. As with many pollutants, the dust the flame retardants are attached to flies up into the air, settles back down, flies back up, and so on, and pretty soon comes to an equilibrium, like the model described later in this chapter. Highest values in the upper right corner of Figure 3.8 come from the urban center and are about three to four times higher than in the rural areas, which are the points in the lower left corner. Different congener distributions are

---

[1] Hale et al. (2002) gives an overview of PBDEs. [2] Law et al. (2006). [3] USEPA (2005b) discusses five ways flame retardants interrupt fires.

Wet deposition of mercury across the U.S.

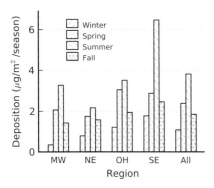

Figure 3.9: Wet deposition of mercury across the U.S. accounts for 50–90% of total mercury deposition (after Prestbo and Gay 2009). For comparison, globally averaged anthropogenic mercury emissions sit around $4\ \mu g/m^2$. MW, Midwest; NE, Northeast; OH, Ohio River Valley; SE, Southeast.

found in the air versus the soil due to differential deposition, with penta-99 having the greatest deposition rate. Most abundant was penta-BDE-47 and penta-BDE-99.

A study from California, where levels are high due to early adoption of flame retardants as a fire safety measure, showed that prenatal and childhood exposures correlated with reduced fine motor coordination, cognition, and attention.[1] As an early adopter of flame retardants, California was the guinea pig for their consequences, and until a better understanding of their problems developed, the worst PBDEs were used there until banned in the mid-2000s. Mothers' prenatal levels of the worst four congeners in the top quartile exceeded 42 ng/g in lipids, and in the children at age 7 exceeded 130 ng/g. For example, a 10-fold increase in child levels associated with 5-fold increase in attention problems. We'll further discuss flame retardants later in Chapter 8 (see Figures 8.8 and 8.10).

Emissions sent into the atmosphere drop back down to the ground attached to either a dry particle or a wet water droplet, perhaps after a chemical reaction or two. The results for mercury in Figure 3.9 separate out wet deposition over various regions of the U.S. Although dry deposition is also important and may be even higher than wet, the wet portion is estimated at 50–90% of total Hg deposition.[2] The highest deposition rates are found in the Midwest and Southeast, ranging up to more than 25 $\mu g/m^2$ in South Florida, whereas the highest atmospheric concentrations are around 16 ng/L in Oklahoma. Fortunately, there are downward trends of about 1–2% per year, in line with Figure 3.4.

---

[1] Eskenazi et al. (2012) examined the effects of PBDE on childhood development. [2] Prestbo and Gay (2009) summarized wet deposition of Hg.

We've already seen that there are many strange and small units, so let's take a moment to put these values into handleable units. Given that 1 acre equals 4,047 m^2, then 1 g/acre is about 0.2 mg/m^2, and depositions rates are micrograms of mercury per square meter per year. We can compare these deposition values to the human emissions shown in Figure 3.2, where we estimated global averages of 10 $\mu g/m^2$. Compare that value to just one season's worth of wet deposition in Figure 3.9, but, of course, emissions aren't evenly dispersed over the globe, so neither is deposition.

Total wet deposition is related to the product of air concentration and precipitation. We saw in Figure 2.3 that climate change increases, or at least alters, precipitation rates and patterns across the country. That means climate change may influence mercury deposition, along with the deposition of other pollutants, as evidenced by historical patterns.[1] Either mercury concentrations will decrease through dilution, or deposition will increase, and it's not clear which process will dominate.

Though small amounts of mercury are deposited, we have to think about what happens to the mercury after deposition. Microbes transform inorganic mercury into methylmercury (MeHg), a neurotoxin. This molecule bioaccumulates up the food chain (see Figure 7.6) and ultimately causes problems for organisms like humans.

For metals more generally, the importance of dry versus wet deposition varies. In southern California dry deposition is much more important than wet deposition, with the latter accounting for 1–10% of metals and dry deposition being responsible for 57–100%, depending on the metal, of the amounts found in stormwater runoff.[2] As could be predicted, proximity to urban areas is the most important factor for high deposition of these metals. Dry deposition of metals ranged from 1 to 100 $\mu g/m^2$/day, with nickel and chromium at the low end and copper and zinc at the high end. Deposition was two to three times as high as urban levels 10 m away from the I-405 Freeway but had decreased to nearly ambient urban levels 150 m away.[3] A comparison of rates in the 1970s and in 2006, showing measurements on the order of $\mu g/m^2$/day, indicates that lead deposition decreased by one to two orders of magnitude—no surprise given the banning of lead additives to gasoline—and chromium levels were about one-fourth the earlier levels.[4] On the other hand, copper and zinc depositions were roughly similar between the two periods. Interestingly, the present major source of lead comes from historically deposited lead resuspended as dust.

Extensive calculations lead to estimates of total airborne copper emissions for cars and medium- and heavy-duty vehicles of 0.58 mg/km, and a roughly

---

[1] Martínez-Cortizas et al. (1999). [2] Sabin et al. (2005). [3] Sabin et al. (2006). [4] Sabin and Schiff (2008) examine dry deposition of metals in southern California.

The sky is falling in New York City.

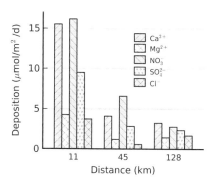

Figure 3.10: Various pollutants deposited with dust decrease as a function of distance from Central Park in New York City (Lovett et al. 2000). These locations correspond to urban, suburban, and rural areas in a northerly direction.

equal amount drops to the ground.[1] Two independent estimates for total vehicle miles in the San Francisco Bay area average to 157 million miles per day, giving an estimate of nearly 54,000 kg/year of copper released to the air. That's a couple hundred kilograms per day via the air, plus add an equal amount for the copper that falls directly to the ground.

Figure 3.10 shows the results for 1997 of the bulk deposition of pollutants in sites located along a line going north from New York City in urban, suburban, and rural settings and clearly shows an urban enhancement.[2] At each site about a dozen oak trees were chosen to hold dust collection plates located midway between the trunk and the edge of the canopy. Of the various types of deposition, bulk deposition means the collector is always open, whereas wet deposition is taken only while it's raining. This study examined throughfall, which involves deposition onto foliage by wet and dry mechanisms, as well as subsequent interactions with the foliage, meaning it includes the collection by leaves as well as exudates. This material then drips from leaves with rainwater, and this water gets collected below trees, and the amounts of pollutants are measured.

Urban dust comes from roadways, excavation, construction, and demolition, and the nitrogen and sulfur from burning fossil fuels. It's usually alkaline, meaning it has the ability to bind with the hydrogen ions that define acidity. As a result, urban dust might be "scrubbing" the urban air of pollutants, collecting acidic gases and causing them to deposit locally, rather than farther downwind. In this way urban dust acts like the powdered limestone blown into coal power plant emissions to scrub away pollutants like mercury.

---

[1] Rosselot (2006) discusses copper coming from autos in the San Francisco Bay area. [2] Lovett et al. (2000).

Herbicide and pesticide deposition in national parks.

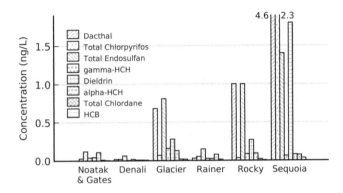

Figure 3.11: Atmospheric deposition of agricultural chemicals onto snow in northwestern North America (Hageman et al. 2006). Several of these pollutants have been banned, indicating legacy depositions from past use as soils erode.

The values in Figure 3.10 for New York City mesh with calculations of emissions and deposition in the Los Angeles basin.[1] In that area the estimate for local deposition of local emissions of pollutants was about 35–45%, an increase over areas without local emissions.

As for long-range deposition, each of the locations sampled in Figure 3.11 for pesticides is a U.S. National Park, and the sample is snow collected from north-facing open areas. Dacthal is a common agricultural and home pre-emergent herbicide that disrupts cell division in root tips. The USEPA health advisory for a 1-day exposure for a young child is 8 mg/L, while drinking water advisories kick in around 1 μg/L.[2] Endosulfan is an organochlorine (a "persistent organic pollutant") that was used solely in agriculture as a sprayed pesticide for fruits, vegetables, cotton, and ornamental plants. Its use was terminated in 2010 due to worker and wildlife risks through its neurotoxicity. Chlorpyrifos is an organophosphate pesticide used in agriculture, horticulture, and turf areas, with recently proposed regulations addressing spray drift exposures. U.S. agriculture uses around 10 million pounds each year.[3]

---

[1] Lu et al. (2003) modeled emission and deposition in the Los Angeles basin. [2] The USEPA provides basic information and health guidelines for all of these chemicals through www.epa.gov.
[3] USEPA, www.epa.gov/oppsrrd1/REDs/factsheets/chlorpyrifos_fs.htm.

Lindane (γ-hexachlorocyclohexane, or HCH) has been used as insecticide on fruit, vegetable, and forest crops, as well as crop seeds, and is used as a prescription lotion for lice and scabies. It's presently used in Mexico as a tick and flea treatment for livestock. Lindane can cause nose and throat irritation at low doses but brings on an array of problems in chronic exposure situations. The α and β isomers of lindane are stable and have toxic effects without pesticidal efficacy. Plus, these isomers wind their way through the ecosystem. The USEPA also considers lindane a suspected carcinogen, and all U.S. pesticide applications have been terminated.

Other historical chemicals (meaning no longer used) include dieldrin, an agricultural pesticide also used against malaria-bearing mosquitoes, banned in 1987, and chlordane, banned in 1988 was an agricultural and household pesticide (including uses as a foundation termite treatment), with nervous system effects; and hexachlorobenzene (HCB), which was widely used prior to 1965 as a fungicide, with other industrial uses, including fireworks. HCB affects the immune system and liver, and causes other widespread problems.

Regarding Denali National Park and Noatak and Gates[1] National Parks in Alaska, all depositions shown in Figure 3.11 are due to long-range transport simply because there are no local pesticide sources.[2] Other parks have varying amounts of regional transport contributing to deposition, and the authors found that regional cropland within 150 km is a very important determinant of concentrations. For example, regional transport accounted for more than half of all pesticides at Sequoia, Glacier, and Rocky National Parks, but it was more variable at Rainier, all of which have nearby agriculture.[3] For example, 98% of dacthal came from regional transport at Rocky National Park.

Thinking about these far-flung, cold wilderness areas brings to mind snow. Snowflakes collect atmospheric pollution better than raindrops: snowflakes have a larger surface to volume ratio and fall at a slower speed, providing more time for pollutant collection,[4] perhaps qualitatively reflecting the deposition velocity differences between forests and grasslands of Figure 3.5. Measurements in Cincinnatti, Ohio, in February 1998 show heavy metal concentrations in snow near highways that were two or more orders of magnitude greater than snow from urban areas that don't have much traffic. A study done during a 47-cm snowfall over 2 days, with sampling continued until snow melted away 4 days later, found lead concentrations of 1–10 mg/L for the snow, compared with a relatively low 90 μg/L for rainfall. Comparable ratios of 100 times greater were also observed for heavy metals Zn, Cu, and Cd.

---

[1] Noatak and Gates National Parks in Alaska are very close, and their data were combined.
[2] Hageman et al. (2006).   [3] Supplementary information provided by Hageman et al. (2006) shows a map of U.S. cropland.   [4] Glenn and Sansalone (2002) and Sansalone and Glenn (2002) studied pollutant deposition from snowfalls.

With winter comes ice-covered roads, and deicing chemicals in runoff correlated with the amount put on the roads. One example shows the scale: 220,000 kg of cyanide-containing deicing salt (28 mg cyanide/kg salt) was applied over about 5 days and yielded 6 kg cyanide in stormwater runoff at an estimated concentration of 0.3 mg/L of snowmelt.[1] Cyanide and lead are both used to prevent clumping in these salts, but safer alternatives are available. These studies agree in their conclusions that highly developed areas had greater pollutant loadings for snowmelt.

Others concur that heavy metals get more particle-bound in snowfall scenarios compared with rainfall situations. Comparing snowmelt and rainfalls with similar total runoff volumes in Colorado,[2] the runoff fraction (see Figure 5.11) for rainfall was nearly double that of snowfall from 100- to 200-acre drainage areas. However, it's important to note that the ground wasn't frozen during the measured events. Peak flow for snowmelt was much lower, meaning fewer large particles get carried off in snowmelt runoff. In this comparison, overall, pollutants were a bit lower for snowmelt, except for oil and grease, probably from snow slush washing cars' undersides, but the pollutants in snowmelt were much more dissolved and wouldn't settle out well in detention ponds.

Perhaps it's no wonder, then, that studies indicate that bacterial populations decrease at the soil's surface when snowmelt takes place. For example, in a worst-case scenario, soil cores from Scotland's highlands exposed to a lab mixture of contaminated snowmelt resulted in a 28-fold decrease in the bacterial population at a depth of 3 cm (in the organic-rich Ah horizon) but increased by 11-fold at a depth of 39 cm (in the BC horizon).[3] Since early snowmelts are the most polluted, researchers also examined effects of a cleaner leached snowmelt. In this case, the bacterial population in the shallow Ah horizon increased 2-fold but decreased 20-fold in the deeper BC horizon. Much of the results were attributed to the greater acidity of the polluted snowmelt (pH 2.3) compared with the leached snowmelt (pH 5.4). The acidic stormwater killed the microbes in the upper soil layer but made more nutrients available to microbes in the deeper layer.

Whether by rain, snow, or dry deposition, pollutants collect onto roads in considerable amounts, as we'll see in the next few figures. The pollution buildup study shown in Figure 3.12 concerned suburban roads in Sydney, Australia, in the early 1990s.[4] The area receives about 1 m of annual rainfall, with 70% coming during January to July, their fall season. The researchers examined a concrete gutter 16.4 m × 0.44 m that edged an asphalt road 8.5 m wide, measured from the median. The road area per meter of curb

---

[1] Sansalone and Glenn (2002). [2] Bennett et al. (1981) measured runoff from 14 storms with snowmelt or rain runoff in Boulder, Colorado, from the mid-1970s. [3] Thompson et al. (1987a, 1987b). [4] Ball et al. (1998).

Roads get dirty.

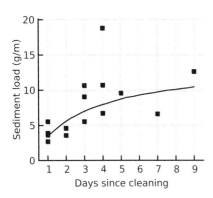

Figure 3.12: Accumulation of sediments onto the roadways around Sydney, Australia (Ball et al. 1998). Levels saturate after a week or so as resuspension from the roadway balances deposition.

was thus 8.5 m^2, and they report collecting 0.85 g/m^2 of sediment in half a year. Accounting for the entire year, double that number to 1.7 g/m^2 annually. That gives 1.7 g/m^2 × 4,047 m^2/acre, which yields about 6.9 kg/acre, or 15 pounds/acre. A handy conversion factor is that 1 g/m^2 is about 4 kg/acre or 8.9 pounds/acre. Traffic counts averaged 8,800 vehicles per day, and the road ran through a medium-density residential neighborhood. Using some initial testing procedures, researchers concluded they collected 95% of particles >75 μm and 82% of particles <75 μm.

The study provides a list of pollutants in the runoff from roads, and it includes the usual suspects (dirt, nutrients, heavy metals) but also includes cyanide from deicing compounds and PCBs from deposition, tires, and roadside spraying. Average amounts of the main pollutants, given in units of milligrams per meter of roadside curb, were sediments, 7,240; iron, 96.0; lead, 3.7; zinc, 1.8; copper, 0.9; chromium, 0.14; and phosphorus, 0.8.

Notice that the sediment load increases quickly and then tapers off to a more-or-less constant value. This saturating dynamic results from a simple process that increases load by a constant deposition of sediments, but a fractional resuspension of the road's load takes place as cars drive over the road. A mathematical representation consists of a single differential equation, $dP/dt = I - aP$, where P is the pollutant load, t is time, I is the constant pollutant input, and a is the fractional resuspension rate. Given that the load reaches its half-saturation value after just 3 or 4 days, the resuspension rate a must define a period of a few days, with that saturation time being proportional to $1/a$.[1] At equilibrium, the saturation load equals $P^* = I/a$,

---

[1] Kim et al. (2006) describe a similar model in detail for their results depicted in Figure. 3.14, and their parameter estimates lead to loss rates of about 2–6% per day for various constituents.

Roads and roofs get dirty.

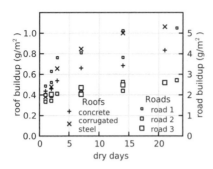

Figure 3.13: Pollutant deposition increases with time on roofs and roads in Brisbane, Australia (Egodawatta et al. 2009). In this situation, saturation of particulate matter takes place after a couple of weeks.

meaning—rather obviously—that higher input and lower resuspension rates result in higher sediment loads until the next rain washes the load away.

Results of Figure 3.13 arise from a two-part study in the city of Brisbane, Australia, that first examined pollution deposition and then examined the subsequent wash-off (see the associated wash-off curve in Figure 4.7).[1] After constructing artificial roofs of 3 m^2, these surfaces were raised to roof height, exposing them to the ambient deposition conditions. After some time, the buildup samples were taken by washing surfaces with a soft brush.

Patterns in the data show quite clearly that the sediment load saturates after about 2 weeks, a little slower than the road data used for Figure 3.12,[2] though no report of nearby traffic counts are given.

The type of roofing is also important, and the differences could influence local water quality through runoff. One such study was performed at an old U.S. Army fort, Camp Mabry in Austin, Texas, where asphalt roofs account for 25% of the watershed area, while metal roofs cover 4%.[3] As one might expect, depositions of PAHs and trace elements were about 50% greater on roofs 12–15 m away from major roads compared with roofs 100 m away, but a comparison between metal and asphalt roofs near freeways demonstrated that metal roofs had greater cadmium, nickel, and zinc yields in their particulates, while asphalt roofs had greater lead yields. This difference between roof types was not true for PAHs. In the end, roofs contribute up to 17% of sediments at the watershed scale (proportionally less than their imperviousness fraction) but contribute 55% of the zinc and significant amounts of cadmium and lead.

---

[1] Egodawatta et al. (2009) included earlier data on road buildup from a conference presentation cited within. [2] Replication of the roof buildup data is not clear, and Egodawatta et al. (2009) connected their buildup curve to a power-law expression, $B = aD^b$, where D is the number of days and B is the buildup. [3] Van Metre and Mahler (2003).

Roofing material itself can also be a pollution source, not just a passive collector. A study of roofs in central Paris, France, showed that neither slate or zinc roofing acted as a source of PAHs.[1] As for heavy metals, however, zinc and titanium came from zinc roofing, and lead, titanium, and copper came from slate roofs. This work introduced the measurement of total aliphatic hydrocarbons (TAHs), which are good old hydrocarbon chains without the aromatic rings of PAHs, but TAHs can include nonaromatic carbon rings. TAH deposition ranged from 140 mg/ha/day in summer to 1,790 mg/ha/day in winter, whereas PAH deposition was less than 6 mg/ha/day every season. Heavy metal deposition[2] hovered around 2.3–2.8 g/ha/day, and other elements[3] constituted 78–116 g/ha/day. Measured export increased with runoff, which would happen if concentrations were constant.

Yet another extensive study of roofing materials, done in Pennsylvania, included everything imaginable: galvanized metal, tar and cedar shingles, plastic panels, fake slate, sealants, and wood with various treatments.[4] Four-foot by eight-foot A-framed roofing sections with each roofing type were set up for 2 years and used to examine runoff for various compounds, all corrected for atmospheric deposition. All of the roofing types released pollutants as a degradation by-product, but in the end, researchers determined that treated galvanized roofing produced the most pollutants of interest, including metals and nutrients. Even 60-year-old sheet metal roofing from a charismatic old barn continued releasing pollutants. Further, and not surprising, pressure-treated wood released a lot of copper, which used to be in the older wood preservation compounds.

Finishing out the examination on the time dependence of accumulation on roads, researchers studied eight highways in Los Angeles, California, over the 2 years 2000–2001 (Figure 3.14).[5] Traffic ranged from 122,000 to 328,000 cars/day, about 20 times more traffic than reported for Figure 3.12.

Accumulation rates during the initial 10 days are 0.544 and 0.0113 g/m^2/day for total suspended solids (TSS) and oil and grease, respectively. Rates decrease by 79% and 61% thereafter. Other numbers cited within this study include deposition values of 0.7 g/axle/km of solids from traffic, which multiplying by an average of, say, 150,000 cars per day (and never mind multiple axles), yields about 150 kg/km of solids deposited on roads each day. That number equals 150 g/m of roadway.

Now, the accumulated masses shown in Figure 3.14[6] are nowhere near that amount of buildup. At a buildup rate of 0.5 g/m^2/day, and assuming a road lane 3 m wide, then a traffic lane accumulates 1.5 g/m/day, or 1.5 kg/km/day.

---

[1] Rocher et al. (2004). [2] Ba, Cd, Co, Cr, Cu, Mn, Ni, Pb, Sb, Sr, V, Ti, and Zn. [3] Ca, K, Mg, Na, P, and S. [4] Clark et al. (2008). [5] Kim et al. (2006) studied highway pollutant deposition rates in southern California. [6] COD, chemical oxygen demand; TSS total suspended solids; TOC total organic carbon; O&G oil & grease; TKN total Kjeldahl nitrogen; TP total phosphorus.

Roads get dirtier and dirtier.

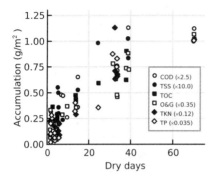

Figure 3.14: Pollutant buildup on highways in Los Angeles, CA (Kim et al. 2006). In this location, it takes more than a month for saturation to take place. (Multiply each data set by the number in the legend to obtain the actual pollutant value; for example, TSS at 40 days is roughly 10 g/m^2.)

There certainly seems to be a wide discrepancy between the various published estimates of deposition, but that might well be expected due to the differences in locations.

Sediment loads in Figure 3.14 mesh well with the numbers in Figure 3.12 since the latter plot accounts for roughly 8.5 m^2 per meter of road. Yet some unknown factor seems different about this high-traffic Los Angeles road, given the long saturation time of a few weeks compared to a few days in Figure 3.12.

Figure 3.15A comes from a study involving stormwater runoff from bridges in Brisbane, Australia,[1] which demonstrates the traffic dependence of pollutant deposition and buildup through their concentrations in runoff. Using bridges for these measurements turned out to be quite convenient. Imagine a bridge, and the bridge having curbs with drains for the rains that fall, and those drains having downspouts where scientists can measure water quality. First-flush samplers collected the first 20 L of runoff from these bridge drains.

Analyzing the data indicated that traffic counts explained 24% and 38% of the variation in zinc and TSS, respectively, but less for other pollutants. Although the correlation of pollutants and traffic isn't terribly strong, with around 70% of the variation unexplained, traffic is certainly an important factor. The results really mean that traffic count shouldn't be the sole basis for which highways get the greatest stormwater treatment attention. Another important point the study revealed was that concrete versus asphalt surfaces showed no significant difference in pollutant concentrations.

A significant automobile-related heavy metal is zinc, which constitutes about 1% of a car tire's mass. Total U.S. zinc emissions were just under 50,000 tons in 1995, most of that coming from zinc production itself (33,000 tons),

---

[1] Drapper et al. (2000).

Roads get dirtier with more traffic.

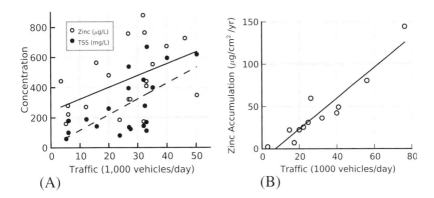

Figure 3.15: (A) Dependence of pollutant accumulation with traffic levels in Queensland, Australia (Drapper et al. 2000). (B) Zinc accumulation increases with traffic counts across the eastern U.S. (Councell et al. 2004).

and the next highest (8,000 tons) from waste incineration.[1] Zinc accumulation in urban areas increases significantly with traffic, according to a study that compiled results from 11 urban and suburban watersheds throughout the eastern US: traffic explained 88% of the variation (Figure 3.15B).

How bad is zinc? Tadpoles of a wood frog, *Rana sylvatica*, were subjected to zinc at the concentration in tire particles and aged sediments, giving water concentrations of tens of micrograms per liter.[2] Tadpoles exposed to tire particles took about 1 week longer to metamorphose compared with controls, which took about 6.5 weeks. Many macroinvertebrate species demonstrate problems when subjected to zinc and other heavy metals, with ecological implications for first-order streams.[3]

On a more general note, metals have a special place in the world of pollutants because, unlike organic compounds, biological processes can't break them down and detoxify them. That said, as with the methylation processes with mercury, biological processes may make them more or less bioavailable. Still, metals are taken up but then released via decomposition and persist in the biosphere until they're locked away permanently through burial in deep sediments.

Cars are important, in part, because when you step on your car's brakes, the metal rotors scrape particles off the brake pads pressed against them as the

---

[1] Councell et al. (2004) discuss environmental sources of zinc. [2] Camponelli et al. (2009).
[3] Beasley and Kneale (2002).

Apply brakes, add pollution.

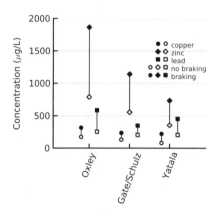

Figure 3.16: A part of the traffic study in Queensland, Australia (Figure 3.15A) compared bridge lanes with and without braking. As indicated by the elevated concentrations, braking vehicles erode brake pads, giving increased copper and zinc depositions (Drapper et al. 2000).

friction between the two slow down your vehicle. Those particles add up—as all car owners know—as evidenced by having to change the brake pads every once in awhile. A small part of the research project in Brisbane, Australia (Figure 3.15A) very cleverly studied isolated bridges with exit ramps in one direction but not the other. That difference provided a test on the influence of braking on the heavy metals (and so forth) deposited as a result of braking from high speed, while controlling for many environmental conditions, including traffic volume (assuming both directions have roughly equivalent counts). Copper mostly comes from brake pads, while zinc comes from tire wear, and both brake pads and tires get worn down as cars slow down (see Figure 3.18). As the study authors note, there were too few results with not terribly strong statistical significance to make any conclusions with high confidence, but pollutants from the "no braking" side of the bridge were always lower than from the "braking" side, a strongly suggestive result confirming what seems reasonable.

Cars release large amounts of pollution into the environment, ranging from oil to heavy elements, well beyond the worries of carbon dioxide released from burning the fossil fuels that power them.[1] Just one example is an estimate of 16,000 tonnes of oil annually dripping from U.K. cars as they travel a total of 500 billion kilometers. Changing that number into more manageable units, that's 16 billion grams, giving about a third of a gram per 10 km. Think of an oil change every 3,000 miles, or about 5,000 km, and that's around 170 gm, or a fifth of a liter, per oil change. Much of that oil drips onto roads and from there goes into streams via stormwater flows. This number shows the reason

---

[1] Napier et al. (2008) describe the heavy elements coming off of automobile wear.

The diversity of automotive pollutants.

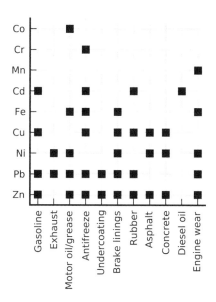

Figure 3.17: Vehicles slough off various elements from different parts, fluids, and surfaces (after Nixon and Saphores 2007). For example, lead and nickel come from exhaust, and asphalt sheds zinc, nickel, and copper.

that used oil and oil filters represent the main contributor of hydrocarbons to surface water pollution.[1]

Another major automotive fluid is antifreeze, with its primary ingredient ethylene glycol. Also used as deicing fluid and brake fluid, ethylene glycol has various noncancerous health risks. Only about 12% of antifreeze gets recycled, compared with 50% of used motor oil and filters. Do-it-yourselfers are mostly to blame for the lack of recycling. Propylene glycol antifreeze is available, which the Centers for Disease Control recognizes as generally safe,[2] but it holds only 10% of the antifreeze market.

The study providing data on automotive pollutants in Figure 3.17 also summarized the large costs of automotive-related water quality issues.[3] Highways contribute half of suspended solids and one-sixth of hydrocarbons reaching streams. Fractions of particulates connected to vehicles come from tire wear (1/3), pavement wear (1/3), engine and brake wear (1/5), and exhaust (1/12). Estimates for tire wear range from 0.0012 to 0.2 g/km per vehicle, with high PAH and zinc content in the released particles. Other important source fractions are shown in Figure 3.18. High-cost-scenario control costs are $100,000 to $300,000 per lane mile, with the low estimate for rural roads and the high

---

[1] Nixon and Saphores (2007). [2] ATSDR, www.atsdr.cdc.gov/toxprofiles/tp.asp?id=1122&tid=240.
[3] Nixon and Saphores (2007).

Car parts parting pollutants.

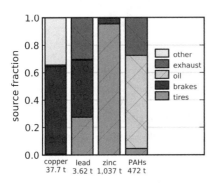

Figure 3.18: Various automotive sources for copper, lead, zinc, and PAHs shown in their relative contributions to surface contamination, as estimated for the U.K. (Napier et al. 2008). All values are in tonnes (metric ton = 1,000 kg) and represent contributions from all passenger vehicles.

estimate for urban roads. These costs represent between 3% and 14% of highway expenditures.

Further, leaking underground fuel storage tanks represent a cost of $1 billion to $2 billion per year over a decade to clean up the 450,000 confirmed leaks in the U.S. Compare this number with the 1.62 million registered underground fuel tanks. Although most leaks tend to be shallow, and passive bioremediation is effective, when groundwater is involved costs range from $100,000 to $1 million. Chemicals of concern are various PAHs: benzene, toluene, ethylbenzene, and xylene.

In finding money to pay these cleanup costs, a one penny increase in gas tax would provide $1.5 billion/year, with the implication that a $0.12 increase could provide enough revenue to fix both tanks and runoff problems. These researchers estimate the total cost of highway runoff at $3 billion to $16 billion per year and remark that, as expected, prevention would be cheaper than remediation.[1]

Another study estimated that, in 1991 dollars, water pollution costs from U.S. automobile use accounted for $400 million to $1.5 billion of a total automobile-related pollution cost estimated between $38.3 billion and $546.6 billion.[2] This total cost includes human mortality from air pollution (the far greatest cost), damage to crops and forests, oil spills, and climate change. This estimate was described as a rough one, and the author argues that noise is likely even more costly.

Fortunately, studies show a dramatic decrease in emissions over the last several decades (see Figure 3.4). Total copper emissions, for example, decreased from 150 tonnes in 1980 to 75 in 2003. Automotive sources became

---

[1] Nixon and Saphores (2007). [2] Delucchi (2000).

the largest source of copper in the environment around 2001, as other emission sources decreased in importance. As a now classic example, lead emissions dropped dramatically from 3,000 tonnes in 1990 to a few hundred in 2003. Meanwhile, zinc emissions roughly doubled from 1970 and strongly correlate with total vehicle miles driven.

Of course, good things come with a price. While catalytic converters reduce NO_x and hydrocarbon emissions like PAHs, catalyst erosion causes increasing metal emissions.[1] Catalytic converters reduce CO emissions by conversion to CO_2, convert reactive nitrogen to N_2, and burn up any remaining hydrocarbons in the exhaust. While reducing these known pollutants, catalytic converters themselves degrade, leaving depositions of platinum-group metals on our roads, with platinum, palladium, and rhodium soil concentrations of 64–73 ng/g, 18–31 ng/g, and 3–7 ng/g, respectively, near South Bend, Indiana.[2] Concentrations in soils 1 km from a road were 3.6 ng/g, 1.54 ng/g, 0.09 ng/g, respectively. The loss of platinum, for example, could be as high as 0.19 μg/km. Some plant uptake of these elements is found, but the consequences are unclear.

Other than simply getting rid of most vehicles, controls on these pollution sources include better traffic flow that reduces braking and acceleration. Since braking is a prime source of metallic pollutants (see Figure 3.16), one suggestion is to replace speed bumps with speed cameras. Ceramic brake pads are also preferable over cheaper standard ones, but those better pads come with greater upfront cost to the car owner, which reduces implementation.

Figure 3.19 lists a few of the many, many sources of PAHs, including pollutants from car exhaust, oils, and tires, erosion from streets, vegetation, and more broadly sourced atmospheric deposition.[3] Yet another source includes parking lot "sealcoats" that protect asphalt from various forms of degradation and enhances appearance. One commonly used type of sealcoat contains crude coal tar, which is the residue that remains after the coking of coal, or coal tar, which is the residue that remains after the distillation of crude coal tar. In contrast, a different asphalt-based sealcoating comes from petroleum. Coal tar can be more than 50% PAHs by weight, and coal tar sealants are typically 20–35% coal tar.[4] The experiments providing these results were done in Austin, Texas, where records show a total application of more than 2.5 million L each year, about a gallon per person per year, prior to a local ban on coal tar sealants in 2006. Sealcoats erode rapidly, with reapplication after 2–3 years.

Experiments were performed after the passage of at least 5 dry days, and then 2 mm (less than one-tenth inch) of distilled water sprayed onto the parking lot was collected as runoff. We've seen how initial abstraction of dribblings

---

[1] Beasley and Kneale (2002) discuss various sources. [2] Ely et al. (2001) discuss catalytic converter wear. [3] Mahler et al. (2005). [4] Mahler, pers. comm.

PAHs from parking lot sealants.

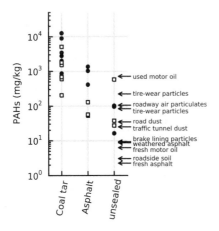

Figure 3.19: PAH concentrations in particulates carried in runoff from parking lots with and without seal coats, comparing coal-tar- and asphalt-based seal coats (Mahler et al. 2005). Squares denote test plots, and circles are samples scraped from parking lots. Also shown for comparison are PAH concentrations of various compounds.

of rain evaporate and seep into crevices: of 100 L sprayed onto the parking lots, the average runoff collected was just 58 L.

Figure 3.19 shows the sum of 10 PAHs and compares that amount with those of other sources scanned from the literature. Overall, unsealed parking lots yield an average of 54 mg/kg of sediment, asphalt-based sealcoated parking lots average 620 mg/kg, and coal-tar-based sealcoated parking lots average 3,500 mg/kg—quite a difference. A subsequent study indicated that the data from asphalt sealcoated lots were anomalously high, probably because the asphalt sealcoating covered prior coal tar sealcoatings. Concentrations on asphalt sealcoated lots are more typically in the 10 mg/kg range.[1] Scrapings from the parking lots showed PAH levels for coal-tar-sealed lots at 40 times greater than asphalt-coated lots; however, abrasion by cars increased erosion on asphalt-coated lots more than on coal-tar-sealed lots.

In four highly urbanized Texas watersheds (>65% urban), PAH fingerprinting shows stream PAH loads come mostly from coal tar sealant exports, despite covering only 1-2% of the watershed. Researchers estimated that 90% of PAH loads from parking lots would be eliminated if sealants were not used. Another study from Marquette, Michigan, demonstrated that commercial parking lots account for 64% of PAH load (not to mention one-fourth to one-third of zinc, copper, and cadmium).[2]

A follow-up study examined the prevalence of coal tar sealant effects across nine U.S. cities,[3] where researchers swept and analyzed dust from

---

[1] Mahler et al. (2010). [2] Steuer et al. (1997). [3] Van Metre et al. (2009).

seal-coated and non-seal-coated parking lots. In six central and eastern cities[1] where coal tar sealcoat use predominates, scientists found a median PAH sum of 2,200 mg/kg and 27 mg/kg, respectively. In three western cities[2] where asphalt-based seal coats are used, they found a median PAH sum of 2.1 and 0.8 mg/kg, respectively. These dust concentrations are reflected in upper sediments in lake bottoms of the two regions. PAHs likely have biological effects at levels exceeding the "probable effect concentration" of about 23 mg/kg dry weight,[3] a level that was exceeded in the top 5 cm of sediment in one lake in New Haven, Connecticut, and two lakes in Minneapolis, Minnesota. Additional studies provide clear evidence of a high quantity of PAHs emanating from seal-coated surfaces both in Austin, Texas, and Madison, Wisconsin.[4]

In contrast, an article written by coal tar industry representatives in the professional stormwater industry magazine *Stormwater* disputes the relevance of coal-tar-based sealants as an important contributor to PAHs,[5] arguing for more research, and argues that coal tar sealant bans are ineffective because PAHs are ubiquitous.

Further studies of coal tar sealants examined the sediments of 40 U.S. lakes sampled between 1996 and 2008.[6] Sediments from the greatest depth, corresponding to the lower portion of reservoirs, and post-1990 sediments were examined for PAH content. Coal tar sealcoats contribute about one-half the PAHs, followed by vehicles and coal combustion, though the most polluted lakes have larger coal tar contributions. These sealants seem to be the larger driver of PAH loads than the urban land use fraction, accounting for roughly three-fourths of total PAH loads. Clear historical trends depict their contribution since these sealants' introduction in the 1960s and 1970s. PAH pollution increases are a problem that can be reversed: a 58% decline in PAHs in area lake sediments in Austin, Texas, followed the 2006 ban on coal tar sealcoats,[7] though disputes with the coal tar industry remain.[8]

Data shown in Figure 3.20 were taken at Spring Creek, Pennsylvania, during 2000–2002, examining land use and pollutants.[9] Here I show results only for chloride and lead. Samples were taken across 10 subbasins ranging in size from 3.9 to 71.7 km^2, with 0.3–53.2% urbanization, 29–78% agriculture, and 11–60% forest. Chloride likely came from salts used for deicing roadways, and the mileage of roads within a subbasin increases with urbanization.[10] Suspected lead sources include roadside sediments from past use of leaded gasoline, as well as paints and roofing materials, though we've seen that lead comes from various car parts (Figure 3.18). More roads in urban areas means more salt to melt road ice, and that means more salt in urban streams. There

---

[1] Minneapolis, MN, Chicago, IL, Detroit, MI, New Haven, CT, Washington, DC, and Austin, TX.
[2] Seattle, WA, Portland, OR, and Salt Lake City, UT.   [3] MacDonald et al. (2000).   [4] Selbig (2009).
[5] O'Reilly et al. (2010).   [6] Van Metre and Mahler (2010).   [7] Van Metre and Mahler (2014a).
[8] DeMott and Gauthier (2014), and Van Metre and Mahler (2014b).   [9] Chang and Carlson (2005) investigated winter water quality in Pennsylvania.   [10] Also see Wilson (2011).

More urbanization, more roads, more pollution.

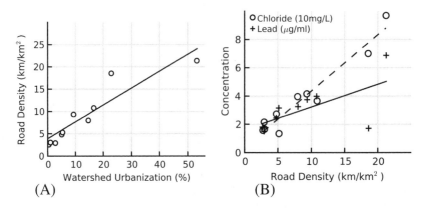

Figure 3.20: (A) Road density increases in urbanized watersheds, and (B) with increased road density come increased concentrations of lead and salt pollution (Chang and Carlson 2005).

was a nearly equally strong negative correlation of salt in forested watersheds, with a Spearman's rank correlation coefficient of –0.72.[1]

One wonders how these results are anything more than obvious, though actually measuring the correlation has great importance! Seeking the mechanism of the correlation, we know that roads collect pollutants and increase runoff, hence polluted runoff. Yet with roads come houses, people, and industries, and it may be difficult to disentangle the many aspects of urbanization that correlate with roads—it is important to find the correct culprit for pollutants.

One type of experiment takes advantage of road construction in relatively undisturbed watersheds, comparing before and after situations. One study took this approach concerning the construction of a four-lane highway in West Virginia, examining nine sites ranging from 33 m to 1,047 m downstream from the highway.[2] Effects were found on turbidity, TSS, and total iron while the highway construction was taking place. Both during and after construction, changes in chloride and sulfate were seen (see, e.g., Figure 11.10). After the construction was finished, changes in acidity, nitrate, and macroinvertebrate index scores were observed.

---

[1] Spearman's correlation should be used with nonlinear monotonic data; Pearson's correlation should be used with linear data.   [2] Chen et al. (2009) studied the effect of highway construction on streams.

Small particles have high zinc concentrations.

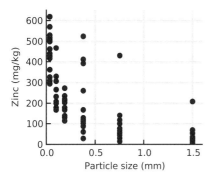

Figure 3.21: Zinc concentration in various particle size classes of the sediments picked up by street sweeping (German and Svensson 2002). Similar results were seen for copper. Street sweeping represents a *nonstructural* stormwater control measure.

Presumably, we would like to remove these pollutants from our roads before they enter streams. In a very detailed experiment in Sweden on the dry version of street sweeping, a 260-m stretch of road had several street sweeping regimens, with a 340 m control stretch for comparison.[1] The road had a traffic level of 11,200 daily vehicles. An initial sweeping performed, hopefully, to get the roads to a "pristine" condition produced 11.1 kg sediment. Once the measurements officially started, 3 weeks of weekly sweeping produced 37.9 kg, whereas 3 weeks of sweeping every workday produced 46.6 kg. Yes, daily sweeping was more effective in picking up sediments than weekly sweeping, but the amount produced clearly tapered off. Effectively, this reflects the saturation taking place in deposition shown in Figure 3.12.

Small particles have a relatively greater zinc concentration, as shown in Figure 3.21. Similar results are seen for other heavy metal concentrations. Despite the fact that the really small particles have higher concentrations, more than half of the metal comes from the fraction of sediment greater than 0.125 mm in size simply because that fraction constitutes more of the sediment mass. Indeed, 80% of the sediment is between 0.075 mm and 2.0 mm, and just one-fifth to one-fourth of the sediment mass comes from particles smaller than 0.25 mm. Sweeping also pulls up the larger particles more efficiently than the smaller ones: before sweeping, 26% of the sediment was finer than 0.25 mm, whereas after sweeping the fraction increased to 40%. Perhaps counterintuitively, sweeping preferentially leaves behind a distribution of smaller-particle sediments with consequently higher concentrations of heavy metals.

The small-particles-are-worst phenomenon is confirmed in Figure 3.22. In this study the runoff was collected, solids were dried and separated into

---

[1] German and Svensson (2002).

Small particles hold heavy metals.

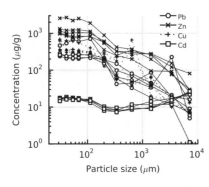

Figure 3.22: Heavy metals in the suspended solids extracted from stormwater runoff in Cincinnati, OH (data from Sansalone and Buchberger 1997). Different lines for each metal represent results from five separate rainfall events.

size classes using sieves, and then each class was analyzed for metal content.[1] Data plotted here come from five rainfalls in 1995 from the I-75 Freeway in Cincinnati, Ohio, having traffic counts around 150,000 vehicles per day. Snowfall data show similar patterns. Precipitation depths for these rainfall events ranged from 0.4 to 25.0 mm. Metals have a habit of attaching to the surfaces of particles, and metal content increases with surface area, which means that metal concentration as a surface-to-volume ratio increases as particle size decreases.

High metal content in small particles crosses international boundaries: a study from Hamilton, New Zealand, where half the mass of the samples passed through a sieve of size 250 μm, demonstrated that this half was the most contaminated by heavy metals.[2] An estimated 0.55 g sediment smaller than 2 mm accumulated on a daily basis for each meter of road curb. Particles larger than 125 μm were trapped by vegetated buffers, but smaller ones were not. This larger size represents 43% of Cu, 54% of Pb, and 36% of Zn.

There are three types of street sweepers, described here in order of providing worst to best pollution removal results.[3] Mechanical street sweepers are those with the prototypical rotating brush that sweeps up dirt and trash into a bin; 90% of street sweepers are of this type. Better than simply sweeping, regenerative air street sweepers blast air at the road just like a leaf blower and then suck up the dusty air like a vacuum cleaner. The top-of-the-line street sweepers use a vacuum filter. They can be a dry type, like a household vacuum cleaner, that uses mechanical brushing, then vacuuming up the dust. A wet type sprays the road and then scrubs, vacuuming up the dirty water like a shop

---

[1] Sansalone and Buchberger (1997). [2] Zanders (2005). [3] Curtis (2002) describes the array of street sweepers.

vacuum. Vacuuming is generally expensive and slow and requires technical training to do correctly.

A comparison of the three types of street sweepers comes from Montgomery County, Maryland. In the year 2000, about 5,600 miles were swept (or reswept), picking up 2,500 tons of solids, working out to 0.17 tons (340 pounds) per mile. This total amount of solids included an estimated 350 pounds of copper, 470 pounds of lead, and 2,400 pounds of zinc. Total cost was about $500,000 for the year, which worked out to about 3.4 pounds per $1.

The study authors concluded that vacuuming performed better than removal by stormwater ponds and filters, and although air-based sweeping does well, it's not as good as these other approaches.[1] In contrast, other researchers concluded that street sweeping was an ineffective approach to improving water quality in Champaign, Illinois,[2] and there are indications that the small particles left behind, with higher metal densities, are those most easily flushed away with runoff.[3] Still, one study coming to that conclusion (using a dry brushing-type sweeper) agreed with the above discussed findings and found that loads decreased for all examined pollutants.[4] The relative utility of street sweeping compared with other approaches to pollution control seems unclear.

Chapter Readings

Egodawatta, P., E. Thomas, and A. Goonetilleke. 2009. Understanding the physical processes of pollutant build-up and wash-off on roof surfaces. *Sci. Total Environ.* 407: 1834–1841.

German, J., and G. Svensson. 2002. Metal content and particle size distribution of street sediments and street sweeping waste. *Water Sci. Technol.* 46: 191–198.

Harrad, S., and S. Hunter. 2006. Concentrations of polybrominated diphenyl ethers in air and soil on a rural-urban transect across a major UK conurbation. *Environ. Sci. Technol.* 40: 4548–4553.

Lindberg, S., R. Bullock, R. Ebinghaus, D. Engstrom, X. Feng, W. Fitzgerald, N. Pirrone, E. Prestbo, and C. Seigneur. 2007. A synthesis of progress and uncertainties in attributing the sources of mercury in deposition. *Ambio* 36: 19–32.

Lovett, G.M., M.M. Traynor, R.V. Pouyat, M.M. Carreiro, W.-X. Zhu, and J.W. Baxter. 2000. Atmospheric deposition to oak forests along an urban-rural gradient. *Environ. Sci. Technol.* 34: 4294–4300.

---

[1] Curtis (2002). [2] Bender and Terstriep (1984). [3] Furumai et al. (2002) plots a cumulative curve calculated from the Sansalone and Buchberger (1997) data, but its origin is not clear. [4] German and Suensson (2002).

McDonald, A.G., W.J. Bealey, D. Fowler, U. Dragosits, U. Skiba, R.I. Smith, R.G. Donovan, H.E. Brett, C.N. Hewitt, and E. Nemitz. 2007. Quantifying the effect of urban tree planting on concentrations and depositions of PM_{10} in two UK conurbations. *Atmos. Environ.* 41: 8455–8467.

Napier, F., B. D'Arcy, and C. Jefferies. 2008. A review of vehicle related metals and polycyclic aromatic hydrocarbons in the UK environment. *Desalination* 226: 143–150.

Nixon, H., and J.-D. Saphores. 2007. Impacts of motor vehicle operation on water quality in the US: Cleanup costs and policies. *Transport. Res.* D12: 564–576.

Pozo, K., T. Harner, S.C. Lee, F. Wania, D.C. Muir, and K.C. Jones. 2009. Seasonally resolved concentrations of persistent organic pollutants in the global atmosphere from the first year of the GAPS study. *Environ. Sci. Technol.* 43: 796–803.

Chapter 4

Imperviousness

Natural soils have high variability in texture, which means rain water infiltrates through channels left behind by rotting tree roots and long-dead rodents, filled with organic debris and loosely filled soils. The process of urban development adds imperviousness—surfaces that prevent rain from hitting the ground—and compacts the soils by trucks and other implements during building construction. Like imperviousness, compaction makes these soils much less permeable, and the harm brought to urban soils takes decades to restore by natural processes.

As a village grows from a few businesses and a neighborhood to a large city, the imperviousness increases and becomes more likely connected to a stormwater system that alleviates the urban problems associated with large water volumes. Eventually, a stormwater system of drains and pipes replaces headwater streams and riparian systems. Septic tanks also disappear, as houses get connected to municipal wastewater systems.

Imperviousness varies with land use type, with commercial areas being the most impervious, at about a 90% level, and more than half the surface covered with streets and parking. Particles that wash off of impervious surfaces make their way into streams when entrained with stormwater runoff. Riparian systems act like coffee filters that keep particles of ground-up coffee beans out of a brewing pot. We see this effect in the particles that wash downstream when riparian systems are removed: higher levels of watershed imperviousness lead to finer, siltier sediments. High imperviousness also leads to higher nutrient levels and higher salt burdens.

Roots and tunnels slice and dice soil profiles.

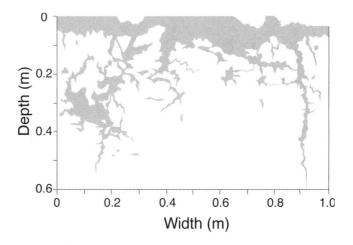

Figure 4.1: Channels, cracks, decayed roots, and rodent tunnels increase the permeability of soils (after Scanlon and Goldsmith 1997). This figure shows the high-permeability regions in a northern Texas playa revealed after excavation of a dye-soaked soil profile.

I have this vision of the impervious surface concept that seems pretty clear: if it doesn't let raindrops hit the ground, then it's impervious. So, here's the rub: just because it's soil and it looks pervious to the naked eye, its history might leave it only partly permeable, and, in part, that's because the "natural" soil infiltration profile is so complicated.

Let's begin the discussion of imperviousness with a quick depiction of "natural" soils. Essentially, the researchers behind Figure 4.1 poured 10–20 cm of dyed water onto a surface, and after letting the water infiltrate for a day or two, they dug a trench to see where the water went.[1] In this image, the top of the soil depicts the vertical profile of a playa (a low spot that collects water on the plains) in northern Texas. Dark areas represent areas where dye went, and light areas were unwetted. Below the mixing zone near the surface where water soaks in quite uniformly, water flows along preferred paths. Obviously, there's high variability in the soil profile caused by things like growing roots that subsequently die and rot, and rodent and ant tunnels, though apparently not in this particular case. These things drill holes through the soil, and either immediately or years or decades hence make easy pathways for

---

[1] Scanlon and Goldsmith (1997).

Building houses compacts the soil.

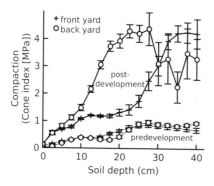

Figure 4.2: Housing construction greatly reduces infiltration rates as measured through compaction (Gregory et al. 2006). These curves compare pre- and postdevelopment compaction profiles measured in the front and back yards of new homes in a Gainesville, FL, subdivision.

water to deeply and/or broadly penetrate dry earth. And that means increased stormwater infiltration.

Again, just because it looks permeable, it doesn't mean it is. Figure 4.2 is a somewhat detailed plot that shows how the very act of building a home, adding clearly impervious surfaces, also unintentionally renders the soils around it partially impervious.[1]

How is soil compaction measured? One common device for measuring infiltration rates is a double-ring infiltrometer. Imagine two concentric rings both kept at constant water level, replenished as water seeps out the bottom and infiltrates into the soil. The volume of replenished water divided by the ring's area measures the infiltration rate over a certain time interval. Only the infiltration rate of the inner ring is important, as the outer ring's infiltration serves to wet the soil around the inner ring, preventing the inner ring's water from seeping outward instead of downward. The outer ring serves to reduce boundary condition effects on the infiltration measurement, and, ideally, one would have a huge outer ring so that the inner ring represents a small interior location completely unaffected by the boundary. Typical values for infiltration rates are, for sandy soils, 414 mm/hr (16.3 in./hr); compacted sandy soils, 64 mm/hr (2.5 in./hr); dry clayey soils, 220 mm/hr (8.8 in./hr); and both compacted and dry or saturated clayey soils, 20 mm/hr (0.7 in./hr). The coefficients of variation (standard deviation divided by the mean) associated with these numbers vary from 0.2 to 0.4 for sandy soils and from 1.0 to 1.5 for clayey soils.

Another device, a penetrometer, produced the data for Figure 4.2 that examines a development in Gainesville, Florida, where the soils were 90%

---

[1] Gregory et al. (2006) studied increased urban soil compaction and infiltration rate reductions.

Digging techniques affect infiltration rates.

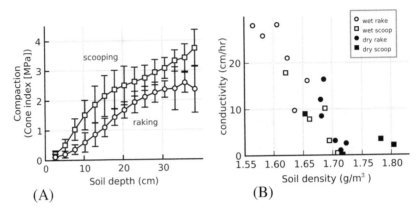

Figure 4.3: Soil properties after using two backhoe digging techniques, scooping and raking (see text), and the resulting differences in compaction profiles (A) and water conductivity (B) versus soil density in wet and dry sandy soils (after Brown and Hunt 2010). Error bars in A show standard deviations.

sand.[1] A penetrometer is a metal rod with sensors that measure depth of insertion and soil strength, measured as a "cone index" that relates to permeability. Low numbers mean greater permeability, leading to higher infiltration rates. Scientists took measurements before the houses were built and then once again after they were built. Comparing the two measurements provides an estimate of the effect of house building on soil compaction. The original plot showed standard deviations, but I'm showing the standard errors, obtained by dividing the standard deviation by the square root of the number of measurements used.

Infiltration rates, averaged from measurements across three housing lots, decreased from 733 mm/hour in the undisturbed condition prior to development to 178 mm/hour after development, which included lot clearing, house building, and landscaping. During this process, the soil density also increased from 1.34 g/cm^3 to 1.49 g/cm^3, representing compaction. A little bit of compaction leads to a great reduction in infiltration (see Figure 4.3B). Infiltration rates for design storms (see Chapter 5) in the region are (or, at least, were) supposed to handle 254 mm/hour, or 10 inches/hour, during a rare 100-year, 24-hour rainfall. Obviously, the construction process isn't doing anything to help achieve those infiltration rates in the front or back yards! One has to wonder whether the infiltration rates used to compute stormwater runoff and

---

[1] Gregory et al. (2006).

infiltration are measured before or after construction. Of course, the whole idea of "low-impact development" includes building processes that protect soils from these kinds of inadvertent harms.

In another experiment, the researchers stripped off the top 10 cm to remove grass and organic matter. Then, infiltration measurements yielded 487 mm/hour for natural forest and 225 mm/hour for pastureland, despite similar soil properties, such as density. Those deep tree roots really fluff up the soils. Next, using a portable soil compactor for 30 seconds, 3 minutes, and 10 minutes, they once again measured infiltration rates. After just half-a-minute of soil compaction, infiltration on a 91% sandy soil dropped to 65 mm/hour! After 3 and 10 minutes of compaction, infiltration was just 30 and 23 mm/hour, respectively, and forest and pasture soils showed no differences.

Yet another experiment measured infiltration rates in pasture soils after nine passes of a pickup, a backhoe, and a dump truck, which reduced infiltration to 68, 59, and 23 mm/hour, respectively, down from a pretreatment rate of 225 mm/hour. Low-impact development involves the choices of vehicles to drive and where to drive them—keeping vehicles off of areas where infiltration should take place helps reduce runoff.

Infiltration rates from various landscapes for the North Carolina Piedmont near Charlotte range from a high of 31.6 cm/hour for medium-age pine and mixed hardwoods, to 11.2 cm/hour for slightly disturbed soils with lawns and large trees, to less than 1 cm/hour for all of the following: previously plowed and cultivated fields, highly disturbed soils with lawns and few trees, and highly compacted soils.[1] In Durham, North Carolina, I can dig a posthole in our clay soils, fill it with water, come back a day or two later and it'll still have water in it. On a larger scale, infiltration variation arises when an undulating topography is flattened for a subdivision—the areas with "cut" soils (where hills once sat) have about one-half the infiltration rate compared to "fill" soils (where the valleys once were).

Chapter 13 describes several stormwater control measures (SCMs) that engineers use to reduce stormwater problems, and many involve infiltration. Many of these approaches involve digging holes and building ponds, which must be done carefully because excavating these pits with heavy equipment can inadvertently compact the soil as with house building (Figure 4.2).[2]

In Figure 4.3A, two digging approaches are compared, scooping and raking. The few times I used a backhoe-like machine, about all I could do was ungracefully scoop soil out of a hole. Put the bucket in, rotate the bucket, pull it out of the hole, and some soil might come out. Professionals can actually control the bucket in different ways: scooping allows the operator to smear

---

[1] Kays (1980) gives infiltration rates in various land use types, along with watershed fractions for Charlotte, North Carolina. [2] Brown and Hunt (2010) compared raking and scooping in sand, loamy sand, and clay soils under dry and wet conditions.

the soil surface at the bottom of the pit, intentionally or not, but the perhaps more elegant approach called raking uses the bucket's teeth to score the bottom's surface, making it jumbled up and broken. Scooping is used for laying sewer and utility lines to deliberately compact the trench to prevent sagging and broken pipes. Raking, on the other hand, would be used for infiltration trenches and septic tank drain field lines.

When one is constructing an SCM specifically designed to maximize stormwater infiltration, the bottom of the pond or trench is critical: that interface needs to conduct water. The results of Figure 4.3 specifically concern excavating the last 30 cm of a pond.

Excavation was tested in dry versus wet soils, and the summary of extensive testing shows that working in wet soils should be avoided. In particular, digging in wet soils and scooping in wet clay soil inhibited infiltration. Figure 4.3B shows conductivity for the sandy soil site in Nashville, North Carolina, as a covariation with resulting soil density. To measure conductivity, roughly speaking, one saturates a sample with water and then pushes water through the sample using a constant pressure. Dividing the volume of water pushed through over time gives a flow rate, and that's the conductivity. Overall, there's a clear pattern of lower conductivity with higher soil density: compaction is bad. As should be expected, raking is better than scooping. A comparison between wet soil and dry soil is complicated by the fact that, despite being separated by just a couple meters, the dry soil was loamy sand (no more than 70–90% sand with just the "right" amount of silt and clay) while the wet soil was just sand.[1]

The overall, general conclusion is that raking while digging is best for enhanced infiltration, especially for wet soils. However, for clay soils the excavation should take place when the soils are dry.

Two additional points are that this study didn't measure the final infiltration rates of these stormwater ponds, and the fine particles in backfill gravel can affect infiltration. Hence, there's a regulation requiring washed gravel for infiltration basins.

Regarding the science behind trying to enhance infiltration, there has been extensive testing on infiltration rates and the effects of compost enhancement of soils in Seattle, Washington.[2] Much of this testing effort attempts to seek solutions that reverse the problems of compaction, a very important inhibitor to infiltration on both sandy and clayey soils (see Figure 4.2). Amending soils with compost helps establish vegetation and reduce runoff through increased storage and infiltration (one municipal study[3] showed a doubled water-holding

---

[1] Complete soil descriptions can be found at Canada's soil science glossary, sis.agr.gc.ca/cansis/glossary/, under "texture." [2] Pitt et al. (1999) is a massively long, data-rich paper on infiltration. Pitt et al. (2008) also discuss compaction's role on infiltration. [3] A report prepared for Redmond, California, by Harrison et al. (1997), cited in Caltrans (2009).

Groundwater recharge rates depend on soils.

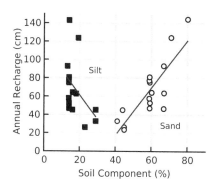

Figure 4.4: Annual recharge rates decrease with increasing silt fraction but increase with increasing sand fraction (Coes et al. 2007). This study, performed on North Carolina soils, confirms our sense that clayey soils (the particle size class below silt) become waterlogged and impenetrable to water.

capacity of soil amended by compost at a two parts compost, one part soil). Tilling 4 inches of compost into the top 12 inches of disturbed urban soils also helps infiltration, increasing it by 1.5–10.5 times more than that of unamended soils, and produces wonderful turf with no need for later fertilization.[1] These features combine to create a greater lag time for runoff (see Figure 13.9). No large differences are observed between the nutrient concentrations in runoff from soils with and without amendments, but the greater water retention of amended soils reduces export.[2]

Results in Figure 4.4, from seven coastal plain sites in North Carolina, confirm what we might think about infiltration and groundwater recharge: the more coarse particles and fewer fine particles, the higher the recharge rates.[3] Scientists here used several different methods, including measuring the water table depth in wells, dating water age using tracers, and measuring stream base flows.

Here in Durham, North Carolina, our soils sit near the lowest point where the silt and sand curves meet. Here's an interesting quote from Elisha Mitchell, for whom North Carolina's Mount Mitchell is named, that comes from an 1822 speech to the North Carolina Agricultural Society:

> The soil of this State [North Carolina] is pronounced, by those who have traveled extensively on both Continents, to be of a middling quality. It is of that kind which seems most to demand the employment of science and skill in its cultivation, and to promise that they shall not be employed in vain. Our grounds are neither so fertile that they will

---

[1] Pitt et al. (1999) [2] A California Department of Transportation study (Caltrans 2009) examined risks and benefits to adding compost to soils. [3] Coes et al. (2007).

Urban soils heal slowly.

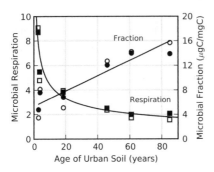

Figure 4.5: Urban soils from the northwestern U.S. recover slowly after urbanization, as indicated by microbial measures (Scharenbroch et al. 2005). Samples were taken in years 2002 (filled symbols) and 2003 (open symbols). Microbial respiration was measured as micrograms of released CO_2 per hour per milligram of microbial carbon.

produce spontaneously what is necessary to the sustenance and comfort of our citizens, nor so sterile that we have reason to abandon them in despair.[1]

A lot of silt, not so much sand, in a wonderful mix of clay— perfect for tracking backyard mud into the house during wet spells and getting one's truck stuck out in the woods, though Professor Mitchell wouldn't have known about the latter issue.

As we've just seen, constructing the urban environment compacts urban soils. However, urban soils can recover from the developmental insult of removed vegetation and topsoil, compression from heavy vehicles, and plantings of grass. Results from two northwestern cities, Moscow, Idaho, and Pullman, Washington, demonstrate that, in response to urbanization, microbial activity was initially quite high, gradually reducing to levels seen in other environments (Figure 4.5).[2] My use of the term "microbial activity" paraphrases their technical term, "metabolic quotient," and their units involve respiration from soil samples and estimates of relative microbial carbon content, two measurements much more complicated than what my simple phrase implies. At the later stage of soil recovery, soil respiration balances net primary production inputs through dead organic matter. Recently urbanized soils have relatively high respiration in response to disturbance. Older urban soils are better developed, with greater microbial carbon content with longer times since urbanization.

---

[1] Printed 60 years later in the North Carolina Department of Agriculture *Monthly Bulletin*, no. 15, 1882. Thanks to Stanley Buol, on whose webpage I found the quote. [2] Temporal development of urban soils in Idaho and Washington is described in Scharenbroch et al. (2005).

Besides a calmed-down microbial activity as urban soils age, organic carbon increases, and the results here show that microbial fraction increases, too.

All in all, humans can't forget soils when dealing with increasing temperatures and CO_2 concentrations: it's estimated that soils could sequester 0.6–1.2 petagrams (abbreviated as Pg and equaling 10^{15} g) of carbon per year with improved soil management, providing at least some offset to our increasing carbon dioxide levels.[1] As an order of magnitude calculation, let's say respiration by urban soils amounts to about 0.25 g of carbon per square meter per hour averaged over the day and year.[2] For the entire year, this value gives a total respiration of about 2 kg/m^2/year. If we were expecting a balance between plant growth and soil respiration, this number exceeds net primary productivity values in Figure 1.2.

Even compared with fossil fuel emissions, land use change contributes a lot of carbon to the atmosphere. Through burning fossil fuels, humans have released 270 ± 30 Pg of carbon over a century or three. At the same time, transformation from rural to urban land use increased soil respiration to the tune of 136 ± 55 Pg. Of this urban respiration, 78 ± 12 Pg has been released from the dark, organic matter stored in the ground, degrading urban soils. Respiration of soil carbon isn't a trifling amount.

Planting trees helps restore urban soils. Tests performed on SCMs designed specifically for enhanced infiltration through compacted urban soils examined how well tree roots performed.[3] Of particular interest were area-saving techniques that cover a deep gravel bed with asphalt to make a parking lot but then plant large trees for canopy shade and deep roots for enhanced infiltration. Greenhouse and field experiments mimicking the conditions within these SCMs showed that black oaks and red maples planted in compacted clay loam soil increased infiltration 153% over unplanted controls, and green ash trees broke through artificial subsoil barriers to increase infiltration rates by 27-fold. Careful design of these structures is warranted, however, because both transpiration and root growth are hindered by too slow an infiltration rate. In other words, the roots benefit from a kick-start in porosity to produce greater infiltration rates down the road.

Let's now consider the location of impervious surfaces. Data in Figure 4.6 originate from a study by the city of Olympia, Washington, and demonstrate that the type and fraction of imperviousness depend on the land use type. This plot compares high-density residential (e.g., single-family housing on small lots, 3–7 housing units per acre), multifamily apartments (7–30 housing units/acre), and commercial areas.[4] Considering nonresidential areas, the imperviousness of industrial areas is 75%, that of commercial areas is 85%,

---

[1] Lal (2003) discusses respiration and sequestration of carbon in soils. [2] Wilson (2011). [3] Bartens et al. (2008, 2009) examined infiltration by tree roots. [4] Data presented in Arnold and Gibbons (1996).

Impervious surface fractions depend on development type.

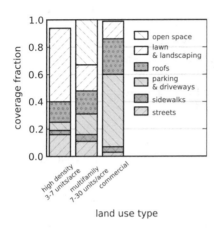

Figure 4.6: Land surface cover changes with land use characteristics (Arnold and Gibbons 1996). Impervious surface fraction is highest for commercial areas, dominated by roofs and parking lots. Coverage fractions do not sum to one.

and for shopping centers it is a whopping 95%, with a large part taken up by parking.

As a consequence of these fractional changes in imperviousness, precipitation has different partitionings in its fate, epitomized by oft-repeated numbers from the USEPA for differing areas.[1] In a natural area, precipitation partitions into 40% evapotranspiration, 10% runoff, and 25% each for shallow and deep infiltration; however, other estimates provide larger evapotranspiration fractions of 60% (see, e.g., Figure 2.5). At the other extreme, with 75–100% impervious the estimates change to 30% evapotranspiration, 55% runoff, 10% shallow infiltration, and 5% deep infiltration. These numbers also contrast somewhat with those of Figure 2.7, which emphasizes runoff rather than broader water cycle partitioning.

We've seen various estimates of how much pollution collects on these impervious surfaces in Chapter 3. Once deposited onto these surfaces, however, those particles don't stay put. Rain comes along, high imperviousness leads to high runoff, and some fraction of those pollutants wash off.

The results of Figure 4.7 considered the wash-off of the pollutants deposited in Figure 3.13.[2] Scientists built a fascinating set of artificial roofs and roads, pictured in their paper. A first, cursory look at the plot shows that the wash-off fractions exceed 1.0, meaning more pollution washes off than

---

[1] For example, Arnold and Gibbons (1996) reproduces a 1993 USEPA schematic partitioning precipitation fates. [2] Egodawatta et al. (2009) examined pollution deposition and subsequent runoff from roofs.

Sediments wash off with heavy rain.

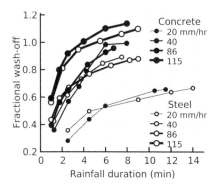

Figure 4.7: Pollutants deposit onto surfaces, and when the rain comes, an increasing fraction washes off with increasing rainfall duration and intensity (Egodawatta et al. 2009). Concrete and steel wash-off acts surprisingly similarly. Deposition variability leads to fractions exceeding one.

was there in the first place. Clearly that can't happen, and the study's authors attribute this absurdity to sampling errors and patchiness in how deposition takes place on the roof surface. In other words, it was not really possible to determine initial particle load accurately for each run, and spots where deposition was measured might have had lesser amounts than the locations where efficient wash-off actually took place.

This problem is a common one, and we encountered a similar situation with the interception of rain by trees (see Figure 2.4). Subtracting two numbers nearly equal in size, with appreciable uncertainties in their values, results in a small number having a large relative error.

One might expect from the slopes of these curves that the runoff from the earlier parts of a rain have more pollution than the later parts. For example, the intermediate intensities (the intermediate-size circles in Figure 4.7) wash off about 60% of the pollutants in the first 2 minutes of the simulated rainfall, but the 2 minutes between 4 and 6 minutes wash off only an additional 10–20% or so. The runoff from later periods of the rainfall should be cleaner. In fact, other studies also found such a so-called first-flush effect (see Figure 5.16)—especially in the first 1 mm of rainfall—and high metal contamination coming off of metal roofs.[1]

An extensive summary of pollutant measures from urban surfaces provides detailed data ranges for pollutant and impervious surface types arising from many papers.[2] Overall, as we've seen, high-traffic areas are worst. However, the data also show that roofs are problematic for many variables compared with rainwater, with metal roofs the worst, but even green roofs have elevated

---

[1] Förster (1996) measured runoff from experimental and real roofs in Bayreuth, Germany.   [2] Göbel et al. (2007) summarized runoff pollutant concentration studies.

Smaller particles and more sediment
with more imperviousness.

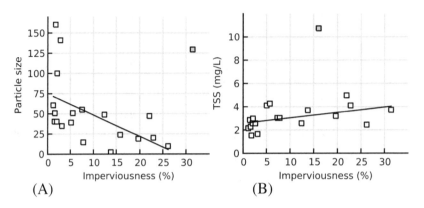

(A)   (B)

Figure 4.8: Sediment characteristics change with imperviousness in southern
and central Maine watersheds (Morse et al. 2003). Outliers excluded from the
regression and marked by gray squares included an airport (A) and an ongoing
highway construction site (B).

heavy metals. Each type of road—freeways, main roads, service roads, parking
lots, even bike paths—has a unique pollutant signature.

Another particle size study arose as a result of the USEPA being tasked
with finding low-cost and fast methods of assessing the biological conditions
of streams and lakes.[1] Rapid bioassessment protocols (RBPs) arose from this
demand, and the USEPA published an extensive manual that serves as a major
source for biological index information and detailed procedures.

Many aspects contribute to these protocols, and Figures 4.8 and 4.9 show
the dependencies of three important factors on imperviousness. A study carefully chose 20 watersheds in south and central Maine from a collection of
300 considered, using as selection criteria various stream attributes, such as
width and depth, flow rate, substrate, and pollution sources.[2] An additional
selection criterion was imperviousness, and sets of comparable watersheds
with low imperviousness (<5%, and 60–90% forest cover), and medium
to high imperviousness (up to 31%), as determined from 1991–1998 aerial
photos, were selected.

---

[1] Barbour et al. (1999). [2] Morse et al. (2003) studied imperviousness in Maine streams, selecting
perennial streams with cobble–riffle habitat, width < 8 m, depth < 80 cm, and gradient < 4%,
among other criteria.

Stream nutrients increase with imperviousness.

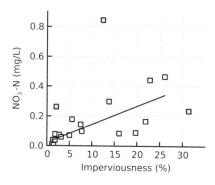

Figure 4.9: Streams from watersheds having higher imperviousness have higher stream nitrate concentrations (Morse et al. 2003). Three sites spanning the range from forest to a severely impacted airport watershed were excluded from the regression line (gray squares).

Researchers performed an extensive and detailed examination during the years 1998–1999 of water quality and insects. Various attributes included specific conductivity (μS/cm), qualitative habitat index, average riparian width, and median particle size (mm) along with biodiversity surveys. Figure 4.8 shows that particle size decreases and total suspended solids (TSS) increase with imperviousness, together meaning that urban streams have lots of silt. Not surprisingly, high imperviousness also caused urban stream syndrome, with narrow stream width and high erosion of stream banks (see Figure 5.14).

Another aspect to these rapid assessments involves land cover correlations, and to that end have arisen state "Gap Analysis Program" (GAP) land classifications—and the meaning of "gap" is important.[1] Imagine all the species out there to conserve in some local area. One hopes that the land conservation programs in that area preserve all of those species, but that's certainly not the case. Thus, there is a "gap" between all of the potential species and those conserved species. GAP land classifications are an attempt to focus on important habitats and lands that fill in the gaps.

As a prelude to the treatment of nutrients in Chapter 6, results of Figure 4.9 depict the concentrations of nitrate in Maine streams,[2] with a very simple overall result: there's more nitrate in the streams of watersheds that have more imperviousness. For rapid assessments, chlorophyll a concentration serves as a proxy for nutrients, though other factors affect that measure, too.

Figure 4.9 shouldn't be taken to mean that factors other than imperviousness aren't important for nutrients. A study from the 1,840-km^2 Cayuga Lake watershed in New York sampled stream nitrate concentration across several sampling periods and found that the most important variable was the

---

[1] See USGS, gapanalysis.usgs.gov. [2] Morse et al. (2003).

Growing cities pipe their stormwater.

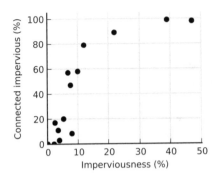

Figure 4.10: As a watershed's impervious surface fraction increases, an increasing percentage of that surface is directly connected to streams, as shown in this example from Melbourne, Australia (Hatt et al. 2004). Most of the variation takes place at 10% imperviousness.

percentage of land in row crops.[1] Land in row crops increased nitrates, but in some cases soils with 10–20% clay correlated with decreased nitrates, along with percent open water and percent developed land. Across the periods, nitrate levels were highly variable in time and space, suggesting a complicated dependence on many parameters. Effects of cropland subsided about 5 or 6 km from the crops, and that reduction demonstrates the importance of riparian buffers, found in agricultural areas with Good Agricultural Practices, (the origins of another GAP acronym), as discussed in Chapter 12.

There's an important historical and urban developmental aspect to imperviousness and stormwater systems. Long before we worried about stormwater and impervious surfaces, people just built structures and roads, and there was no stormwater system. Rain was just rain, and what little runoff there was just infiltrated into nearby natural cover. But as a city gets bigger and denser, it gains more impervious surface, and people learn that even small rains lead to large volumes of runoff. That runoff starts causing floods and water damage and represents a really big problem.

With eyes on the symptom and not the problem, the initial focus centers on getting all that collected water out of spaces where it does harm. That concern with massive water volumes dictates developing stormwater infrastructure for the express purpose of quickly eliminating a high volume of water. Figure 4.10 shows the development of a stormwater system as watersheds become urbanized, with the data coming from 15 small watersheds in the Dandenong Ranges east of Melbourne, Australia.[2]

This study quantified impervious surface using the two different measures, total and connected, discussed in Chapter 2. Data in Figure 4.10 demonstrate that, at least in that region, once a watershed gets to 20–30% imperviousness,

---

[1] Golden et al. (2009). [2] Hatt et al. (2004).

essentially all of that impervious surface has been connected to a storm-water system, with all the water volume flowing through pipes. An additional analysis of streamflow data and pollutant concentrations demonstrated that connected impervious surface provided the strongest correlation with TSS, phosphates, and nitrogen, all of which increased dramatically over this range of imperviousness.[1]

This study concludes that the disconnection of imperviousness is a great solution to stormwater problems, and rainwater harvesting provides one pos-sibility. Of course, disconnection also requires a solution that doesn't get us back to the original problems cities faced with flooding.

Many cities treat stormwater separately from sewage, while others have combined sewer systems; for example, in New York City stormwater and sewage go through a combined water treatment system. Here's the problem: no one can reasonably expect wastewater treatment plants to handle a couple inches of rain over a few square kilometers of impervious surfaces during a 2-hour rainfall—the water volumes are just too immense. The result is "com-bined sewer overflows," (CSOs) where diluted, untreated sewage discharges from the wastewater treatment plant. That's problematic, but the dilution itself is considered a type of sewage treatment. One aspect of a combined system is helpful, however: the first 0.1 or 0.25 inch of runoff (called the first flush) has higher pollution levels and warrants special treatment, perhaps by passage through wastewater treatment plants.

Chapter Readings

Arnold, C.L., Jr., and C.J. Gibbons. 1996. Impervious surface coverage: The emergence of a key environmental indicator. *J. Am. Plann. Assoc.* 64: 243–258.

Gregory, J.H., M.D. Dukes, P.H. Jones, and G.L. Miller. 2006. Effect of urban soil compaction on infiltration rate. *J. Soil Water Conserv.* 61: 117–124.

Hatt, B.E., T.D. Fletcher, C.J. Walsh, and S.L. Taylor. 2004. The influence of urban density and drainage infrastructure on the concentrations and loads of pollutants in small streams. *Environ. Manag.* 34: 112–124.

Kaushal, S.S., P.M. Groffman, G.E. Likens, K.T. Belt, W.P. Stack, V.R. Kelly, L.E. Band, and G.T. Fisher. 2005. Increased salinization of fresh water in the northeastern United States. *Proc. Nal. Acad. Sci. USA* 102: 13517–13520.

---

[1] Walsh et al. (2005a, 2005b) provide in-depth coverage of urban stream health.

Scharenbroch, B.C., J.E. Lloyd, and J.L. Johnson-Maynard. 2005. Distinguishing urban soils with physical, chemical, and biological properties. *Pedobiologia* 49: 283–296.

Schueler, T.R. 1994. The importance of imperviousness. *Watershed Protect. Techn.* 1: 100–111.

Part II
Environmental Harms

Chapter 5

Urban Runoff

Impervious surfaces prevent the infiltration of precipitation into soils, diverting the water downhill. Large fractions of impervious surfaces in a watershed result in high flows of short duration, called "flashy flows," causing damage to streams and riparian areas. Along with imperviousness and flashy flows come reduced water storage throughout the watershed, which reduces the replenishment of a stream's base flow between rainfalls.

Differences in streamflow patterns are observed across watersheds with different land use types, and within single watersheds as urbanization takes place across several decades. In contrast, peak flows decrease and base flows increase with the amount of natural cover in a watershed.

Runoff models predict that the peak flows of fully urbanized watersheds are roughly six times greater than watersheds in natural land cover, and these high peak flows occur with higher frequency, in response to relatively small rainfall events. High peak flows cause the erosion of streams simply because a larger volume of water requires a larger stream to carry that water. Stream erosion leads to downstream sediments that harm ecosystems of streams, lakes, and oceans. In response to greater erosion problems, urban streams have much greater sedimentation problems than in rural watersheds. The faster flows also carry these sediments farther downstream, and small urban watersheds have greater sediment export than do large rural ones.

Urban streams are flashy.

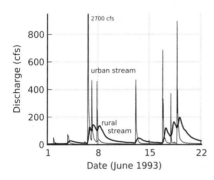

Figure 5.1: Imperviousness increases flashiness (S.R. Corsi, pers.
comm, after Masterson and
Bannerman 1994). Very different
streamflow regimes take place in
nearby urban and rural watersheds
around Milwaukee, WI, during the
same series of rainstorms.

Let's begin our examination of urban runoff by realizing how the time
course of stormwater runoff changes with imperviousness. Figure 5.1 shows
the flows in two nearby, but different, streams that result from a series of
rainfall events. The two streams drain nearby urban and rural watersheds
in and around Milwaukee, Wisconsin.[1] For these 1993 data, the rural stream
data come from measurements at Jackson Creek at Mound Road near Elkhorn,
Wisconsin, whereas the urban stream data come from Lincoln Creek at 47th
Street in Milwaukee. The latter stream sits in the middle of a greatly urbanized
watershed, while the former drains a mostly agricultural area about 40 miles
to the southwest.

Figure 5.1 shows spiky profiles from the urban watershed but muted
profiles from the rural watershed. We'll see the same thing in Figure 13.4
for low-impact developments that seek to ameliorate this flashy-flow situation. When a drop of rain falls in a rural watershed, it has a long flow path,
with infiltration starting right where it hits the ground. It takes time for the
water to make its way over land or through soils and into the stream. Just as
important, a raindrop that falls far from the stream spends a much longer time
getting to the stream than one that falls close by.

In the urban watershed, however, a raindrop falling on a parking lot connected to a storm drain connected to a pipe that smoothly and quickly transports
the water to the stream takes no time at all. All that rain falling on connected
impervious surfaces, both near and far from the stream, quickly gets to the
stream, and that rapid removal results in a large volume of water flowing into
urban streams. Indeed, stormwater systems were designed to rid cities quickly
of stormwater. Figure 5.1 shows how successful that task is achieved!

---

[1] Masterson and Bannerman (1994).

Urbanization reduces base flows.

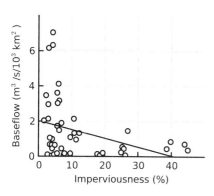

Figure 5.2: Base flows in streams decrease as a watershed becomes more impervious (Wang et al. 2001). These data come from a survey across 47 small watersheds in southeastern Wisconsin. Flows are scaled by watershed size.

The word to describe this large, transient water volume in urban streams is "flashy." We'll see further on in this chapter some of the consequences of flashy flows, including degraded streams, high sedimentation, and poor water quality.

The flip side of flashy flows, or the "direct flows" more generally,[1] is a lack of "base flows," or the water that runs in a stream between storms. Base-flow reductions are about as bad for streams as flashy flows, and we'll spend time examining that issue as well. As an example, Figure 5.2 depicts the results of a study of 47 small southeastern Wisconsin watersheds ranging in size from 10 to 100 km^2.[2] In this flat, fertile area of Wisconsin, 28 of the studied watersheds flowed to Lake Michigan, and 19 drained to the Mississippi River. Land use was primarily either urban or agricultural, with a little bit of forest and wetlands, but both natural areas constituted less than 10% of the area on average. Results shown here are but a small part of a study that examined the effect of land use type on biological integrity in streams. Detailing land use cover into 16 types, the most important type by far was connected imperviousness (see the discussion of Figure 11.15) in explaining biodiversity-related features, and a number of development-related land use features hindered base flows.

Figure 5.2 links flashy flows with lower water tables. The flashy flows of developed watersheds, depicted in Figure 5.1, successfully move water downstream very fast but reduce infiltration through the soil and into the groundwater. Since there's less water in the ground, less water moves from the ground into streams during the dry periods. Hence, reduced base flow occurs with increasing impervious surface fraction.[3] On the other hand, features that

---

[1] Meyer (2005). [2] Wang et al. (2001). [3] Wang et al. (2001) more carefully examined nonlinear regressions of the 90% levels of variation, which showed a stronger correlation with connected imperviousness than the simple linear regression shown in Figure 5.2.

Roads reduce wet-season base flows.

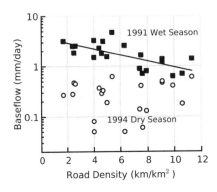

Figure 5.3: An increasing density of roads in a watershed decreases the wet-season base flow in the Puget Sound of Washington State (Konrad and Booth 2005). During the dry season, however, base flow depends on groundwater seepage, which appears uninfluenced by road density.

increased base flows included wetlands, surface water, riparian forests, and vegetated land, meaning that a watershed with a lot of water storage capacity preserved the base flow of streams.

In summary, with a rural stream we have a reasonably sustained base flow from water trickling out of the soils of a pervious watershed, and when a rain comes along it recharges the soils with water while overland flows might bump up streamflows. An urban stream, however, has very little base flow, and when a rain comes, the huge volumes of water go directly from stormwater systems into streams, not into the soil, and then shortly after the rain ends, the flow stops.

As we know from Figure 3.20, roads and imperviousness are correlated as attributes of urbanization. More than 20 subbasins of Puget Sound in western Washington representing a range of urbanization levels and road densities were measured in rainy and dry seasons as part of an extensive monitoring program for the watershed of Puget Sound near Seattle, Washington.[1] However, these researchers argue that the roads affect shallow subsurface groundwater flows but not the recharge from deeper-level groundwater. This idea is borne out from Figure 5.3. During the wet season, more roads mean lower base flow, just like we see with imperviousness (see Figure 5.2). What happens is twofold: first, stormwater is converted to flashy peak flows, and second, the road cuts interrupt flows along the shallow pathways to the streams and force water to percolate into deeper groundwater depths. In the dry season base flows are determined by groundwater seepage into streams, independent of storm-flow hydrology.

[1] Konrad and Booth (2002). Konrad and Booth (2005) summarize many aspects of the data set, including various statistical properties of streamflow.

Chicago's base-flow fraction decreased over the years.

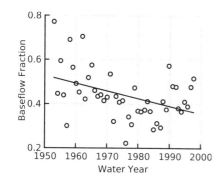

Figure 5.4: In response to flashy peak flows and reduced base flows with urbanization, the McDonald Creek watershed of Chicago, IL, shows a decreasing fraction of total stream-flow occurring as base flow (Meyer 2005). Two other nearby water-sheds demonstrated similar base-flow changes.

Don't let the rather muted logarithmically scaled axis in Figure 5.3 detract from the numbers: wet-season base-flow reductions go from levels of 2–3 mm/day with few roads down to around 0.8 mm/day, a roughly 60% decrease.

Perhaps there's an analogy to climate change: yes, average temperatures are warming, but there are many aspects to changing climate, like the date of first frost and average winter temperature minimums (see Chapter 2). Like-wise, urbanization affects many aspects of water flow in watersheds, more than just annually averaged peak flow and base flow in streams. Effective urban stormwater management replicates the volume of water storage repre-sented by the original soils and land cover, distributed widely across the urban landscape. That replication helps restore the streams to their "natural" flow patterns.[1] Roads affect one particular stormwater attribute differently from general imperviousness.

Figure 5.4 shows results from one of three watersheds studied in Cook County, Illinois, which is better known as metropolitan Chicago.[2] In these watersheds, urban land cover went from about 20% in the 1950s to over 90% by the mid-1970s, with impervious fractions hovering around 45% in the late 1990s. In this work, the flow immediately after a rainfall event was termed "direct flow," but it means the flow resulting directly from the storm as opposed to indirectly via groundwater.

Using flow data like those shown in Figure 5.1, the researcher devel-oped a computer algorithm separating those plots into direct flow and base flow. Figure 5.4 shows that a greater fraction of the streamflow took place as

---

[1] Konrad and Booth (2002). [2] Meyer (2005).

direct flow as the city developed over the decades, leaving behind less base flow. However, results also show that the annually averaged base flow hasn't changed, but the mean base flow during dry months and the median base flow *have* increased. These results indicate that water is constantly added to streams via sewer leaks, outfalls, irrigation, and so on, even during the summer months when no to little base flows are expected. It's also possible that some water release from detention ponds was classified as base flow and that stream deepening has increased groundwater inputs by draining riparian areas. Still, the fraction of the streamflow characterized as base flow decreased by 0.3% per year. That means the streams became flashier. And, of course, the increased precipitation with climate change shown in Figure 2.3 could increase average annual flows, but no such increase was observable in precipitation at this local scale.

Another study involving base flows was done in western Georgia in 18 watersheds ranging in size from 500 to 2,500 hectares.[1] Watersheds were categorized as urban, developing, pasture, managed forest, and unmanaged forest, with three or four watersheds in each category. Urban watersheds range from 25% to 42% imperviousness, with less than 4% in the others.

Researchers used a very long time series of streamflow from each of these watersheds, much like the one shown in Figure 5.1. A rather complicated procedure determined baseflow index values, which represents the fraction of the total streamflow coming from groundwater. A value of zero means the stream discharges only surface runoff, whereas a value of one means all of the flow comes from groundwater-sourced base flow, with no precipitation-induced surface flow. As we have come to expect, urban watersheds have low base flow indices, and at the opposite end, forested and pasture watersheds have high values (Figure 5.5). One of the forested watersheds had been recently clear-cut, leading to higher runoff values (see, e.g., Figure 2.5) and a low base flow index value.

Along with the base flow in the Georgia watersheds of Figure 5.5, extreme streamflow events were measured. Using the time-series data, researchers first determined the median streamflow (50% of values sit above and below the median value) and then counted the number of events exceeding *seven* times the median flow. Thus, "extreme" means those flashy, high-volume runoff events.

What Figure 5.6 demonstrates is that hand in hand with a low base-flow index comes high flashiness. And as we've seen, high flashiness and low base flows occur in urban streams. In pastures, by contrast, precipitation infiltrates into the ground, producing few extreme events but sustaining base flows through groundwater discharge.

---

[1] Schoonover et al. (2006) studied watersheds in Georgia.

Urban streams trickle.

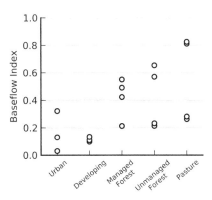

Figure 5.5: Base-flow index refers to the fraction of streamflow that comes from groundwater, as opposed to stormwater runoff. Urban streams are mostly flashy runoff, but watersheds with forests and pastures infiltrate precipitation, which emerges elsewhere as the groundwater discharge that supports base flows (Schoonover et al. 2006).

Flashy streams are trickling streams.

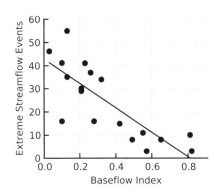

Figure 5.6: Like the paired urban–rural Wisconsin watersheds in Figure 5.1, Georgia watersheds demonstrate that flashy streams create low base-flow conditions (Schoonover et al. 2006).

Not all urban areas show decreased base flow with increasing urbanization. A study across North Carolina comparing 30 years of streamflow data spanning 1960–1990 for closely situated urban and rural stream base flows was rather inconclusive.[1] This particular examination of a dozen sets of paired urban and rural streams didn't bear out the hypothesis that the urban streams would have a greater decreasing base flow trend with time than the rural streams.

The results of Figure 5.7 are quite old, from 1968, with even older citations, but that just demonstrates how long stormwater has been considered within the scientific literature (compared with the onset of significant stormwater

---

[1] Evett et al. (1994) studied urbanization and base flows in North Carolina.

Urbanized watersheds reach sixfold higher peak flows.

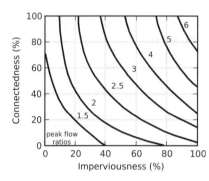

Figure 5.7: Expected urbanized-to-natural peak-flow ratios for differing amounts of imperviousness and connectedness of that imperviousness to stormwater systems (after Leopold 1968). Tabulated data were transformed into approximating curves.

regulations).[1] The data come from a government publication by the U.S. Department of the Interior, which might be one of the best basic discussions of urban stormwater and provides an excellent discussion on the implications of urban land use on hydrological flows, with a special emphasis on sediments. Indeed, the most important concern of the publication was the flashiness caused by impervious surfaces, and it took great care to provide a very clear explanation of the decreased lag time (like that shown in Figure 13.9 for rain gardens) and higher peak flows (like that of Figure 5.1) of urbanized watersheds.

The particular set of curves in Figure 5.7 considers the peak flows observed over a 1-year period for a stream in a watershed with varying imperviousness and connectedness. The curves represent constant ratios for the peak flow of the developed condition to natural condition for a watershed. For example, Figure 5.1 shows peak flows for the rural stream hitting 150 cubic feet per second (cfs), whereas the urban stream usually hits around 600 cfs, giving a rough factor of 4. Those two watersheds were roughly equal in size, and to facilitate comparisons across differently sized watersheds, these curves normalize flows relative to a one-square-mile drainage area, along with some other important factors like average slope and the volume of rain falling across an entire watershed.

Figure 5.8 is closely related to Figure 5.7. Curves here depend on the same variables, the amount of impervious surface in a watershed and its connectedness to stormwater systems, given in each curve's pair of numbers. The first number represents the percentage of the watershed's impervious surface

---

[1] Leopold (1968).

Side effects include frequent discharges.

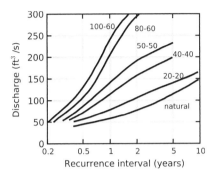

Figure 5.8: Frequencies of peak stream-water discharges, scaled to a 1-km^2 watershed, given the fraction of connected impervious surface and the watershed imperviousness. These results arose from studies in Maryland but generalize to other regions (Leopold 1968). Recurrence interval means the expected average time between events of a certain size.

connected to storm sewers; the second number is the percent imperviousness of the watershed.

Now imagine measuring streamflows coming from a set of watersheds. How often do these different flow (or discharge) values occur? Just appeal to your own experiences, or perhaps look ahead to Figure 5.10: small storms are frequent, and because they're small, they have low discharges. Big storms come along every once in awhile and have really high discharges, and we call them floods, thus the basic shape of the curves in Figure 5.8. Small storms with low discharge have short recurrence intervals and extend into the lower left corner; large storms with high discharge have long recurrence intervals and reach into the upper right corner.

Indeed, the points at a recurrence interval of 1 year match the curves of Figure 5.7. Discharge for a natural stream is about 50 cfs, and for the 80–60 curve the discharge is 200 cfs. These values give a ratio of 4, right about where the 80–60 point sits in Figure 5.7.

Even back in 1968 there was a big emphasis on "storage," just like the goal of modern goals for low-impact development. Rivers rise and flood, and you can think of that situation as adding water storage at a downhill location. To prevent rivers from flooding, we've added reservoirs that fill up, replacing flooded river storage a bit farther upstream. But damage is done by the time the water gets to the reservoirs, so we can add ditches, cisterns, check dams, artificial ponds, and so on, which all add more and more storage even farther upstream. All of that upstream storage helps decrease peak streamflows.

Figure 5.9 shows the net effect of increased runoff coefficients with increasing imperviousness. Figure 5.9A shows a comparison of two nearby drainage areas within 1 km of one another, one undeveloped with a watershed of size 3.3 hectares with woods, fields, and an intermittent stream with riparian

Developed land has more runoff.

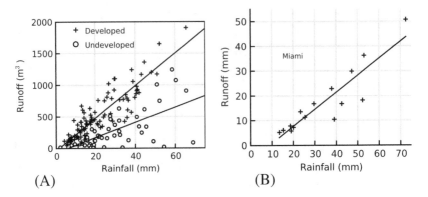

Figure 5.9: (A) Runoff from a developed watershed in North Carolina is high compared with a paired undeveloped one (Line and White 2007). (B) Total runoff increases with total rainfall for a small watershed in urban Miami, FL (Lee and Heaney 2003).

buffers >30 m wide, while the other drainage area covered 4 hectares with 32 single-family homes.[1] The latter bunch of homes is a small suburb called Carpenter Village, in the Piedmont area of North Carolina, just outside Raleigh. The study period lasted nearly 6 years, and one-quarter of the developed drainage was built during that period. Once construction finished, total imperviousness was 53%, with 31% directly connected to stormwater systems.

Average peak runoff from the developed area, 5,573 L/min, was high compared with that from the undeveloped area, 1,321 L/min. Highest peak discharge rates observed over the study period were 28,290 L/min from the developed area, compared with 4,600 L/min from the undeveloped area. Overall, the developed area had 68% greater runoff volume than the undeveloped area, and 95% more total suspended solids and 66% more nitrogen were exported. Most of the suspended solids export took place during the clearing and grading just prior to building, which emphasizes the need for sediment controls during the housing development process.

This connection between runoff and imperviousness was used to great effect in a study that focused on the Kings Creek area in Miami, Florida, a nearly 6-hectare watershed with high-density residential housing. In this development, the directly connected impervious fraction was 44%, but this

---

[1] Line and White (2007) compared runoff from developed and undeveloped drainage areas.

Runoff increased while precipitation was unchanged.

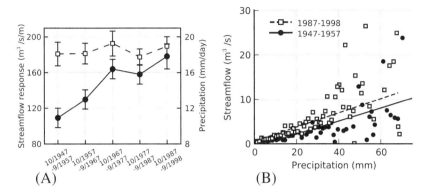

(A)   (B)

Figure 5.10: (A) As a watershed near Annandala, VA, became developed with housing over the decades, streamflow response to rainfall increases (filled circles) despite relatively constant precipitation (open squares). Data include only days that have more than 6 mm of precipitation (Jennings and Jarnagin 2002). (B) Streamflow response to precipitation increased after four decades of urbanization, with a higher frequency of high-streamflow events (T. Jarnagin, pers. comm.).

part of the impervious surface disproportionately contributed 72% of the total runoff.[1] Researchers used a set of storms to parameterize a runoff–rainfall model, using data from Figure 5.9B, and this model was then used with 52 years of rainfall and runoff data to produce Figure 2.7 depicting the various fates of rain. Initial abstraction, or "depression storage," was always 1/10 inch (2.54 mm), and about 43% of storms were below this rainfall depth, meaning there was no runoff at all.

Figure 2.3 shows increased precipitation as a result of climate change. On the basis of that plot, some folks might propose the idea that climate change leads to more rain, which leads to more runoff, and that's the source of stormwater problems. Figure 5.10A shelves that potential argument against the responsibility of urbanization actually leading to our stormwater problems. Using aerial photography spanning 1949–1994, researchers connected impervious cover, streamflow, and precipitation in the Accotink Creek sub-watershed near Annandala, Virginia.[2] During that time, total impervious area

---

[1] Lee and Heaney (2003) studied the runoff–rainfall relationship from the Miami, Florida, data.
[2] Jennings and Jarnagin (2002).

increased 10-fold, from 3% to 33%. In response, normal and extreme flow events increased by 50% and 100%, respectively. In Figure 5.10A, streamflow response measures stream discharge (m^3/sec) per meter of precipitation on the days that precipitation exceeds 6 mm, which avoids rainfalls below initial abstraction, for which we don't expect any runoff. Overall, precipitation shows no change over the five decades, but stream discharge increased greatly.

Figure 5.10B gives a feel for the scatter and distributions of precipitation and streamflow. I plotted the first decade (1947–1957, circles) and fifth decade (1987–1998, squares) and averaged all flows in 1-mm-precipitation bins.[1] Some data sit outside this plot area, but they're pretty sparse, and the lines are a simple linear regression excluding bins with single events, which are mostly very large rainfalls. Large rainfalls are quite rare: for the latter decade, nearly 1,000 precipitation events sit below 10-mm depth, but fewer than 40 events exceed 40 mm. The latter decade's regression line has a sharply steeper slope when the very large rainfall events are included. Comparing the two decades, development over this time span produced much larger streamflows coming from the large precipitation events. For example, the number of flows exceeding 14 m^3/sec increase more than threefold between the first decade and the fifth decade. As we'll see elsewhere (e.g., Figure 5.14), these flushing events cause great harm.

Further demonstrating the increase in runoff, Figure 5.11A comes from a survey of 125 neighborhoods with 16 major land use areas in the Little Shades Creek watershed, near Birmingham, Alabama, holding more than 5,000 acres.[2] A detailed analysis of land cover types folded together runoff features and several decades of precipitation data to produce the estimated runoff fractions shown here. The greatest portion of the watershed, by far, was single-family residential, at 3,611 acres, followed by vacant land at 989 acres. There was seemingly little difference between sandy and clayey soils in terms of runoff fraction, except at the very low impervious surface fractions. That simply means that once impervious surfaces dominate runoff, it doesn't matter whether a parking lot sits on top of sand or clay.

Data in Figure 5.11B come from more 40 sites across the U.S.[3] There's a stark difference in these numbers once you take a moment to calculate the consequences. For example, if we assumed a 1-inch rainfall, the runoff from, say, at the low end of imperviousness, a 1-acre meadow with runoff fraction of 0.06, would produce a water volume of 218 cubic feet[4], which could flood a 10 × 10 foot room with more than 2 feet of water. This is bearable compared with the runoff from a parking lot at high imperviousness, with runoff fraction

---

[1] Taylor Jarnagin kindly provided their flow and precipitation data from which I made the plot.
[2] Bochis-Micu and Pitt (2005) examined runoff characteristics from neighborhoods in Birmingham, Alabama. [3] Schueler (1994) provides a nice and clear discussion of imperviousness and its effects. [4] $(43,560 \text{ ft}^2) \times (1/12 \text{ ft}) \times (0.06) = 218 \text{ ft}^3$.

Runoff fraction increases with imperviousness.

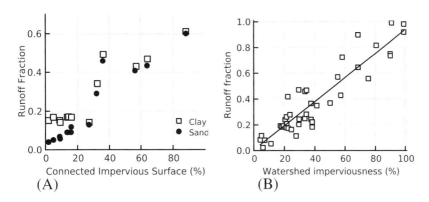

(A)   (B)

Figure 5.11: (A) Fraction of precipitation that runs off a site increases with connected impervious surface fraction, measured across various land use categories in Birmingham, AL (Bochis-Micu and Pitt 2005). Soil type (sandy vs. clayey) affects runoff fraction only for low imperviousness. (B) Similar runoff versus imperviousness results are observed across watersheds spanning the U.S. (Schueler 1994).

---

of 0.95, which would produce more than 3,400 cubic feet of water, filling that room 34 feet deep! It's this high-volume flush of stormwater that erodes streams (see Figure 5.14).

Another consequence of increased runoff is decreased groundwater recharge. A study of the urbanized Peachtree Creek watershed situated in the Piedmont around Atlanta, Georgia, brought together all of the issues discussed so far.[1] This urban watershed experienced peak flows 30–100% higher than other less urban or nonurban streams. Peak flows also drop off 50–100% quicker (meaning they're flashy, like in Figure 5.1), and base flow values were 25–35% lower. Disregarding the seasonal variation in groundwater levels, urban groundwater levels decreased over 30 years compared with nonurban groundwater levels. However, these reductions are small compared with some locations. Because of high withdrawal rates in Delhi, India, groundwater levels have dropped 8 m in 20 years, and in the central portion of Dhaka, Bangladesh, levels have dropped as much as 3.5 m/year.[2] In addition to drastically lowered water tables, aquifers in these cities are experiencing higher chloride levels due to sea water intrusion.

---

[1] Rose and Peters (2001) used USGS data spanning 1958–1995 for Atlanta area streams. [2] Haque et al. (2013).

More rain, more runoff.

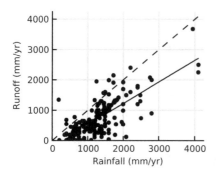

Figure 5.12: Runoff from rivers across the globe increases with regional rainfall (after Alvarez-Cobelas et al. 2008; M. Alvarez-Cobelas, pers. comm.). The dashed line represents runoff equal to rainfall; the slope of the fitted line is 0.71, meaning that 71% of the rainfall leaves a watershed as runoff.

Figure 5.12 presents rainfall–runoff data collected from papers involving 946 rivers worldwide, a massive amount of data.[1] Shown here is a subset of that published data, specifically the original data compiled by the authors themselves. A simple linear regression gives a slope of 0.71, meaning that 71% of the rainfall exits the watershed. This value does not comport well at all with the 60% evapotranspiration estimate (Figure 1.2). Indeed, some catchments even show more runoff than rainfall. Within a catchment, runoff amounts can exceed rainfall for a couple of reasons. First, the groundwater flows might not reflect the surface elevations that characterize the watershed boundaries, transferring water originating as precipitation in one watershed as the water that appears in the stream of another watershed. Second, municipal water systems might import water from one watershed, and that water ultimately exits as wastewater, treated and released in another watershed's streams. Finally, though not affecting excessive runoff estimates, some of these watersheds are on the order of tens of thousands of square kilometers, and precipitation might be double-counted. Perhaps some precipitation lost through evapotranspiration at one location once again falls within the same watershed in a follow-on precipitation event.

At least some current analysis of urban runoff is based on a simple mathematical device called "curve numbers," abbreviated CN.[2] Figure 5.13 shows these curves as calculated rainfall–runoff relations that mimic real data, like that observed in Figure 5.9. The curves arise from a phenomenological equation,[3]

---

[1] Alvarez-Cobelas et al. (2008). Supplementary data provided online by the journal and by Miguel Alvarez-Cobelas. [2] Current version of USDA Technical Release 55 (TR55); it and other tools from the USDA available at go.usa.gov/KoZ. [3] Plummer and Woodward (1998).

Rainfall–runoff curves.

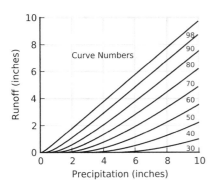

Figure 5.13: Rainfall–runoff curves identified by a so-called curve number allow simple characterization of the runoff consequences of a multitude of watershed features (after USDA 1986). Forested watersheds have low numbers, whereas developed watersheds have high values.

$$Q = \frac{(P - I_a)^2}{(P - I_a) + S}, \tag{5.1}$$

where Q is the runoff, P is the precipitation, I_a is the initial abstraction, and S is the maximum storage of the watershed. Two limits are revealing. First, if the rainfall P is less than the initial abstraction, I_a, then there simply is no runoff, so we start with $Q = 0$ for $P \leq I_a$. Next, for very heavy rainfalls where $P - I_a$ greatly exceeds S, Q becomes linear and approaches $P - I_a$.

The parameters I_a and S are not so simple and contain many assumptions.[1] One is that the initial abstraction (including interception and other factors) accounts for 20% of the watershed's storage. That assumption simplifies the relation to

$$Q = \frac{(P - 0.2S)^2}{P + 0.8S}. \tag{5.2}$$

At this point, professionals rewrite the storage variable in terms of CN using the equation

$$S = \frac{1000}{\mathrm{CN}} - 10. \tag{5.3}$$

If we think of S as inches of stormwater storage in the watershed, we get 0 inches with CN $= 100$, 1.11 inches for CN $= 90$, and 90 inches for CN $= 10$.

---

[1] Plummer and Woodward (1998).

By definition, fully impervious areas have CN = 98, which implies 0.2 inches of storage. In this way a single parameter, the curve number, can wrap together and describe many factors, including soil type, wet or dry soils, and imperviousness.

Another aspect of characterizing stormwater processes is the concept of a "design storm." Because all storms differ in when, where, and how fast precipitation falls, there is a need to standardized the performance of stormwater control measures against some benchmark storm. Here is where a design storm comes to play.[1] One could imagine a storm building up from a slow rain to a deluge and then dwindling back down to a drizzle, or a huge downpour that quickly disappears, or a constant, slow rain for several days. What sort of assumptions about an area's storms should be made when designing stormwater control measures (SCMs)? Using historical precipitation patterns, the statistical information can be pulled out to predict general runoff patterns and specific features like peak flows. Given the complications of a watershed, and how brief, local rainfall events within broader watershed-scale events can produce localized high flows, different assumptions for constructing design storms lead to different watershed-scale flow results. These results often determine which SCMs are chosen and the specific design criteria.

I've mentioned that flashy flows harm streams. Figure 5.14A depicts an example of stream incision (washout formation) below a suburban area in King County, Washington.[2] High flows eroded other streams in the area by several feet per year for at least three decades, leaving channels tens of feet deep cut into the downstream areas. Besides the damage to areas nearby, that incision means a tremendous amount of sedimentation.

You can't expect a big stream to fit in a small channel, and the incision process simply equilibrates a new, higher streamflow due to urbanization with this expectation, as shown in Figure 5.14B. Measured stream cross-sectional area increases with 2-year discharge rates as estimated under present land use conditions (or conditions from 20-some years ago, anyway).[3] At that time discharges were expected to double as development continued, resulting in an additional 75% increase in stream channel size, meaning tens of centimeters in depth, and 1–2 m or two in width.

These results highlight the damage that's been done to streams from the high flows that come about from urbanization. Thoughts have turned to restoring these streams, and that restoration process must take note of three different aspects.[4] First, as shown previously, what was once a natural stream becomes hydrologically and geomorphically simplified, which includes a straightening

---

[1] Nnadi et al. (1999) discuss design storms. [2] Booth (1991) examined stream channel erosion. [3] Booth (1990) examined stream channel size, W, versus 2-year discharge, Q_2. The fitted expression given in the paper is $W = Q_2^{0.68}$. [4] Bernhardt and Palmer (2007) describe what happens to urban streams and ingredients of stream restoration.

Higher flows, bigger stream channels, and fast erosion.

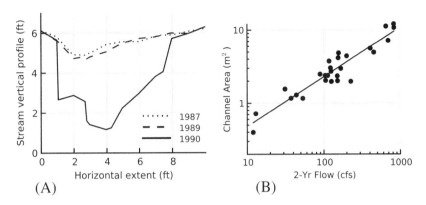

(A)   (B)

Figure 5.14: (A) When new urban development in King County, WA, first generated flashy streamflows, the water overwhelmed a small stream, quickly eroding its channel (Booth 1991). (B) This erosion creates a strong dependence between channel area the maximum 2-year flow rate (Booth 1990).

of the channel while also making it deeper and narrower. That deeper channel acts much like a drainage ditch, drying out the riparian soils of the stream corridor. Those changes affect a multitude of ecosystem services, including regional cooling, nutrient processing, and groundwater storage capacity. Second, ecological simplification takes place alongside the hydrological simplification. As shown, for example, in Figure 11.14, urban streams lose species richness, and with that loss again follows the loss of important ecosystem services like nutrient processing. The third aspect involves the fact that these are *urban* streams, existing where people live and work, and the very degradation of these streams makes them unattractive, unused, and unimportant to voters and taxpayers.

Attempts as stream restoration should try to remediate these three features as much as possible, given the space and financial constraints in urban areas. As a source of encouragement, not all reaches of urban streams get dramatically undercut by flashy flows.[1] Figure 5.15 shows, essentially, that the stream bed height in a low-slope portion of the Soos Creek in King County, Washington, dropped about 0.3 m over a time interval with a four- or fivefold discharge increase. Many processes or features can reduce erosion. For example, large woody debris (covered in Chapter 10) helps increase stream roughness, which

---

[1] Booth (1990).

Not all discharge increases produce deep incisions.

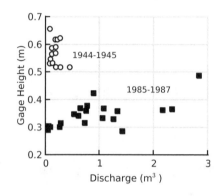

Figure 5.15: Streams that flow through areas with low slopes are much less susceptible to erosion, as shown for a low-slope section of the Soos Creek in King County, WA (Booth 1990). While discharge doubled due to intervening urbanization, the stream bed dropped only about 1 foot.

helps dissipate the kinetic energy of fast-moving water, which decreases erosion and lets streams retain higher gradient slopes without incision.

Moving from water volume to water quality, with runoff comes pollution, but not all runoff is equally polluted. Imagine a storm with a total runoff volume V_T and some total pollutant load of mass M_T. That simple idea leads to a simplistic event mean concentration (EMC), EMC $= M_T/V_T$:[1] *event* because we're talking about a single storm, *mean* because we're talking about a value averaged over the duration of the event, and *concentration* because it's a total mass of some pollutant divided by a total volume of water. Comparisons between various measurement approaches show that a careful sampling procedure results in EMC values roughly three times higher than other, less careful procedures.

Unfortunately, storms aren't so simple that they can be characterized by a single average pollutant value, and Figure 5.16 demonstrates that early water is dirty water.[2] These results are for a small roadway, specifically a 544-m^2 concrete "catchment" of an I-10 Freeway bridge at City Park Lake in Baton Rouge, Louisiana, for a single rainfall.

Let's think more carefully about a storm lasting a time interval, $0 < t < T$, where T is the total duration of the storm. Represent the fraction of the total pollution and runoff water volume, M_T and V_T, respectively, delivered up to time t as $M(t)$ and $V(t)$ and at the end of the storm as $M(T) = 1$ and $V(T) = 1$. The first-flush-is-dirtier concept means high concentrations early on and lower concentrations later, meaning that $M(t) > V(t)$. All that says

---

[1] Athanasiadis et al. (2010) provide an excellent mathematical definition of event mean concentration, and Maniquiz et al. (2010) compare several EMC formulations, finding general agreement among them all. [2] Sansalone and Kim (2008) have good first-flush data.

First flush carries more sediments.

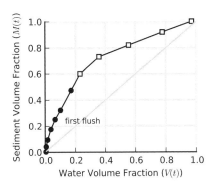

Figure 5.16: Covariation of water and sediment volumes running off a freeway bridge in Baton Rouge, LA, show that the initial, first flush of stormwater runoff carries a great majority of the sediments (Sansalone and Kim 2008). For example, the initial fifth of runoff water volume carries nearly 60% of the total sediment volume.

is that in the first, say, 10% of the runoff volume, we'll have seen more than 10% of the total pollutant load. That's precisely what we see in Figure 5.16 in this specific example, once the water volume fraction has reached 0.2, the pollutant volume fraction has reached about 0.5, so half the pollution was delivered in the first 20% of the runoff.

A bit more technically, the first, dirtier runoff phase of the event can be considered over when the slope of the ratio $M(t)/V(t)$ falls below 1, denoted as the point in the plot where the plotting symbol changes. Beyond that point in time, the runoff is cleaner than its average value.

Runoff carries sediments, but, sadly, that's not such a simple statement either, and there are at least a couple terms to characterize those sediments.[1] The first is total suspended solids (TSS), a measurement that was originally designed for wastewater, which really, truly emphasized suspended material, being the amount of sediment remaining after the heavy particles had settled out. In contrast, suspended sediment concentration (SSC) includes both the TSS portion and the sediments of a water sample that settled to the bottom of a still container.

Problems come in with the application of TSS to natural water samples (and stormwater runoff), where the heavy sediments are an important part of the process. Increasing amounts of sand-size material (more than 25%) increasingly throws TSS measurements off because sand settles so quickly. Since the SSC measurement uses the entire sample, it includes the sandy parts. A careful comparison of the two measures shows that TSS underestimates total sediments by around 20% compared with SSC. Furthermore, TSS variances were double that of SSC variances. Of course, the flow rate scales

---

[1] Gray et al. (2000) compares TSS (total suspended solids) with SSC (suspended sediment concentration).

Faster flows carry more sediments.

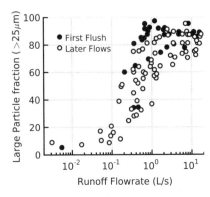

Figure 5.17: Faster flows carry larger particles along with all the smaller particles, producing more downstream sediments (Sansalone and Kim 2008). These results describe the sediments and flows from a small 544 m^2 section of an I-10 Freeway bridge in Baton Rouge, LA, and include eight different rainfall events, with 14 to 17 samples per rainfall event.

with catchment area, and larger basins will produce bigger flows and entrain larger particles.[1] In summary, SSC measurements are more reliable, but both measurements are found in the literature.

The first flush also comes as a big "whoosh." Flow tends to increase quickly to high rates, carrying large particles. This tendency is shown in Figure 5.17, where the filled circles of the first flush congregate at the higher flow rates and large particle fractions. These circles correspond to the early interval marked by the circles in Figure 5.16. Once the surfaces become a bit cleaner with the passage of the first flush, high flows remain for the later flows, and the large particle fraction drops a bit (shown by the open circles). Once the storm nears its end, flow rates decrease, large particles can't be washed away by these slow flows (see Figure 4.7), and the large-particle fraction greatly decreases.

The first-flush phenomenon is complicated. One study of the copper roof on the Academy of Fine Arts in Munich, Germany, showed that 40% of rainfall events showed elevated copper concentrations early during a storm, although 37% of the time the opposite effect, early dilution, was found.[2] Results came from multiple roof sections and different seasons, and the important factors promoting first flush were season, and a roof pointing into the wind, thereby increasing runoff.

A comprehensive examination summarizing data from several U.S. states demonstrated the strength of a first-flush effect by land use type, where commercial areas have the greatest incidence, followed as a group by residential, institutional, and mixed use.[3] Industrial land use has the next lowest, and

---

[1] Sansalone and Kim (2008). [2] Athanasiadis et al. (2010). [3] Pitt et al. (2004) reports on the National Stormwater Quality Database, with extensive data on land use and pollutant levels.

open space has no first-flush phenomenon. Components showing a first flush included chemical oxygen demand (COD), total dissolved solids, total Kjeldahl nitrogen (TKN), and zinc, but excluded turbidity, pH, fecal features, phosphorus, and total nitrogen.

Regarding the nitrogen measures, TKN adds up the ammonia and organic nitrogen in a sample, in contrast to total nitrogen, which adds together TKN and nitrates. There are many different ways to examine both nitrogen and phosphorus, and care needs to be taken when interpreting results because they can depend on which way one measures nutrients. Arguments can be made that the best measures are total nitrogen and total phosphorus, but that's not always possible for technical or financial reasons.[1]

Another first-flush study involving various roofing materials also clearly showed higher concentrations early in storms, including electrical conductivity, zinc, and various PAHs, but not copper.[2] In fact, PAHs also showed a "later flush" that was attributed to a later, increased rainfall intensity that helped dislodge particulates.

In the best of conditions, it seems a first flush depends on many factors. In part, observing a first flush requires a simple stormwater path from the source to the measurement point. The combination of arrival times at a sampling station from many different locations, mixing some first flushes with a long flow path with later flushes arriving via a short flow path smears out concentration variations and eliminates such a phenomenon.[3]

The remaining few pages of this chapter examine sediment export increasingly downstream. Sediment production data shown in Figure 5.18 comes from an urban watershed, Little Falls Branch (4.1 mi^2) near Bethesda, Maryland, and a rural watershed, the Watts Branch basin (3.7 mi^2) near Rockville, Maryland.[4] Data were taken in the early 1960s. The rural watershed performed well in holding back sediment, and stormwater produced very little erosion. The urban watershed, on the other hand, produced much more sediment for identical discharges, but the processes described in Figure 5.7 generated much higher discharge rates that led to even higher sediment loads.

In western Washington near Seattle, sediment export from the Issaquah Creek watershed increased from a preurban rate of 24 tonnes/km^2/year to 44 tonnes/km^2/year under urbanization.[5] Presently about 20% of the sediment comes from channel erosion due to increased stream discharges from impervious surfaces, as depicted in Figure 5.14A. Another 12% includes the disturbed surfaces of the usual suspects of construction, roads, and so on. Since 44 tonnes equals 44 million grams, and there's a million square meters in a square kilometer, that's a soil loss of 44 g/m^2 each year. That's around 0.1 mm of soil in depth. An interesting and sobering comparison is that 7

[1] Dodds (2003). [2] Förster (1996). [3] Pitt et al. (2004). [4] Leopold (1968) calculates sediment production values. [5] Nelson and Booth (2002).

Urban watersheds produce high sediment loads.

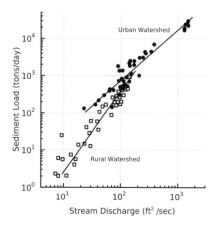

Figure 5.18: Urban streams have higher flows and more sediments than rural streams in Maryland (after Leopold 1968). The rural (as defined in 1968) stream drained a 3.7-square-mile watershed near Rockville, MD; the urban stream drained a 4.1-square-mile watershed near Bethesda, MD.

inches of topsoil has been eroded by cotton farming in regions of Georgia over the decades and centuries.[1] Sediment export varies quite a bit within a watershed. The lower portion of Issaquah Creek exports 180–320 tonnes/year, while 20 km upstream the number reaches 700 tonnes/year, and that means 700 g/m^2!

Sediment loss implies deposition somewhere else; rates of 6–9 mm/year are seen in the intervening stream bed, and deposition at bridge crossings was measured at 7–30 mm/year.[2] The deposition of 10 mm over a square meter gives a volume of 10,000 cm^3, or 10 L, which represents a soil mass of about 25 kg. The delta formed where the creek flows into Lake Sammamish, Washington, grew by 2,600 tonnes/year from 1944 to 1995. Total export from the entire watershed of 144 km^2 is around 4,000 tonnes/year of fine sediments (<2 mm) and 2,300 tonnes/year of coarse sediments (>2 mm), for a total of 6,300 tonnes. In this watershed, with about 73% forested land cover, about half of the sediments come from landslides, 20% from stream erosion, and 15% from forest roads.

The construction practices that take place in urbanizing areas unleash lots of sediments in small areas, an estimated 20,000–40,000 times more sediments than farms and forests, as shown in Figure 5.19. Compare this result with those from Figure 10.12, which shows the spacing of deep pools in rivers with the amount of large woody debris in streams, which greatly differs between urban and forested locations. Flashy flows in urban areas dislodge sediments, eliminate many pools in the streams, and flush out much

---

[1] Schoonover et al. (2006).   [2] Nelson and Booth (2002).

Small urban watersheds produce sediments.

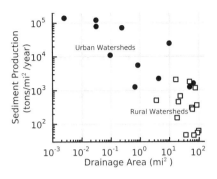

Figure 5.19: Urban and rural stream sediment loads for watersheds all across Maryland (after Leopold 1968). Urban and urbanizing watersheds produce 20,000–40,000 times more sediments than rural ones.

of the woody debris. Indeed, a simple calculation folding together peak flow increases (Figure 5.7) and channel size versus flow information (e.g., Figure 5.14B) suggests that flashy urban flows cause stream channel erosion that alone accounts for five times the amount of sediment production that comes from forested watersheds.[1]

Sediment may sound like a minor concern, but a number of problems come from high sediment loads.[2] First, sediment acts like a cloud, and the decreased light transmission through the water column reduces photosynthesis, yielding a lower primary productivity. As a result, turbid lakes have fewer aquatic plants, and that means fewer resources going into the herbivore trophic level, producing an overall reduced ecosystem diversity.

Sedimentation can also either increase or decrease temperatures, depending on lake characteristics, which can affect algal dominance (see Figure 9.8). Low light itself results in flagellate phytoplankton dominance, in part because clay binds to algal cells and makes them sink.

With greater turbidity comes fewer zooplankton with smaller body size and lowered reproduction, an effect seen in, for example, *Daphnia ambigua*. Suspended sediments inhibit larger cladoceran filter feeders that dominate clear water, favoring smaller, more selective rotifers. Sediments can also act as food supplements to large filter feeders, which then dominate their trophic levels.

Turbidity also reduces predation by fish and alters daily migration, which alters competitive outcomes. Detecting changes in predation is more complicated since a bit of turbidity can enhance the contrast between prey and their background for visual predators, but it can also reduce a predator's reactive

---

[1] Leopold (1968). [2] Donohue and Molinos (2009) discuss impacts of sediment loads.

distance. For their part, while being hidden in a cloud, prey reduce predator vigilance and enhance foraging and reproduction, which can alter trophic
dynamics.

Sediments falling to the stream bottom affect benthic invertebrates, with
fine sediments being more harmful. "Benthic" simply means the bottom of
a stream or lake, and the region includes a bit of the sediments. Fine sediments can clog pores, affecting water and oxygen penetration into mud (see
Figure 10.3).

Sediments also deoxygenate water through a rather complicated process.
Adsorption of dissolved organic carbon onto clay particles feeds bacteria,
enriching the microbial ecosystem but reducing oxygen for other organisms
with this decomposition. In one example, when suspended sediments were
100 mg/L, the organic carbon on sediments provided 30 times more resources
than phytoplankton. Some planktivores eat the particles directly, bypassing
their usual protist prey, while feeding by others is disrupted by the sediments,
and many planktivores starve. These changes can disrupt the ecosystem.

As with all of ecology, the disruption arising from increasing sediments
is complicated, and there are no hard-and-fast rules, with patterns being seen
that are opposite of what's expected. Even the type of clay is important:
montmorillonite (also called bentonite) versus kaolinite, which differ in their
aluminum and silica fractions, as well as the size of their particles. Overall,
though, increasing turbidity reduces primary productivity and changes species
composition, along with a change in ecosystem function.[1]

Integral to erosion and sedimentation is a process called "saltation."[2]
Imagine a particle in a blowing wind that lands on the ground or stream
bed. It doesn't settle calmly like the slow descent of a parachutist; rather, it
drops violently like a little meteorite. When it hits the surface, what happens
next depends on a few features. First, a basic principle is that the horizontal
air (or water) speed at a surface drops to zero, no matter what the air speed is
at a meter above the ground. The "shear" measures how quickly the wind increases with height above the ground: high winds mean high shears at ground
level. Now, when the particle hits the ground it bumps into other particles,
and they may get knocked off the ground into the air. If the shear is high,
then some of these particles get caught in the wind like a kite. On the other
hand, if the shear is low, these particles just drop back to the ground with a
tiny thud. This saltation process is sustained like a little nuclear reaction when
the wind shear is high, with one particle hitting the ground leading to several
more being ejected, and erosion takes place.

The process is relevant to the formation of sand dunes and deposition
velocities, and to stream bed erosion as well.[3] An important conclusion is that

---

[1] Donohue and Molinos (2009). [2] Kok and Renno (2009) describe the mathematics behind
saltation. [3] Sklar and Dietrich (2004) apply saltation to stream bed erosion.

Sediments settle at the coast.

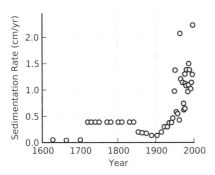

Figure 5.20: Sedimentation rates in Pamlico Sound on the North Carolina coast increased greatly over the last half century (after Cooper et al. 2004). Durham County's Eno River and Falls Lake reservoir, as part of the Neuse Watershed, ultimately empty into Pamlico Sound.

sediment supply features prominently in bed erosion rates. At low sediment concentrations, erosion increases with increasing sediment concentration because there are more particles to dislodge bottom sediments. At really high concentrations, however, sediment blankets the river bottom (called "fully alluviated"), protecting it from erosion. As a result, erosion rates are greatest for an intermediate sediment concentration.

The Issaquah Creek watershed discussed earlier was studied extensively for sediment transport.[1] One can speak of the distinction between bed load and suspended sediments. In reality, particle size is continuous, with big particles bouncing along the bottom (the bed-load fraction) and fine particles never touching the bottom. The ratio of suspended load to bed load varies from river to river, and it changes throughout a system as water flow varies. The binary classification fails; nonetheless, it's used, and the bed-load transport rate is just 300 tonnes/year, while suspended sediments are about 10 times greater.

Let's review the numbers, then, for this one watershed. The total sediment production from the watershed, of which one-third is coarse and two-thirds is fine, runs at about 6,400 tonnes/year. Of the fine sediments that make their way to Lake Sammamish, about two-thirds of it deposit onto the delta, while the remaining one-third washes out into the lake. Most of the coarse sediments, constituting one-third of the total sediments, never reaches the delta—it gets deposited in the lower and middle reaches of the channel. As a result, the delta grows by about 2,600 tonnes/year.

Once the sediments wash into the streams, they're transported by the streams. Wherever the water flow decreases, the heavier-than-water sediments fall out, just like the sand being carried in high winds drops to the ground when

---

[1] Nelson and Booth (2002). Many thanks to Derek Booth for clarifying discussions.

the air calms. One such example comes from my city, where half of Durham sits in the Neuse River basin, meaning that half of the city's stormwater empties out into Pamlico Sound at the North Carolina coast.

Sediment cores taken in 1997 show that as much as 2 cm of sediment, nearly 1 inch, drops into the sound each year (Figure 5.20). That's at least 10 times more than three centuries ago, with most of that increase taking place in the last half century. Pity the critters living at the sound's bottom, evidenced by measured changes in the ecological community through several centuries as the stormwater sedimentation accumulated.[1]

Chapter Readings

Allan, J.D. 2004. Landscapes and riverscapes: The influence of land use on stream ecosystems. *Annu. Rev. Ecol. Evol. Syst.* 35: 257–284.

Jennings, D.B., and S.T. Jarnagin. 2002. Changes in anthropogenic impervious surfaces, precipitation and daily streamflow discharge: A historical perspective in a mid-Atlantic subwatershed. *Landscape Ecol.* 17: 471–489.

Konrad, C.P., and D.B. Booth. 2002. Hydrologic trends associated with urban development in western Washington streams. USGS Water-Resources Investigations Report 02-4040 (Tacoma WA).

Lee, J.G., and J.P. Heaney. 2003. Estimation of urban imperviousness and its impacts on storm water systems. *J. Water Resources Plann. Manag.* 129: 419–426.

Leopold, L.B. 1968. Hydrology for urban land planning: A guidebook on the hydrologic effects of land use. USGS Circular 554 (Reston, VA).

Meyer, S.C. 2005. Analysis of base flow trends in urban streams, northeastern Illinois, USA. *Hydrogeol. J.* 13: 871–885.

Schoonover, J.E., B.G. Lockaby, and B.S. Helms. 2006. Impacts of land cover on stream hydrology in the west Georgia Piedmont, USA. *J. Environ. Qual.* 35: 2123–2131.

---

[1] Cooper et al. (2004) present the work on sedimentation rates in Pamlico Sound.

Chapter 6

Nutrients

Plants and animals need nutrients, but urban and agricultural land uses contribute nutrients, specifically nitrogen and phosphorus, to surface waters. At low terrestrial delivery rates, microbial populations react with higher metabolic rates, much like fertilizing garden plants. In oxygen-depleted soils, certain microbes use nitrate in a way similar to oxygen while consuming organic matter. This helpful denitrification process converts nitrate to nitrogen gas, removing it from the biosphere. When deposition levels reach a certain level, uptake simply can't deal with it all, and nutrient export to streams begins.

Agricultural crops need high amounts of fertilizers for high yields, and some fraction of that fertilizer finds its way into waterways. Of agricultural practices, cropland exports the greatest amount of nutrients, particularly corn. Watersheds with a high diversity of crops have a much lower nitrate export than do single-crop watersheds. However, urban land use remains an important source of nutrients, both nitrogen and phosphorus. For many land use types, total nutrient export depends greatly on precipitation and its runoff. Flashy flows export the greatest share of these nutrients, and nutrient export also increases with imperviousness. Forest and pastures have particularly low nutrient concentrations in their stormwater runoff. For comparison, pristine forested watersheds export a fraction to a few kilograms of nitrogen per hectare each year, but agricultural and urban areas export tens of kilograms of nitrogen per hectare.

Nitrogen transformations: Nitrification and denitrification.

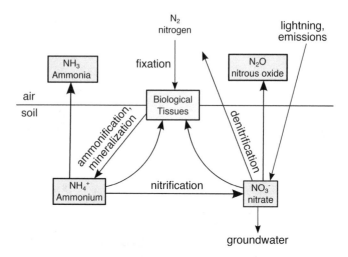

Figure 6.1: Generalized components and pathways of nitrogen. Synthetic fertilizers represent an external source of ammonium and nitrate.

High amounts of nutrients entering surface waters lead to algal blooms. When the algae in those blooms die, bacteria surge in response, consuming to their little vacuoles' content, depressing oxygen levels to below those suitable for fish.

Biological molecules, the proteins and nucleic acids, hold nitrogen atoms, and availability of biologically accessible nitrogen often limits plant growth. Figure 6.1 depicts the connections between important nitrogen-based molecules, including processes involving nitrogen that build up biological molecules, and the steps that break them down. When herbivory or plant death happens, the subsequent decomposition leads to ammonium ions (NH_4^+) and ammonia (NH_3) released into the soils. That breakdown process is called mineralization. Nitrification transforms ammonium into the nitrate ion (NO_3^-) and many bacteria and fungi make their living using the energy released in this transformation. To complicate matters, some microbes process ammonium to the intermediate nitrite (NO_2^-) which other microbes then transform to nitrate. Plants can take in nitrate and/or ammonium and by doing so incorporate nitrogen back into their tissues.

There are many sources of reactive nitrogen.[1] We've seen that humans have greatly increased the amount of nitrogen available to the biosphere (Figure 1.4). These numbers vary greatly, but preindustrial additions included roughly 120 Tg N/year through biological fixation plus another 5 Tg N/year through lightning. Humans now add 45 Tg N/year through biological fixation with legumes, 100 Tg N/year with synthetic production, and 25 Tg N/year from burning fossil fuels.[2]

Putting these numbers in context, somewhere around 100,000 Tg N resides in terrestrial ecosystems and soil biomass, meaning that once a nitrogen atom is fixed from the atmosphere, it spends roughly 800 years on land before moving back to the atmosphere or the ocean.[3] In contrast, the air holds about 4 billion Tg N, and above an acre of land the air holds about 35,000 tons of nitrogen gas. We have a seemingly infinite supply of N_2.

Back to the function of nitrogen, excess applied nitrogen results in the eutrophication of estuaries, and that fertilization leads to algal blooms, which lead to anoxia (absence of O_2) or hypoxia (low O_2) upon microbial decay. Costs of eutrophication from excessive nutrients are estimated at $2.2 billion annually, though the variance in these estimates, like all estimates of ecosystem services values, is large.[4] Just a few of these costs include reduction in property values that range from $300 million to $2.8 billion, and recreation-based economic losses ranging from $370 million to $1.2 billion. A further $800 million is spent on bottled water due to taste and water-quality concerns linked to eutrophication.

Atmospheric deposition of nitrogen changes soil chemistry, leading to acidification, reduced plant growth, and tree mortality. It can also lead to biodiversity loss, with dominance of strong N-competitive grasses. Nitric acids leach calcium from soils, and that mobilizes aluminum, which is toxic in aquatic systems. Uptake of ammonium (NH_4^+) acidifies soils because a hydrogen ion gets left behind in the soil, whereas uptake of NO_3^- alkalinizes soils because it pulls a hydrogen ion out.[5] Especially high surface nitrogen levels are found in winter and early spring, before leaf-out, due to reduced uptake by vegetation.

Unlike with plant uptake, the deposition of nitrogen can lead to the loss of NO_3^-—which is soluble in water—from the watershed, and chronic loss leads to acidification because the H^+ is left behind.[6] This leaching tends to take place outside the growing season, when the demand for nitrogen is low. Similarly, the loss of NO_3^- as it flows out of a water body results in lake acidification because it leaves positive hydrogen ions behind and results in

---

[1] Vitousek et al. (1997) review changes in N cycle by humans. [2] Galloway and Cowling (2002), Galloway et al. (2003), and Schlesinger (2009) provide estimates. [3] Schlesinger (2009). [4] Dodds et al. (2009) estimate eutrophication costs. [5] Stoddard (1994) provides a great discussion of the nitrogen cycle, including the basic chemistry involved. [6] Stoddard (1994).

the reduction in a lake's acid neutralizing capacity (see Figure 8.6). However, lake acidification usually comes about because of the deposition of acids, like sulfates (SO_4), more so than the loss of nitrates. Furthermore, freshwater lakes are usually limited by P, not N.[1]

As much as we rely on the use of nitrogen to grow crops, we also rely on the ecosystem service of wetlands and riparian areas to remove nitrogen before it enters our lakes, reservoirs, and groundwater. The next few pages discuss the chemistry that underpins this removal. First I'll review photosynthesis, which transforms CO_2 into biological molecules, and then the breakdown of these molecules in oxygen-rich environments. This example shows what happens to electrons in each process and provides the context for understanding how nitrogen, in the form of nitrate, substitutes for oxygen in anaerobic environments.

We're probably all familiar with the carbon dioxide (CO_2) that makes up a small portion of our atmosphere, the primary greenhouse gas we've been increasing with our fossil fuel emissions. We also know about photosynthesis, the process by which light energy is captured by specific molecules in plants, energizing electrons, and ultimately used to convert CO_2 and H_2O into CH_2O retained by the plant and O_2 released into the atmosphere. These ingredients are further connected into carbon chains, rings, and myriad carbon-based molecules. This assembly of basic ingredients into biological molecules points out the need to consider both energy *and* electrons.

One of the first steps in photosynthesis breaks apart water to liberate electrons and hydrogen atoms, $2H_2O \rightarrow 4e^- + 4H^+ + O_2$, and produce an oxygen molecule.[2] In this reaction the oxygen atoms in the water molecules release the electrons that had been ripped away from the hydrogen atoms. The molecular changes that take place in the reaction are nicely depicted by Lewis dot diagrams:

$$
\begin{array}{ccc}
\text{H} & \text{H} & \\
{}^{x\,a}_{a}\text{O}{}^{a}_{a} + {}^{x\,b}_{b}\text{O}{}^{b}_{b} \rightarrow & {}^{a\,a}_{a\,a}\text{O}{}^{b\,b}_{a\,a}{}^{b\,b}_{b\,b}\text{O} & + 4e^- + 4H^+, \quad (6.1) \\
\text{H} & \text{H} &
\end{array}
$$

where a and b indicate the oxygen atoms' electrons, and the released energetic electrons on the right-hand side ($4_e{}^-$ are noted by x in the water molecules on the left. Photosynthesis breaks apart the water molecules to make O_2, but four electrons and hydrogen ions get left behind.[3] It's the electrons that carry the energy to power the synthesis of biological molecules.

Recognize that most reactions take place in the water of cells—certainly the reactions we care about here—and water molecules often mediate the

---

[1] Schlesinger (1997).   [2] Discussion here follows Bohn et al. (2001).   [3] You might think that these leftover bits might form two H_2 molecules, but it doesn't work that way.

transfer of electrons. This transfer takes place as hydrogen atoms constantly flit between H_2O molecules, such that two H_2O molecules can become one H_3O^+ and one OH^-, although these configurations have only a very short, 10^{-13} sec lifetime.

It's these electrons that become highly energized and transfer that energy in the complicated conversion of the molecule adenosine diphosphate (ADP) into adenosine triphosphate (ATP). Using that energy, transported safely through a cell as chemical potential energy in ATP, plants fix carbon using the stripped-off electrons and hydrogens via $CO_2 + 4e^- + 4H^+ \rightarrow CH_2O + H_2O$. Again using a Lewis dot diagram to show where the electrons go,

$$
\overset{b\,b}{\underset{b\,b}{O}}\overset{o\,o}{\underset{b\,b}{_{b\,b}}}C\overset{o\,o}{\underset{a\,a}{_{a\,a}}}\overset{a\,a}{\underset{a\,a}{O}} + 4e^- + 4H^+ \rightarrow \overset{H}{\underset{H}{\overset{x\,o}{\underset{o\,x}{C}}}}\overset{a\,a}{\underset{a\,a}{_{o\,o}}}\overset{a\,a}{O} + \overset{H}{\underset{H}{\overset{x\,b}{\underset{b\,x}{_{b}O_{b}}}}}, \qquad (6.2)
$$

where the CH_2O is but one small part of a to-be-assembled glucose molecule.

I use the chemical reaction in Diagrams 6.1 and 6.2 to show that the processes involving nitrification and denitrification are more complicated than just electron energy levels. One needs to think about not only energy, which can be transformed from electric potential energy, heat, and light, but also conserving electrons across a wide variety and number of molecules.[1]

These electron movements are relative to the pairs of molecules giving up and accepting the electrons. For example, in Diagram 6.1, water loses electrons, while in Diagram 6.2, the carbon dioxide gains electrons. In this context are a couple of important terms, "oxidation" and "reduction." In these pairs, electrons move from reductants (electron donors) to oxidants (electron acceptors). These descriptors represent a more nucleic-centered view of the world, where the molecule losing the electron is the reducing agent, and it "reduces" the molecule that gains the electron. The molecule gaining the electron is the oxidizing agent, called the oxidant, and it oxidizes the molecule that loses the electron.[2]

Once the more complex biomolecules implied by the production of CH_2O in Diagram 6.2 have been built up, the unbending path toward greater entropy sends these electrons back to their lower energy states. Actively guiding that path are the many respiring animals and microbes, "burning" the carbon-based molecules in oxygen-rich, aerobic environments. In aerobic respiration, sugars are broken down while oxygen molecules are split into oxygen atoms, which then take up electrons as they reform water molecules:

---

[1] I found a couple of sources quite helpful: Bartlett and James (1993) and Schlesinger (1997). [2] A fine explanation can be found at www.webassign.net/-question_assets/-ncsugen chem102labv1/-lab_11/manual.html.

$$
\begin{array}{c}
\text{H} \\
\overset{x\,o}{\underset{o\,x}{\text{C}}}{}^{a\,a}_{o\,o}\text{O} \\
\text{H}
\end{array}
+
\begin{array}{c}
\text{H} \\
{}^{b}_{b}\overset{x\,b}{\text{O}}{}^{b}_{b} \\
\text{H}
\end{array}
+
\overset{z\,z}{\underset{z\,z}{\text{O}}}{}^{z}_{z}{}^{z}_{z}\overset{z\,z}{\underset{z\,z}{\text{O}}}
\;\rightarrow\;
\overset{a\,a}{\underset{a\,a}{\text{O}}}{}^{o}_{a}{}^{o}_{a}\text{C}{}^{o}_{b}{}^{o}_{b}\overset{b\,b}{\underset{b\,b}{\text{O}}}
+2
\begin{array}{c}
\text{H} \\
{}^{z}_{z}\overset{x\,z}{\underset{z\,x}{\text{O}}}{}^{z}_{z} \\
\text{H}
\end{array}
\qquad (6.3)
$$

or $CH_2O + H_2O + O_2 \rightarrow CO_2 + 2H_2O$. Oxygen serves the critical role of electron acceptor in this aerobic respiration process, taking up the x electrons.

This chapter involves stormwater problems dealing with nitrogen and the important nitrogen transformations taking place in anaerobic environments. Although the reduction–oxidation reactions described in Diagrams 6.1–6.3 involve carbon in aerobic environments, my motivation for the carbon–nitrogen bait-and-switch is threefold: (1) it's important to explain the need for microbes to transfer electrons while gaining energy from biomass; (2) using either oxygen or nitrate as an electron acceptor concerns microbes consuming organic matter; and (3) denitrification chemistry is vastly more complicated than aerobic respiration, with details too messy to discuss here.

Pivoting now to denitrification, we're interested in the anaerobic process taking place in boggy, water-logged wetland soils and in lake-bottom sediments rich in organics and fine soils. The lack of oxygen leads to denitrification, which transforms nitrate (NO_3^-) into nitrogen gas (N_2). Microbes have such an ability, dumping electrons into nitrate and using it as the oxidizing agent much like O_2 in aerobic environments. The sacrifice on the part of the microbe is that the anaerobic process of denitrification provides less energy to the microbe, and if oxygen is around, the aerobic option wins because more energy can be extracted from the organic material. In the anaerobic reactions, NO_3 takes up the electrons in the reaction, $2NO_3^- + 10e^- + 12H^+ \rightarrow N_2 + 6H_2O$, which corresponds to the appropriate portion of Diagram 6.3. The end point here is that the nitrogen escapes as a gas, making its way back into the air and leaving streams and lakes. Other anaerobic reactions can convert the nitrate (NO_3^-) into ammonia or nitrite (NO_2).

Other related denitrification steps end with nitrous oxide (N_2O), a greenhouse gas with some 300 times more heat-trapping strength than carbon dioxide. It also stays in the atmosphere 120 years, but from the perspective of aquatic nutrients, it's taken away from the nutrient pool.[1] N_2O in the atmosphere is increasing by 0.2–0.3% per year, and the molecule also catalyzes stratospheric ozone loss.

Denitrification increases in finer soils when the soil pores are filled with water.[2] For example, when nitrates were added to fine, wet soils containing carbon, upward of 90% of the nitrogen was lost through denitrification. However, sands experimentally amended with carbon—meaning glucose was added—also produced high denitrification rates when relatively drier. High

---

[1] Schlesinger (1997). [2] Weier et al. (1993) found this increase in denitrification.

NO_3 inhibited the important transformation of N_2O to N_2, and in the fine soils greater nitrogen content decreased denitrification. Researchers concluded that perhaps the water saturation inhibited oxygen diffusion, resulting in anoxic conditions that favor denitrifying bacteria. Sufficient water and carbon were the most important factors for increased denitrification. Field predictions are difficult for the N_2:N_2O ratios coming out of denitrification because of high variability in the natural conditions, and because outcomes depend sensitively on details. These concerns make it an important factor in stormwater treatment that control measures carefully account for the production ratio of N_2O versus N_2 when dealing with denitrification solutions.

Finally, sulfate (SO_4^{2-}) represents yet another electron acceptor in anaerobic environments, and sulfate-reducing bacteria are the same ones that methylate mercury.[1] That unfortunate connection links the wetlands that eliminate nutrients through denitrification to the conditions that can lead to high mercury concentrations in fish (see Figure 7.6).

Regarding nitrogen as a promoter of primary productivity, recall that corn yields increase when intentionally provided nitrogen fertilizers by their farmers (Figure 1.3). Peat bogs also need nitrogen, and they get *all* of their nutrients from atmospheric deposition.[2] In this case, though, it's not plants that use the nitrogen, it's the microbial population that decomposes the dead plant material.

Under natural conditions, various biologically produced chemicals inhibit microbial decomposition. *Sphagnum* produces something called "uronic acids and polyphenols" that inhibit microbes and vascular plants from growing.[3] The atmospheric deposition of nitrogen, however, changes that chemistry, leading to high decomposition of the stored plant material (carbon) and the release of carbon dioxide.

In the study represented in Figure 6.2, samples were collected from 12 bogs in nine European countries and studied back in the lab. As the plots show, CO_2 emissions saturate with increasing nitrogen deposition. That saturation exemplifies that ecological processes can be limited by many factors, and in this case it reflects an increasing limitation by phosphorus as the soils are released from nitrogen limitation. Further, higher nitrogen deposition results in lower polyphenol production by living *Sphagnum*, further promoting microbial activity. The biology of the bogs also changes because increased microbial decomposition releases even more nitrogen.

Astonishingly, peat bogs hold about *one-third* of the world's soil carbon, but account for only 2–3% of the land surface. Hence, what happens to peat bogs has important implications for climate change.

Atmospheric nitrate deposition takes place throughout a stream's watershed, but does that deposited nitrogen make its way into the stream? Results

---

[1] Jeremiason et al. 2006. [2] Bragazza et al. (2006) studied the effect of atmospheric nitrogen deposition on peat bogs. [3] Bragazza et al. (2006).

Microbial respiration increases with nitrogen deposition.

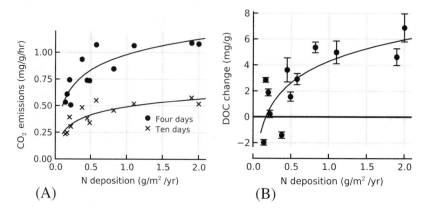

Figure 6.2: Atmospheric nitrogen deposition alters the respiration (A) and dissolved organic carbon (DOC;B) within northern European peat bogs (Bragazza et al. 2006). Note that 1 g/m^2 is about 8.9 pounds/acre, meaning that the highest deposition level here reaches almost 18 pounds/acre.

shown in Figure 6.3 come from forested northeastern U.S. watersheds where researchers measured the nitrate export from draining streams.[1] These data come from studies performed from West Virginia to Maine, involving some 83 data sets that had measured appropriately export values over the appropriate time periods.

As a comparison for these numbers, a map of the northeastern U.S. from 1993 showed nitrogen deposition up to 12.7 kg/ha/year, though fine-scale features such as the enhanced levels of urban areas weren't detailed.[2] Note that 10 kg/ha equals 1 g/m^2, or about 8.9 pounds/acre; thus, these deposition values match those of the independent variable in Figure 6.2. Broader-scale anthropological effects were seen in concentration of several pollutants in precipitation, including NO_3, SO_4, Ca, K, and H ions, all of which increased two- to threefold moving from east to west across the study region. These longitudinal changes were attributed variously to industrial activity, animal feedlots, and fertilizers. Deposition rates at high elevation sites, for example, can be up to 35 kg/ha/year and even higher, but these high values aren't represented

---

[1] Aber et al. (2003). [2] Ollinger et al. (1993) produced color maps estimating early 1990s deposition of N and other pollutants.

Nitrogen uptake reaches a limit, and then export happens.

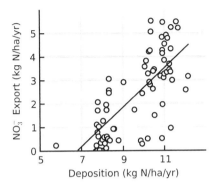

Figure 6.3: Atmospheric deposition of nitrogen in northeastern U.S. lakes results in very little export until a critical loading value around 8 kgN/ha/year, or about 7.1 pounds/acre (Aber et al. 2003).

in Figure 6.3. Considering dry deposition alone, latitude best predicted the variation, while longitude was relevant only for SO_4.

Within these forested watersheds, tree species react differently to nitrogen deposition: soils beneath sugar maples showed increases in nitrification, but beech soils did not. Logging, plowing, and fires all produce spikes in NO_3^- stream exports, followed by a sharp, many-year decrease and century-long soil nitrification signal.

Several interesting points from the study included the idea that deposition generally increases with elevation, meaning deposition is underestimated by several kilograms. The estimate of wet deposition increased only indirectly through increased precipitation with elevation, and for dry deposition no account for higher deposition velocity for increased conifer representation in forests was included. In the end, nitrate deposition rates could be quite high, around 30 kg/ha/year.[1]

The overall conclusion is that nitrogen deposition is altering the nitrogen status of northeastern forests, affecting hardwoods the most.[2] Increasing nitrogen deposition decreases the C:N ratio in biomass; that decline fosters increased nitrification (the conversion of ammonia into nitrate), and the nitrates leak from the watersheds into streams and lakes.

In forests, these processes result in the leaching of nitrate, which acidifies soils. Nitrogen content increases, and that changes the herbivore community, in particular, given that insects prefer plants with higher nitrogen content, giving a boost to a couple of problematic ones: the hemlock woolly adelgid and the beech scale. Calcium, magnesium, and aluminum all come from a

---

[1] Ollinger et al. (1993). [2] Aber et al. (2003).

steady weathering process, but the deposition of sulfur, nitrate, and acidic compounds subverts this process to leach them out of the ecosystem.[1]

Chronic deposition of nitrogen has extensively changed forest ecology rather than led to a fertilization of forests.[2] These various studies examined sulfur, nitrogen, ozone, and mercury deposition in eight ecosystems. Acid deposition includes $SO_4{}^-$ and $NO_3{}^-$. Bogs are highly nitrogen limited, as discussed in Figure 6.2, and nitrogen deposition alters competitive outcomes, displacing low-growing species like *Sphagnum* by taller, light-capturing species. Deposition of sulfur to bogs promotes the anaerobic bacteria that also methylate mercury, endangering fish in downstream lakes.

Chronic nitrogen deposition also leads to nitrogen saturation, leading, ironically, to soil fertility reductions.[3] Forests are usually nitrogen limited, and although a small amount of deposition has a fertilizing effect, chronic deposition has toxic effects through the aftereffects of nitrate leaching. Sulfur deposition is usually high relative to biological demand, and it also leaches out of forest ecosystems but takes calcium and magnesium with it. Soils of unglaciated regions have a greater capacity to store sulfur, with potential for longer-term effects. Resulting acidification causes aluminum to move around in soils, and that has toxic effects on roots. After aluminum makes its way to streams, it kills fish by interrupting respiratory processes. Acidification can even bring greater clarity to lakes by reducing turbidity, but that lets light reach the bottom, causing weeds to grow and changing ecosystems by altering predator–prey interactions.

The importance of these problems combines with harmful air quality effects of atmospheric reactive nitrogen[4] to make nitrogen emission reductions critical. Fortunately, emission reduction approaches have seen success over the past decades (see Figure 3.4).

Nitrates coming from atmospheric deposition and microbial nitrification have different ^{15}N and ^{18}O isotope fractions, and that serves as a fingerprint to source stream nitrates. The isotope differences arise from the different processes that create the nitrates. In the atmosphere, nitrate comes from lightning and the burning of fossil fuels, which pulls its ingredients from the air. In contrast, microbially produced nitrate pulls the three oxygen atoms in a two-to-one proportion from the surrounding water and air, respectively. Those different pathways create different isotope signatures for both atoms.

Figure 6.4 shows results from one study using such a fingerprinting approach, and it confirmed the importance of catchment-scale microbial processes in forming nitrate in the White Mountains of New Hampshire.[5] Studied watersheds were completely forested, one with an old-growth forest and the other last logged in the first decade of 1900. Another study of Catskill

---

[1] Lovett et al. (2009) has a helpful soil interaction flowchart. [2] Lovett et al. (2009) examined chronic deposition of nitrogen in forests. [3] Lovett et al. (2009). [4] Wilson (2011). [5] Pardo et al. (2004).

Nitrogen isotopes in streams fingerprint sources.

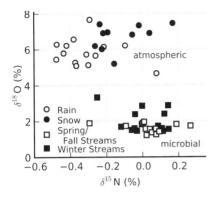

Figure 6.4: Isotopic signatures in precipitation and streams around the White Mountains of New Hampshire help identify nutrient sources (Pardo et al. 2004). Lower percentages of ^{18}O indicate microbial sources, whereas higher percentages indicate atmospheric sources.

Mountain streams of New York[1] also showed that stream nitrates came mainly from microbial nitrification, except during periods of high flooding. In addition, groundwater storage moderated precipitation pulses in exported nitrogen and, presumably, increased the time for microbial interactions to occur.

It might not be such a clear distinction, however. An examination of the isotopic signatures in fertilizers available in Spain[2] showed that the isotope signature of the nitrogen in dissolved nitrates ($\delta^{15}N_{NO_3}$) ranged between–0.16% and +0.56%, and the oxygen signature ($\delta^{18}O_{NO_3}$) ranged from +1.8% to +2.51%. One outlier revealed a Chilean geological nitrate source with an oxygen signature ($\delta^{18}O_{NO3}$) near 5.0%. These values sit right in the cloud of points from streams rather than atmospheric sources.

Up to this point this chapter has discussed background issues surrounding nitrogen, and we now pivot toward the nutrient-related consequences of human land uses of agriculture and urbanization. As discussed in Chapter 1, we have a lot of people in this world, and to feed them all we need good agricultural lands. Too much water makes portions of this area unsuitable for farming, and to make these lands workable—in other words, to keep tractors from sinking into muddy fields—a large area covering Alberta, Saskatchewan, Manitoba, Montana, the Dakotas, Minnesota, and Iowa has agricultural lands that are drained through tiling (explained below).[3] Draining the water exposes these rich soils to food production, and up to 80% of some small watersheds is

---

[1] Burns and Kendall (2002) studied nitrate sources in the streams of the Catskill Mountains of New York. [2] Vitòria et al. (2004) discuss isotopic fractions in nitrates in many sources. [3] Blann et al. (2009) review and discuss tile drainage in agricultural lands and provide a good overview of pesticide and herbicide effects on organisms.

Tile-drained agricultural runoff exports nitrogen.

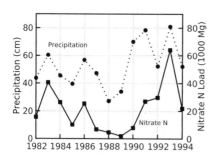

Figure 6.5: Nitrate export from tile-drained agriculture over a 12-year span closely reflects precipitation (Randall and Mulla 2001). Thus, reducing runoff reduces nutrient export. Data come from the Greater Blue Earth River watershed in southwestern Minnesota.

drained by tiling. For example, the percentage of drained land in Indiana rose from 0% in 1870 to 40% in 1920, tapering off to 45% in 1960.[1]

Land drainage is a very old practice, dating from at least the digging of ditches in Roman times. Today, it's a process called "tiling." Think of the black plastic piping available at many big-box hardware stores, but in rolls hundreds of feet long, buried three feet deep. Water drains in from the soil and flows out the end of the pipe into a large drainage ditch. It's an efficient practice, too, without the need for lots of ditches that displace rows of crops.

Subsurface draining of soils costs about $300–600 per acre but increases yields by about 5–25%. If corn yields are 200 bushels/acre, that's 10–50 bushels more per acre per year. At $5/bushel, that's a yearly return of $50–250. Drainage also helps prevent the salt buildup in irrigated soils by drawing water deeper into the ground and out the pipes.

Compared with surface drainage approaches, subsurface drainage also increases the water storage capacity of the land. Along with increased storage, subsurface drainage decreases peak flows and increases base flows, whereas surface drainage increases peak flows. However, on permeable soils, subsurface drainage can increase rapid drainage, and therefore increase peak flows. Drainage only weakly affects total runoff, with an increase only in the neighborhood of 10%.[2]

These pipes also drain away nitrogen, as Figure 6.5 shows for the 3,540-sqaure-mile Greater Blue Earth River watershed in southwestern Minnesota, with a clear correlation between precipitation and nitrogen export in the runoff.[3] Evidence shows that the nitrogen builds up in dry years, even in the biomass of unfertilized farmland, and drains away during wet years. Fields that had corn or corn–soybean rotations yielded a bit more than 50 kg/ha/year,

---

[1] Kumar et al. (2009).   [2] Blann et al. (2009).   [3] Randall and Mulla (2001).

whereas alfalfa exported less than 2 kg/ha/year. Low spring fertilizer applications produced the least nitrate exports, while fall applications had a 36% higher export. Comparing applications of 134 and 202 kg/ha, 15% of the nitrate contained in the runoff for the lower rate came directly from the fertilizer, but the number was 65% from the high rate. The authors recommend stopping the tile-draining practice, but in the face of that unrealistic hope, they recommend construction of control systems like wetlands for drainage water. Among additional concerns, better measurements of available nitrogen would help minimize application rates.

Nutrient export from abandoned fields with tile drains is minimal, just 0.1 kg/ha/year.[1] Export can be reduced, however, when clover is planted with corn or when orchard grass forage is planted with clover, but there is a yield sacrifice, more than half in the case of clover and corn[2]—in other words, if you want corn, don't plant it with clover. The best suggestion was a cover crop of rye to reduce nitrate export.

Another problem with tiling is that the very removal of the water, and the nutrients it carries, alters the natural wetlands and streams and delivers that water to rivers. Nutrient export increased by a factor of 2-7 over the last century, estimated as 2–3 kg/ha across the watershed but up to 50 kg/ha directly from agricultural lands. Nutrients exported with these waters are said to be the key water-quality problem in the Mississippi River basin and the primary cause of low oxygen levels in the Gulf of Mexico.[3]

Agriculture in the Midwest isn't alone in its connection with stream nutrients. On the eastern coast of the U.S., agriculture also affects stream nutrient levels, shown in Figure 6.6 for 15 subbasins of the Choptank River that flows into Maryland's Chesapeake Bay.[4] Nutrient concentrations were measured during base-flow periods.

One study summarized 40 other studies of nutrient runoff from agricultural land,[5] including 15 states of the U.S. and two Canadian provinces. Agricultural subtypes were crops, pastures and hay, and rotation crops, meaning that multiple crops were planted over the years on the same fields. That rotation makes it impossible to point the "blame" on any single crop. Sediment-associated particulate nutrients and dissolved nutrients were the two forms examined. Though the study was primarily an exploratory run for a large database of studies, some initial analyses were presented. Only dissolved nitrogen depended on applied nitrogen rates, and then only weakly, showing the great importance of erosion and such on nutrient transport. Phosphorus loads (both particulate and dissolved) depended on the soil phosphorus levels, but with lots of scatter. No-till farming practices had the lowest total nitrogen and phosphorus loads, except for pasture/hay fields, as one might expect, though

---

[1] Burton and Hook (1979) measured nitrogen coming off of an abandoned farm field. [2] Qi et al. (2005). [3] Blann et al. (2009). [4] Sutton et al. (2010). [5] Harmel et al. (2006) summarized nutrients from agricultural lands.

Nutrient concentrations increase with agriculture.

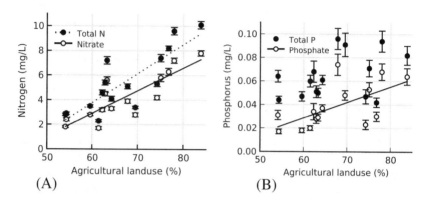

(A)   (B)

Figure 6.6: Agriculture in the Chesapeake Basin adds nitrogen (A) and phosphorus (B) to streams (Sutton et al. 2010). All quantities, except for total phosphorus, have significant correlations.

dissolved nutrients were highest, but not by much, for no-till agriculture. It was thought that the higher level arises because fertilizers aren't tilled into the soil, so there's greater nutrient-laden surface runoff. What was unexpected, however, was that the addition of various conservation practices, such as terraces, buffers, and filter strips, showed an increase in loads over conventional agricultural practices. The major suspicion for that result is that the practices were added in places that had high soil erosion potential in the first place, meaning the set of studies with various conservation practices is biased toward high nutrients. That suspicion is supported by looking at the individual studies that show decreases in nutrients when the conservation practices are added.

Figure 6.7 presents data taken in the 404-km^2 Ipswich River basin in northeastern Massachusetts. The complete data set spanned a broad variation in land use types, from 94% forest to 93% urban to 25% agricultural.[1] These plots come with new nitrogen units. Figure 6.7A lists micromolar nitrate concentrations; 1 μM means a concentration of 10^{-6} moles/L, where a mole is an Avogadro's number (6.02 x 10^{23}) of molecules. As we know, a nitrate molecule (NO_3) has one nitrogen atom and three oxygen atoms. A mole of nitrogen atoms has a mass of 14 g, and a mole of oxygen has a mass of 16 g, meaning a mole of nitrate has a mass of 14 + (3)16 = 62 g. Thus, a concentration of μM equals a concentration of 0.062 mg/L, and 100 times that amount equals 6.2 mg/L, the nitrate concentration found at about

---

[1] Williams et al. (2005).

Urban and agricultural land use increases nutrients to Plum Island Sound.

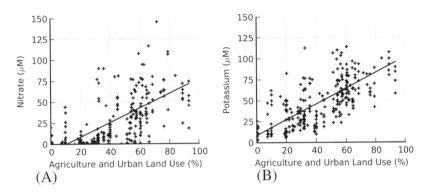

Figure 6.7: Nitrate (A) and potassium (B) flowing from the Ipswich River basin in Massachusetts to Plum Island Sound increase with urban and agricultural land use density (Williams et al. 2005). A 100 μM solution equals 6 mg/L for nitrate and 4 mg/L for potassium.

70% agricultural and urban land use. Potassium has a molar mass of 39.1 gs, meaning a concentration of μM equals a concentration of 0.04 mg/L.

A more localized examination of two headwater streams in this basin compared low and high levels of urban land use, 16% and 79%, respectively, with impervious surface fractions of 8.2% and 24.6%, also respectively.[1] Not surprisingly, runoff from the more urbanized area was about 30% greater, and its retention of nitrogen was lower (78–85% vs. 95–97%), compared with the lower-density subwatershed. Granted, the nitrogen loading of the more urbanized subwatershed was about 45% greater, too. In the end, however, the urban stream exported more than six times more nitrogen per square kilometer than the less urbanized one.

Interestingly, the greatest source of nutrients to the river system, after atmospheric deposition, was determined to be the import of food to feed humans. As we all know, the mouth is not the end point of nutrients, and many nutrients make their way down to the stream after passing through a few biological processes, one of which involves septic tanks.

If you live in a rural area, you might well have a home with a septic tank, into which the wastewater enters upon exiting a house. In urban areas this wastewater flushes into a municipal wastewater treatment system, but with a septic system all of this material goes into the septic tank, usually including

---

[1] Wollheim et al. (2005).

Septic tanks disappear with urbanization.

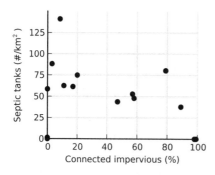

Figure 6.8: Septic tank density varies with impervious surface fractions in Melbourne, Australia (Hatt et al. 2004). Failing household septic systems lead to nutrients and pathogens in runoff, but expansions of municipal sewer systems replace these systems.

the washing machine water, sink drains, shower drains, and, of course, the toilets. Some homes have gray-water systems that divert wastewater from washing machines and, perhaps, showers into an alternative set of pipes that bypass the septic tank and go directly into a drain field or landscaping system. This diversion can extend the useful life of the septic system and make water available for secondary, nonpotable purposes like subsurface irrigation. Within the septic tank, all of the biological material gets broken down through microbial action, and the water flows out of the tank, through drain pipes and into a drain field. There the water infiltrates into the soil and groundwater with further microbial decomposition.

Sometimes septic systems fail, leading to seepage onto the ground and potentially into streams, and in past times these septic systems intentionally discharged very closely or directly into streams and lakes. These near-surface discharges lead to a short-circuit of the microbial action that removes the nitrogen and pollute waterways with excess nutrients.

As cities grow with higher housing densities, it's a good thing to get rid of unreliable septic systems, and their potential for leakage, and connect these older homes to more reliable municipal sewer systems. Figure 6.8 demonstrates how septic tank density decreases with increased connected imperviousness in Melbourne, Australia.[1]

An important component to help us understand nutrient export is an examination of watershed-level imports. Important quantities in this direction are the net anthropogenic phosphorus and nitrogen inputs, which estimate the input and output of nutrients from a watershed and include fertilizers, detergents, and atmospheric deposition. Fertilizer coming in gets added, crops

---

[1] Hatt et al. (2004). Chris Walsh kindly provided the original data.

Phosphorus inputs increase with agriculture and urbanization.

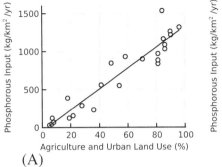 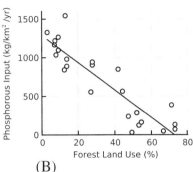

(A)   (B)

Figure 6.9: Phosphorus inputs measured between 1972 and 1992 increases with the sum of agriculture and urban land use fraction (A), and decreases with the forest land use fraction (B) (Han et al. 2011). Watersheds include 18 around Lake Michigan and 6 around Lake Erie.

leaving get subtracted, and food coming in gets added. However, nutrients in locally produced food that gets eaten, "processed," and treated in septic systems or wastewater treatment plants are an example of internal cycling within the watershed rather than a source or sink, and researchers must very carefully ensure these nutrients aren't double- or triple-counted.

Figure 6.9 shows curves for phosphorus coming into 18 Lake Michigan and 6 Lake Erie watersheds from 1974 to 1992 against natural and managed land use types.[1] Here the Lake Erie watersheds were more than 70% agricultural land use, with much greater variation for the Lake Michigan watersheds, having a broad range of agriculture, forest, and urban land use. (Figure 6.22 shows the curves of phosphorus export from these same watersheds.)

Input calculations used a watershed-scale estimation procedure folding in land use types and practices, including details such as county-level fertilizer use, per capita phosphorus use (0.6 kg/person/year), and changing dishwasher use patterns (from 7 times/week in 1974 to 5.6 times/week in 1992). Detergent, for laundry and dishwashing combined, was the dominant source of phosphorus in the most urbanized watersheds, reaching nearly 650 kgP/km^2/year for a population density of 360 people/km^2. In contrast, atmospheric deposition contributes about 6–14 kgP/km^2/year.

---

[1] Han et al. (2011) studied the phosphorus budgets in Lake Michigan and Lake Erie watersheds.

Monoculture crops add nitrates to streams.

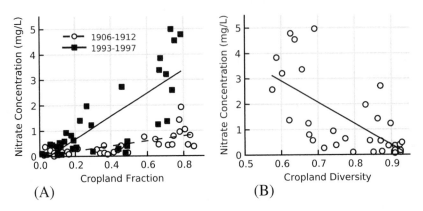

(A)   (B)

Figure 6.10: (A) Nitrate levels in runoff from agriculture land increases with the land use fraction in crops, measured across a large fraction of the continental U.S. Levels greatly increased during the 20th century. (B) Nitrate concentration decreases with an increase in the diversity of crops in a watershed (Broussard and Turner 2009).

In summary, the phosphorus inputs for the Lake Michigan and Lake Erie watersheds, respectively, equaled 31 and 99 kgP/km^2/year from detergents and 501 and 1,310 kgP/km^2/year from fertilizers. After subtracting exports, the Lake Michigan and Erie watersheds gained an averaged 558 and 1,112 kgP/km^2/year, respectively. Of this amount, rivers exported roughly 10%, meaning an overall increase in watershed nutrient pools.

Figure 6.10A shows the increase in stream nitrate concentration over the span of nearly a century, between 1906–1912 and 1993–1997.[1] The early data come from a 1924 paper, and the later data come from studies published by the USGS. The Haber-Bosch process for producing synthetic nitrogen (see Figure 1.4) was invented right in the midpoint of when the earlier data was taken, and the consequences of that invention were in full effect when the later data were gathered.

These plots come from the agriculturally heavy watersheds of the continental U.S., plus a touch of Canada, and each point represents a single watershed. Watersheds include urban and industrial land uses, but the agricultural dependence still comes through. No change in export was seen for watersheds with

---

[1] Broussard and Turner (2009) studied U.S. land use change over a century and water quality effects.

Agriculture avoids hydric soils.

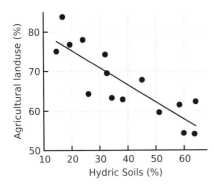

Figure 6.11: Agriculture avoids hydric (water-saturated) soils in the Choptank River basin (Sutton et al. 2010). This correlation between water saturation (hence denitrification potential) and agriculture obscures the connections among soils, agriculture land use, nutrient levels, and infiltration.

less than 30% cropland during the entire time span. Effectively, the relation is driven by increased nitrate export since the 1970s in watersheds with 30–60% cropland.

Also interesting is that total farmland area increased only 11.5% over the century spanning 1900–2002, having already been dominantly agricultural, and much of that small increase must have taken place during the first half of the century. The more striking change in agriculture is reflected in the 63% decrease in the *number* of farms and the 65% increase in the number of farms greater than 1,000 acres (405 hectares). These large farms now account for about 67% of all U.S. farmland.[1]

Cropland diversity is calculated from the acreage planted in different crops within a watershed and really just gives the chance that two different acres of cropland, chosen randomly within the same watershed, are growing two different crops.[2] A single crop gives a diversity value of 0, and two equally abundant crops would give a diversity value of 0.5. Although these researchers considered nine crops—barley, corn, cotton, hay, oats, rice, sorghum, soybeans, and wheat—an increasing cropland diversity index in the Mississippi River watershed also reflects a decreasing amount of acreage planted in corn. In sum, to get lower nitrate levels we need less corn, more perennial crops, and more sustainable agriculture.

One concern about blaming agriculture or urbanization unfairly is that, perhaps, these land uses have chosen specific soil types that export nutrients differently than other soil types. Figure 6.11 addresses this point. "Hydric" means water saturated, and hydric soils are those that have developed under

---

[1] Broussard and Turner (2009) and Wilson (2011). [2] Technically, Broussard and Turner (2009) used a modified Simpson's diversity index.

really wet conditions or are permanently or occasionally experiencing water saturation. Essentially, at least during some portion of the growing season, these soils are water saturated enough that anaerobic processes take place.[1] Agricultural land set-aside programs like the USDA's Conservation Reserve Enhancement Program (CREP) set their sites on tree planting for these soils, which are often associated with headwater streams. I always associate clayey soils with hydric soils, but that connection isn't necessarily so. It's likely that this impression comes from the reduced infiltration due to high clay content. The real feature of hydric soils is that the water table comes very close to the soil surface. For the most part, these soils can't be worked with heavy tractors without them sinking into deep mud.

Figure 6.11 comes from the study of nutrients coming from the Chesapeake basin (Figure 6.6) that sought an explanation for a lack of effect from restoring stream buffers. Correlations like this one, where the water-saturation conditions for anaerobic denitrification occur and the application of fertilizers is less likely, make it more difficult to make conclusions surrounding nutrient concentrations.

The results in Figure 6.12 are important, but somewhat complicated, too.[2] Imagine, if you will, going down to a stream and measuring its flow rate (expressed in Figure 6.12 as water depth per day). After a year's worth of measurements, you might reasonably expect to find a flow frequency distribution shaped like a normal distribution, with an average value and a standard deviation (even if not a normal distribution, it doesn't matter to this discussion). We know that the area under such curves equals 1, because every measurement taken is included under the curve. Another equally valid way to plot the same information is to first line up all of the measurements from the lowest value to the highest value and then plot the fraction of all measurements that are less than or equal to a particular flow rate as a function of the flow rate. More technically, that prescription plots the integral of the distribution against the distribution's variable. The resulting curves can be cleaner than the original distributions, making it easier to distinguish results from different streams.

What do we see from these plots? First, flow distributions change with land use, shown in Figure 6.12A for proportion runoff versus runoff flow rate. The two broader-sloped curves for the watersheds with higher imperviousness—Glyndon and Dead Run—have a much greater range of flashy flows and low base flows than the other three steeper curves that represent more consistent flows—not too fast and not too slow (note that the flows are shown on a logarithmic scale). Since these two urban watersheds have greater fractions at lower flow rates, it means the rivers spend more time with reduced base

---

[1] See the Kenai Watershed Forum page on hydric soils at www.kenaiwetlands.net/NRCS HydricSoilsIntro.html. [2] Shields et al. (2008).

More nutrients flow during urban streams' flashiness.

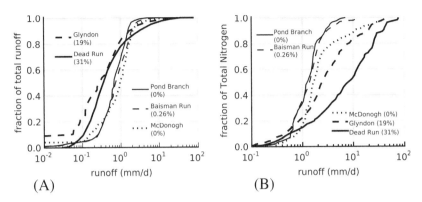

(A)   (B)

Figure 6.12: (A) Cumulative runoff fractions at increasing flow rates, for several watersheds around Baltimore, MD, show high base flows for rural streams and flashy flows for urban streams. For example, Dead Run sits above the rural streams at low flows but below at high flows, meaning rural streams have sustained, moderate flows. (B) Urban streams' nitrogen export curves mostly sit below the rural streams' curves, meaning most urban nutrient export occurs at relatively high flow rates (Shields et al. 2008). Watershed imperviousness is in parentheses.

flows, consistent with the results of Figure 5.2. Furthermore, since the curves for these watersheds drop below the others at high flow rates, it means they also spend more time with high flow rates. In other words, the three more natural catchments have stable base flows without much flashiness from stormwater runoff.

Figure 6.12B shows that for the two watersheds with more imperviousness, Glyndon and Dead Run, more nitrogen is exported at the high flows. We see this relation, for example, by focusing on a flow rate, say, of 10 mm/day, where the Dead Run watershed has exported only about half of its total nitrogen, but the natural streams have exported all of theirs.

Now compare the 60% total nitrogen levels of the different curves in Figure 6.12B. The flow rate for the high-imperviousness Dead Run is much greater than, for example, the flow value for the natural-condition watershed Pond Branch. Higher imperviousness leads to more nitrogen export at higher flow rates. Figure 6.13 plots the 75% levels of total export from the curves in Figure 6.12B against the imperviousness of the watershed. Indeed, more

More imperviousness produces more nitrate at greater flow rates.

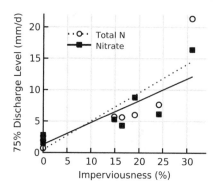

Figure 6.13: Flashy urban flows carry more nitrogen from urban watersheds of Baltimore, MD (Shields et al. 2008). Figure 6.12B shows that the flow rate at which 75% of total nitrogen (and similarly for nitrate) has been exported increases with imperviousness, meaning that reducing nitrogen export from urban watersheds must address high flow rates.

nitrogen leaves at higher flows, meaning that controlling nitrogen export from urbanized watersheds means removing nitrogen from high-flow scenarios. This result is important because control measures like stream restoration work only for low-flow situations: if streams are restored but maintain their high flow rates, then restoration doesn't work as well to remove the nutrients.

While nutrients might increase in urbanized watersheds, their removal doesn't. In the watershed of Chattahoochee River near Atlanta, Georgia, a study examined several aspects of ecosystem function, including leaf breakdown rate, gross primary productivity, community respiration, net ecosystem metabolism, and nutrient removal.[1] These "functions" are processes that take place alongside species interactions, or consumer–resource interactions, that we might think of as the foundation of ecology. Thinking about functions rather than species reorients our thoughts about ecosystems not as collections of organisms but, perhaps, more like a car with many parts. Somehow all of the parts, when put together, work to move the vehicle, each part somehow contributing through some specific function. In this automotive analogy, the important function is the car's movement. When something breaks down, a good mechanic can rummage through a parts store and find a suitable, if not identical part, that restores the function of the broken part. Similarly, the biosphere exists and persists, with species as parts, each species contributing to the flow of energy and nutrients within the system.

The study demonstrated higher nutrient levels in the streams of more urbanized watersheds, but really focused on measuring the subsequent nutrient uptake (Figure 6.14). Uptake was measured by adding nutrients, in the forms

---

[1] Meyer et al. (2005).

Urbanization reduces nutrient uptake.

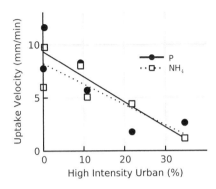

Figure 6.14: The uptake velocity of nitrogen and phosphorus in streams draining watersheds around Atlanta, GA, decreases with increasing fraction of high-intensity urban land use (Meyer et al. 2005).

of NH_4 and PO_4, for an hour or two at one spot and then measuring concentrations 200–750 m downstream. These biologically active nutrients were paired with bromine, a nonreactive tracer used to scale loss rates via nonbiological processes like water exchange between the stream and groundwater. Knowing the initial amount added, and measuring the concentration at a downstream location, one can calculate the uptake rate per unit distance.

Nutrient uptake rate is simply the vertical movement rate of the nutrients through the water column toward the benthos, or the stream bottom. This nutrient uptake is precisely analogous to the deposition velocity for air pollutants discussed in Chapter 3 (see Figures 3.5 and 3.6). Just as some pollutant in the air settles out or gets captured by the tree canopy as the wind blows, nitrogen and phosphorus in a stream get drawn out of the water column by interactions with the stream bottom. Again, the analogy is an eraser made of nutrients that slowly wears away as the water flows downstream, and the uptake velocity describes how fast the eraser wears down.

Figure 6.15 shows measurements of nitrate levels in streams and how they vary across different land use types in different geological areas.[1] The Potomac River basin examined here feeds the Potomac River, which drains into Chesapeake Bay, and has overall land use fractions of 35% agricultural, 10% urban, and 50% forest. Geological areas listed on the figure are in the Valley and Ridge province and the Piedmont province of the watershed. The Valley and Ridge province has two subunits, one called the Valley and Ridge subunit that's primarily forested and has steep ridges and deep valleys. The Great Valley subunit has carbonate bedrock, with limestone and dolomite (calcium magnesium carbonate), and has mostly agricultural lands.

---

[1] Miller et al. (1997) examined nitrate concentrations in the Potomac River basin.

Nitrate levels in Potomac watershed streams vary with land use type and soils.

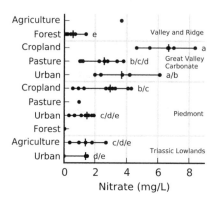

Figure 6.15: Nitrate concentrations in Potomac River streams vary across the different land use types and different subunits of the Potomac River watershed (after Miller et al. 1997). Letters refer to statistically different groupings, and the small bars depict median values.

The Piedmont province has a Piedmont lowlands subunit with "fractured siliciclastic or crystalline rocks" which are granites, gneisses (coarsely layered rocks), and metaquartzites (cemented sandstone) and with a mix of agriculture, urban, and forest land uses. The triassic lowlands subunit has siliciclastic rocks and diabase soils that originated as volcanic intrusions into the surrounding geology. These lands are mostly agricultural. The various subunit basins average 25 km^2, ranging from 6 to 95 km^2. Water samples were collected during base flow, with the requirement that there be no rainfall for at least 3 days. Note that this study took place in the mid-1990s, and the USEPA introduced stormwater management practices for agriculture around that time.

Row cropping resulted in the highest stream nitrate levels. In cases where row cropping and pastures couldn't be separated, both were swept into the term "agriculture." The worst situation was agriculture on the Great Valley carbonate bedrock, where stream nutrient transport was efficient. Urban land use on the Great Valley subunit also produced high nitrate levels, in part because portions so designated had a higher mix of agriculture than the urban land use areas on the other subunits. Within agriculture, corn, soybeans, and grain all correlated with increased nitrate levels, but pasture correlated with decreased levels.

In the several geological regions of the Chesapeake Bay watershed, water chemistry depends on the amount of carbonate rock more than on land use type.[1] These chemical measures included concentrations of calcium and magnesium, the pH level, conductivity, and stream alkalinity. A similar dependence

---

[1] Liu et al. (2000) correlated stream chemistry to carbonate rock.

Agriculture and urbanization reduce stream water quality.

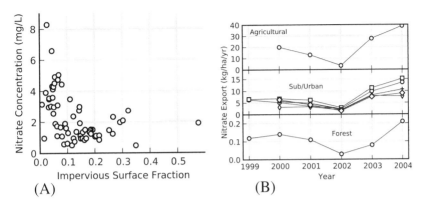

Figure 6.16: (A) Streams with low impervious-surface fractions, draining agricultural areas around Baltimore, MD, have high nitrate concentrations that level off in high-impervious urban areas. (B) These levels reflect nitrate export levels in agricultural and urban areas that far exceed forested watershed levels (after Kaushal et al. 2008).

was seen in the Potomac River watershed.[1] Other regions with less carbonate rock demonstrated that nitrate, sodium, and chlorine concentrations were all connected to land use type, not geology. Geology had such an important effect that, for example, dissolved silicate (quartz plus water) had opposite dependencies on land use type in the coastal plain (increasing concentrations with cropland) and piedmont (increasing concentrations with development). Indeed, with all the added nitrogen, coastal ecosystems might be experiencing silica limitation, changing the balance between ecological competitors, resulting in a decrease in diatoms and increase in algae.[2]

Figure 6.16 shows the connection between nitrogen and land use in the Baltimore area of Maryland.[3] True, agriculture adds nitrogen to streams: farmers apply roughly 60 kgN/ha/year (recall the importance of nitrogen fertilizers on crop yields in Figure 1.3). One can certainly question whether farmers apply more nitrogen than necessary; if so, overapplication leads to more runoff than necessary. On the other hand, suburbanites in these watersheds apply an average of 14.4 kgN/ha/year to their lawns simply for the looks, which, some might argue, serves no useful purpose. Some of this applied fertilizer

[1] Miller et al. (1997).   [2] Justić et al. (1995) looked at silica limitation.   [3] Kaushal et al. (2008).

Nutrient export varies with land use type.

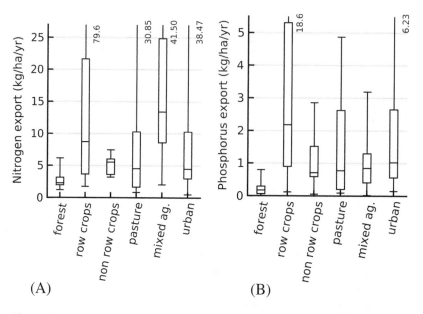

(A)   (B)

Figure 6.17: Nitrate (A) and phosphorus (B) export from different land use types scattered across the midwestern and southeastern U.S. (Beaulac and Reckow 1982). The bar within the box marks the median value, box ends mark the 25% and 75% values, and the whisker bars (or numbers) denote extreme values.

also enters waterways when it runs down ditches and into streams instead of beautifying lawns. In Figure 6.16A, results from Baltimore, Maryland, show that high nitrate concentrations in streams occur in areas with lower fractions of impervious surfaces, reflecting greater runoff from agricultural land. However, the levels remain high for higher impervious-surface fractions due to lawn applications.

Total export equals concentration times volume, and dependent on the information available, a study publishes one, the other, or both. Figure 6.16B shows that agricultural areas export around 20 kgN/ha/year, and urban and suburban areas export 5–10 kgN/ha/year, while forests export very little. Note the marked export decrease during the 2002 drought: no rain, no runoff.

The results of Figure 6.17 are more than 30 years old now but still useful.[1] Data used here came from slightly earlier literature reviews, with the original

---

[1] Beaulac and Reckow (1982) examined nutrient export from various land use types. K. Reckow, pers. comm.

studies performed between the 1970 and 1980, covering the midwestern and southeastern U.S. Here the results were reexamined for their dependence on land use types. Distributions of samples are highly skewed, with lots of low values but a long tail of high values, and this situation exists for both phosphorus and nitrogen export. Overall, we learn that nitrogen and phosphorus export is very high from feedlots, high from row crops, low from forests, and intermediate from pastures. Urban export is similar to that of pastures, but an important note is that these urban values are low, likely because they involve old residential areas with low imperviousness.

Compare these numbers with export from natural watersheds (see Figure 6.20). Numbers run about 100 kgN/km^2/year, or 1 kgN/ha/year, whereas numbers in Figure 6.17 are 5–10 times higher or more. Phosphorus levels show similar relative values. It's also interesting that since these numbers were taken, the production of synthetic nitrogen has more than doubled (see Figure 1.4).

As with the results in Figure 6.15, conclusions from old data like these need a degree of caution. Many new agricultural practices have come along since the 1970s, include stream buffers, low-till planting, and contour farming, not to mention the high-tech approaches to fertilizer application that have come up since the 1990s. These approaches include satellite and computer-controlled application of fertilizers based on nutrient levels, soil properties, and moisture, as well as prior yields.[1] Sensors can also identify weeds during herbicide treatment and grain pathogens to bypass harvesting to avoid toxins and weed seed contamination.

We've just seen that pastures seem pretty good from a water quality perspective. My late father used to call golf by the pejorative "pasture pool," so we might now examine golf courses within the framework of pastures. Golf courses appear—to use another Minnesota term—not so bad from a nutrient export perspective. A study from Austin, Texas, examined a 76-hectare golf course with a river running through it and compared inflow nutrients with outflow nutrients.[2] The course was constructed in 1963, and the study was performed 35 years later during 1998–2003. Export from the golf course amounted to 1.2 kg/ha/year of nitrate, 0.23 kg/ha/year of ammonium, and 0.51 kg/ha/year of phosphorus. These numbers are extremely small compared with the annually applied fertilizer: 36.5 kgN/ha and 8.2 kgP/ha. Two important points are that grass clippings are recycled back into the turf, and discharged runoff represents just 22% of precipitation. From this one study, anyway, it seems that a well-maintained golf course, including fertilization, irrigation, and overseeding, does not export great amounts of nutrients.

Healthy lawns also provide better water-quality benefits than a dead and dying lawn. A study performed in St. Paul, Minnesota, examined turfgrass

---

[1] Gebbers and Adamchuk (2010).   [2] King et al. (2007).

under four fertilizer treatments: no fertilizer, nitrogen (N) and potassium (K) with no phosphorus (P), N+K and one unit of P; and N+K and three units of P.[1] The phosphorus unit was 21.3 kg/ha for the first year and 7.1 kg/ha thereafter. Treatments also included either removing the lawn clippings or letting them remain on the lawn. Results were very interesting. First, whether clippings stayed on the lawn or not didn't matter for nutrient export, but there were three times more clippings with fertilization. Second, adding phosphorus didn't matter as long as nitrogen and potassium was applied. Third, after the first year of fertilizing the lawns, more phosphorus was exported, and more runoff occurred, from the unfertilized lawns than from the fertilized lawns because the healthier lawn retained more nutrients than the unhealthy lawn. That's despite the fact that no phosphorus was added to the unfertilized lawns! There was also the issue that frozen soils led to a greater export of phosphorus from the Minnesota lawn/tundra—indeed, 80% of the phosphorus export took place while the soil was frozen.

In my region of North Carolina, a late 1990s to early 2000s survey of homeowners in the Neuse and Tar-Pamlico watersheds indicated that, across five communities, 54–83% of homeowners apply fertilizers to their lawns.[2] Households using lawn care services ranged from 16% to 43%, while households performing soil testing ranged from 16% to 35%. Two-thirds of the homeowners watered their lawns during dry periods, ranging from 54% to 89% across the five communities. Misinformation abounds: across the four communities where the question was asked, 35–91% of households used pesticides on their lawns, but the primary reason for using them, ranging from 45% to 69%, was to combat weeds. I'm certain the weeds were grateful for the help in removing their pests! Regarding nitrogen use, the survey results indicated an estimated annual average application of 111 kgN/ha, a very high amount compared with golf course applications of 77.6 kgN/ha and comparable to those needed for high corn yields (see Figure 1.3).

Relatively little is understood about phosphorus and turf, but the annual application ranges from 2 to 10 kg/ha, while the annual "harvest" of lawn clippings contains 0.4–7.5 kg/ha.[3] Losses from minerals and gains from atmospheric deposition seem to be about equal to the amount that's exported, 0.2–0.7 kg/ha. It's thought that fertilizers applied to lawns might not lead to massive phosphorus runoff.

Finally, from the "people study things I never thought of" department: People hire lawn companies to reduce soil compaction by pulling a coring machine around their yards that plunges little pipes into the ground and pulls up plugs of soil that get dropped on the grass. An alternative approach is a machine that plunges a solid tine into the ground, making a hole but pushing

---

[1] Bierman et al. (2010) examined nutrient export from lawns. [2] Osmond and Hardy (2004). [3] Soldat and Petrovic (2008).

and compacting the soil around the hole. Golf courses use one or the other of these approaches in seeking turf perfection, and one study compared the effects on nutrient runoff from a Minnesota golf course.[1] Two days after treatment, the hollow tine approach had 55% less runoff than the solid tine approach, but 63 days after treatment, this runoff reduction had dropped to 10%. Similarly, nutrient losses were 39–77% less with hollow tines after 2 days and 5–27% less after 63 days. There was no control approach of doing nothing, and despite these approaches, turfs exported more than 10% of the applied nutrients for both treatments, quite a bit higher than the few percent export quoted for the Texas study described previously.

These differences simply demonstrate that nitrogen retention varies. One important factor is altitude.[2] In the Catskills Mountains of New York, nitrogen export in forests decreased from 25% in the lower watersheds (800 m altitude) to 7% in the highest ones (1,275 m). This result was found when comparing nitrogen export from streams at different altitudes in search of an effect of an atmospheric deposition that increases by 40% from the lowest to highest elevations. Likely, retention increases with elevation because the thickness of the "Oa" soil horizon[3] also increases with elevation. The lower temperature at higher elevations leads to slower decomposition, a thicker Oa layer (about a 6-cm increase over a 200-m elevation change), and more nitrogen retention in the organic matter. Increased abundance of conifers at higher elevations also likely increased deposition and decreased decomposition.

We've seen that plant growth requires nutrients, nitrogen in particular for terrestrial ecosystems. Algae are primary producers in aquatic and marine ecosystems, turning the energy of light into biomass, and high production of algae goes along with high levels of nutrients, much like farmers fertilizing fields for high crop yields (see Figure 1.3). Figure 6.18 shows precisely this high level of productivity from data taken along five large rivers at multiple locations all over the state of Minnesota that pass through a variety of ecosystem and land use types. Chlorophyll a provides a measure of the suspended algal biomass in the water column of these streams.[4]

Although the data in Figure 6.18 imply a simple dependence of algal biomass on nutrients, the relationship is complicated. Even the ratio of total nitrogen to total phosphorus can affect the results, and one must carefully interpret productivity on this basis.[5] Furthermore, higher turbidity and total suspended solids limit light availability, reducing algal biomass, and algal species composition also changes with many factors, including the nitrogen to phosphorus ratio, with cyanobacteria (blue-green algae) dominating when biomass is high.[6]

---

[1] Rice and Horgan (2011). [2] Lawrence et al. (2000) found that nitrate retention increases with increasing altitude. [3] "O" stands for organic, "a" refers to the decomposed part, and this soil sits just below the leaf litter. [4] Heiskary and Markus (2001). [5] Downing and McCauley (1992).
[6] Smith (2003).

High nutrients produce algal blooms.

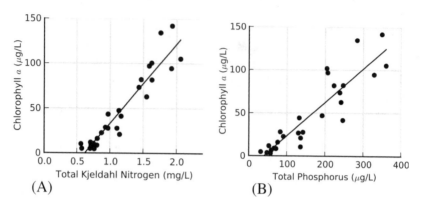

(A)   (B)

Figure 6.18: High concentrations of either nitrogen (A) or phosphorus (B) produce algal blooms in Minnesota rivers (Heiskary and Markus 2001). Total Kjeldahl nitrogen (TKN) includes ammonium and organic nitrogen but excludes nitrates.

Still waters run green.

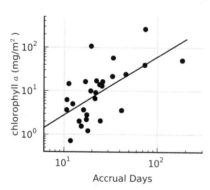

Figure 6.19: The longer collected water sits without mixing(accrual), the greater the abundance of phytoplankton, as measured by chlorophyll a in New Zealand streams (Biggs 2000). Low-accrual days would be like a small pool refreshed by a large stream, and the reverse for high-accrual days.

Finally, simply the number of so-called accrual days that water sits around with reproducing algae plays an important role in algal abundance, as shown in Figure 6.19.[1] These data come from a set of 25 streams and rivers in New Zealand, implying that the flush of runoff coming from precipitation sweeps

[1] Biggs (2000).

Even from pristine watersheds, more runoff means more nutrients.

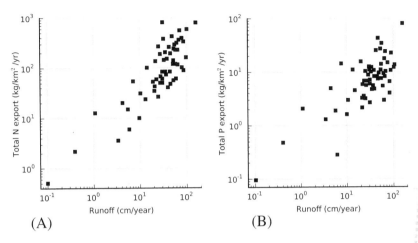

Figure 6.20: Nitrogen (A) and phosphorus (B) runoff from 63 pristine U.S. streams (Smith et al. 2003). For comparison, 100 kg/km^2 equals 1 kg/ha and 0.9 pounds/acre. In large part, more runoff means more nutrient export.

the waterways clean of algae. Flushing can be thought of as cropping the biomass, reducing its total abundance.

The water residence time for a water body, be it lake, stream, or reservoir, is an incredibly important variable for algal blooms.[1] In an extensive survey covering lakes in North America, Europe, Africa, and Australia, phosphorus input and water renewal time combined within the framework of a model for lakes explained 88% of the variation in a lake's primary productivity.[2] Residence time also plays a role in the primary productivity of estuaries (see Figure 12.17).

An examination of nutrient export scaling up to continent-sized watersheds began with measurements of systems as close to natural as possible. The plots in Figure 6.20 use USGS data from 63 minimally impacted U.S. streams, which we'll call "pristine."[3] All of these reference sites are small watersheds because large ones always have at least some amount of development, even if limited. Nonetheless, all watersheds, no matter their size, suffer from atmospheric deposition, which researchers estimated increased exports by 15–100% over truly pristine systems.

---

[1] Søballe and Kimmel (1987). [2] Schindler (1978). [3] Smith et al. (2003) provide nitrogen export data from pristine U.S. streams. The data are in the paper's supplementary information.

This work provides some base level of export for natural systems, with stream concentrations running at 0.02 mg/L in the West and 0.5 mg/L at the Southeast coast. Compare these numbers with the human-impacted concentrations of Figure 6.15, which are at least twice as high. This particular study, which was quite extensive, states that present nitrogen concentrations are about 6.4 times higher than these background levels, and phosphorus levels are about doubled. It's important to emphasize that nitrogen concentrations are rather independent of runoff amounts, which means that total nitrogen export depends more on the *volume* of runoff than on the concentration,[1] where runoff volume explains 70–80% of the variation in nutrient export for the data shown in Figure 6.20. Furthermore, as shown in Figure 4.9, runoff depends greatly on the amount of imperviousness: no impervious surfaces, very little runoff and very little nutrient export.

Speaking of concentrations, let's connect the units of Figure 6.15 to those of Figure 6.20. At high runoff amounts, say, around 100 cm/year, nitrogen export sits at a few hundred kg/km²/year. Let's convert 100 kg/km²/year into pounds/acre/year. Since there's 100 hectares (a hectare is 100×100 m) in kilometer, 100 kg/km²/year equals 1 kg/ha/year. Since 1 kg weighs 2.2 pounds,[2] and 1 hectare equals 2.47 acres, then 1 kg/ha/year equals 0.9 pounds/acre. Hence, the nutrient exports shown in Figure 6.20 aren't terribly high: a few hundred kg/km²/year of nitrogen export translates into just a few pounds/acre/year. Compare this amount with the amount of nutrients exported from human-altered landscapes in Figure 6.17.

Figure 6.21A demonstrates that, overall, about 20–25% of the nitrogen humans add to the landscape (including contributions to wastewater systems) makes its way to rivers, and the remainder is retained (perhaps in vegetation) or denitrified.[3] These data involve 16 watersheds from Maine to Virginia, and data taken from 1988 to 1993. As a fraction of anthropogenic inputs, export from drier northeastern watersheds amounts to 10–15%, but in the wetter watersheds the amount is greater, around 35% (Figure 6.21B). The thinking is that in wetter watersheds water flows are greater and heavier, and thus in situ denitrification has less opportunity to take place (see Figure 6.20). Researchers emphasize that the increase with precipitation doesn't come about by flushing nitrogen built up during dry years because the pattern arises across watersheds and comes from averages over six years, a long enough period for intra-annual variability to be averaged out. As climate change leads to increased precipitation (see Figure 2.3), this leads to more discharge, and increased discharge means we can expect more nitrogen export in the future.

---

[1] Smith et al. (2003) state that, although atmospheric deposition did not predict total nitrogen export, it was important for nitrate, ammonium, and dissolved organic nitrogen. [2] Recall that a kilogram is a unit of mass, while pounds is a unit of weight. [3] Howarth et al. (2006) summarized N export data from the late 1980s to early 1990s.

Put more nitrogen and water in, get more nitrogen out.

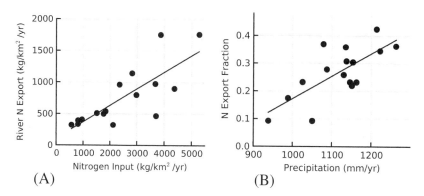

Figure 6.21: For U.S. rivers that drain into the North Atlantic, nitrogen flux increases with human inputs into the watersheds (A) and watershed precipitation rates (B) (Howarth et al. 2006).

Phosphorus input also determines, in part, phosphorus export (Figure 6.22). Export measurements come from the product of monthly total phosphorus concentrations and daily streamflow measurements, which yields a total mass, and phosphorus inputs come from Figure 6.9. In these particular watersheds, agricultural and urban land use together explained 67% of the variation in phosphorus export.[1] Fertilizer represented the largest source and was strongly correlated with the amount of land planted in corn and wheat, explaining 95% of the variation in phosphorus export. Export values varied by three orders of magnitude across the watersheds. Net anthropogenic phosphorus input (NAPI) differs from net anthropogenic nitrogen input (NANI) in that fertilizer dominates phosphorus everywhere, but NANI was determined more broadly by fertilizer, human food, and atmospheric deposition in agricultural, urban, and forest land uses, respectively.

Expanding our view even further, Figure 6.23A displays data on the human influences on nutrient budgets compiled from 14 regions of the North Atlantic Ocean.[2] Of the nitrogen inputs, sewage and wastewater account for 12% going into the North Atlantic averaged across all source regions, with the highest input being 26% from the Northeast U.S. coast and the lowest being 7% for northern Canadian rivers. Only 25% of nitrogen and phosphorus goes

---

[1] Han et al. (2011). [2] Howarth et al. (1996) examined nutrient budgets.

Put more phosphorus in, get more phosphorus out.

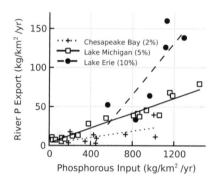

Figure 6.22: Phosphorus exported from watersheds increases with varying levels imperviousness and developed land use fractions that lead to varying levels of anthropogenic phosphorus inputs (Han et al. 2011).

All around the world, nitrogen export increases with human inputs.

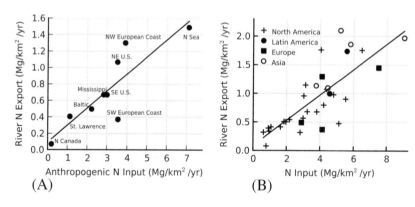

Figure 6.23: (A) North Atlantic nitrogen export from listed watersheds increases with human inputs (Howarth et al. 1996). (B) Nitrogen export increases versus human and natural inputs for selected regions across the world (Boyer et al. 2006). Mg-megagrams (10^6 g).

downstream, as shown by the relations in Figures 6.21 and 6.22.[1] Only about 1% of anthropogenic nitrogen makes its way into long-term groundwater storage. These researchers believe denitrification in wetlands and aquatic systems dominates the loss of nitrogen.

---

[1] Howarth et al. (2006).

Nitrogen export increases with an area's human population.

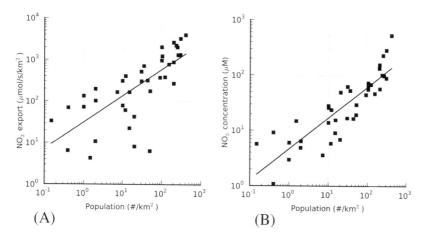

(A)   (B)

Figure 6.24: Both nitrate export from (A) and nitrate concentration in (B) the rivers draining watersheds increase with human population density, showing that it serves as a good proxy for watershed nitrogen input (Peierls et al. 1991).

Comparing the two nutrients, nitrogen to phosphorus input ratios (N:P) vary from a high of 271:1 in the Gulf of Mexico to a low of 4.8:1 in the Amazon basin.[1] Compare these numbers with the North Atlantic Ocean's average N:P input ratio of 13:1. However, these ratios sit around 200:1 for forested watersheds, with agriculture ranging from 30:1 to 300:1 and urban areas ranging from 5:1 to 40:1. These researchers conclude that N:P ratio isn't a great measure of disturbance, though disturbance can produce extreme values and highly eutrophic coastal waters, as seen at the mouth of the Mississippi River.

Another summary spanned river systems across the world, showing that rivers transported 11 teragrams (10^{12} Tg) of nitrogen to terrestrial and aquatic systems, and 48 Tg to coastal areas, giving the export/input plot shown in Figure 6.23B.[2] Given that Earth's land surface measures 150 million km^2, or 15 billion hectares, that total export of 59 Tg, or 59 billion kg, represents roughly 4 kg/ha. Since the slope of the line in Figure 6.23B is about 0.2, that means the net nitrogen export is about one-fifth of the input nitrogen, which implies an input of about 20!

---

[1] Howarth et al. (1996). [2] Boyer et al. (2006) looks at nitrogen exports from worldwide watersheds and compares them with net imports.

Sources of the nitrogen exported by streams around the world.

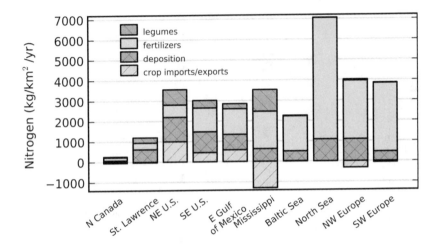

Figure 6.25: The relative contributions of four nitrogen sources for the river exports across the North Atlantic shown in Figure 6.23A (Howarth et al. 1996). Export of crops from the Mississippi and northwestern Europe basins lowers the nitrogen exported via rivers.

Previous plots show that nitrogen export depends on a complicated summation of human nitrogen inputs. Skipping the complicated details, Figure 6.24 plots nitrate export and concentration for 42 rivers around the world directly against the respective watershed's human population density.[1] This nitrogen could come from sewage, atmospheric deposition, fertilizers, and deforestation, and each component has a reasonably strong connection to human density. As shown, there's a similar increase in total export and concentration.

Figure 6.24B lists nitrate micromolar concentrations (1 μM nitrate equals 0.06 mg/L; see Figure 6.7 discussion), setting the bottom of the nitrate concentration scale. One hundred times that amount (100 μM) equals 6 mg/L, the concentration found in areas with population densities of about a few hundred people per square kilometer. Similarly, the export scale can be converted by noting that, again, a mole of nitrate has a mass of 62 g and a year equals

---

[1] Peierls et al. (1991) showed the connection between human population and nitrate export.

31.5 million sec (give or take a few hours), meaning that 1 μmoles/km^2/sec (after multiplying by 62 g/mole and 31.5 x 10^6 sec/year) equals very nearly 2 kg/km^2/year. Again, that number sets the bottom of the nitrogen export axis of Figure 6.24A, and we find a nitrogen export of 1,000 kg/km^2/year at a population density of a few hundred people per square kilometer. Realize that 1,000 kg/km^2 equals 10 kg/ha, or about 10 pounds/acre. Many of these values then hover around a total export of about 30 pounds/acre. This amount compares well with the range shown in Figure 6.23.

As mentioned earlier, net anthropogenic nitrogen inputs include human-related atmospheric deposition, fertilizers, and crop nitrogen fixation, as well as the overall nitrogen import or export contained in foods and feeds.[1] Figure 6.25 shows how these various components vary among different geographical areas in their contribution to nutrients in the North Atlantic Ocean. Several different approaches went into making these detailed estimates, sometimes resorting to just using per capita estimates. Of note, sewage inputs represent a dominant source of nitrogen, and about 12% of the export from North American rivers, compared with 25% for western Europe, 6% for northwestern Africa, and 3% for Central and South America.

Chapter Readings

Aber, J.D., C.L. Goodale, S.V. Ollinger, M.-L. Smith, A.H. Magill, M.E. Martin, R.A. Hallett, and J.L. Stoddard. 2003. Is nitrogen deposition altering the nitrogen status of northeastern forests? *Bioscience* 53: 375–389.

Bernhardt, E.S., L.E. Band, C.J. Walsh, and P.E. Berke. 2008. Understanding, managing, and minimizing urban impacts on surface water nitrogen loading. *Ann. N Y Acad. Sci.* 1134: 61–96.

Bragazza, L., C. Freeman, T. Jones, H. Rydin, J. Limpens, N. Fenner, T. Ellis, R. Gerdol, M. Hájek, T. Hájek, P. Iacumin, L. Kutnar, T. Tahvanainen, and H. Toberman. 2006. Atmospheric nitrogen deposition promotes carbon loss from peat bogs. *Proc. Nat. Acad. Sci. USA* 103: 19386–19389.

Galloway, J.N., J.D. Aber, J.W. Erisman, S.P. Seitzinger, R.W. Howarth, E.B. Cowling, and J. Cosby. 2003. The nitrogen cascade. *Bioscience* 53: 341–356.

Galloway, J.N., and E.B. Cowling. 2002. Reactive nitrogen and the world: 200 years of change. *Ambio* 31: 64–71.

Han, H., N. Bosch, and J.D. Allan. 2011. Spatial and temporal variation in phosphorus budgets for 24 watersheds in the Lake Erie and Lake Michigan basins. *Biogeochemistry* 102: 45–58.

---

[1] Howarth et al. (1996) estimated nutrient inputs.

Pardo, L.H., C. Kendall, J. Pett-Ridge, and C.C.Y. Chang. 2004. Evaluating the source of streamwater nitrate using δ_N^{15} and δ_O^{18} in nitrate in two watersheds in New Hampshire, USA. *Hydrol. Proc.* 18: 2699–2712.

Smith, R.A., R.B. Alexander, and G.E. Schwarz. 2003. Natural background concentrations of nutrients in streams and rivers of the conterminous United States. *Environ. Sci. Technol.* 37: 3039–3047.

Stoddard, J.L. 1994. Long-term changes in watershed retention of nitrogen: Its causes and consequences. In Environmental chemistry of lakes and reservoirs, L.A. Baker, ed. Advances in Chemistry 237. American Chemical Society (Washington, DC), USA; 223–284.

Chapter 7

Mercury and Other Metals

Mercury and other metals enter ecosystems as a result of atmospheric deposition and automotive sources. Heavy metals, including copper, lead, and zinc, found in lake sediments closely match the emissions of regional sources, showing large increases since the mid-1900s. Land animals eat the vegetation that takes up heavy metals, and, in a positive sign, mercury and lead content in moose reflect the decreasing levels seen over recent decades with the introduction of emissions regulations.

For the most part, mercury enters aquatic organisms through phytoplankton, which are then eaten by predators, biomagnifying mercury concentrations as the biomass progresses up the food chain. Mercury concentrations within fish alone increase fourfold from detritus feeders to fish-eating predatory fish. Furthermore, the bigger a fish is, the higher mercury concentration it has from a longer lifetime of feeding, as well as a recent diet of bigger fish also containing high levels of mercury.

In a fascinating ecological complication, algal blooms that result in depleted oxygen levels also dilute mercury within the bloom, resulting in lower concentrations of mercury in the trophic levels above phytoplankton. Thus, reducing nutrient loads without addressing mercury levels may be hazardous to fish health and, ultimately, the humans that eat fish.

Mercury and lead in moose teeth.

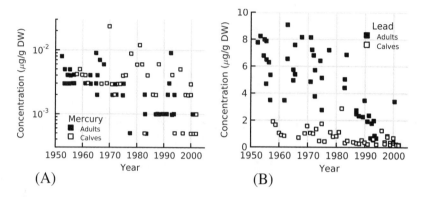

(A)   (B)

Figure 7.1: Mercury (A) and lead (B) in moose teeth collected from Isle Royale National Park that sits in Lake Superior (Vucetich et al. 2009). Atmospheric deposition of these heavy elements reaches this isolated wild population, and the decreasing concentrations demonstrate the positive consequences of emissions regulations. DW, dry weight.

Emissions, long range transport, and deposition lead to heavy metals distributed across the terrestrial landscape. Evidence of this distribution can be seen in the mercury and lead uptake in moose teeth collected from 1952 to 2002 shown in Figure 7.1.[1] These teeth were collected from Isle Royale National Park, Michigan, an isolated island in Lake Superior very near Thunder Bay, Ontario. The reduction seen over this interval comes from terminating lead additives to gasoline and reduced coal emissions and took place alongside a similar reduction seen in gulls.[2] Isotopic signatures even point to increased use in the 1980s of coal in Missouri and Tennessee, as well as the removal of lead additives in gasoline.

There are some surprising sources of mercury that lead to localized pollution, including mercury-coated seeds for antifungal purposes—occasionally these seeds are ingested.[3] Mercury once was a component of latex paints, playing an antifouling role, but that use was terminated around 1991. Neurological effects have been the primary focus for understanding and preventing mercury contamination. With reduced production and emissions, bioaccumulation is now the main source for contamination issues.

---

[1] Vucetich et al. (2009). [2] Koster et al. (1996). [3] Tan et al. (2009) provide an excellent medically focused overview of mercury.

Heavy metals affect the endocrine system, which involves various glands, including the thyroid and gonads,[1] that release hormones into bloodstream. Across many species mercury has an affinity for testes, affecting male reproduction. Females also have a variety of reproductive problems: altered menstrual cycles, reduced ovulation, and spontaneous abortions. These reproductive concerns have been long known: mercury was used to induce abortions as far back as the 1400s.[2] Reproductive problems have been seen in female dentists and their assistants, but problems with the heart, hypertension, and atherosclerosis (hardening of the arteries) has also been seen in humans due to mercury. Across species, females tend to accumulate mercury in the brain, and males in the kidneys; however, males seem to have more neurological and behavioral problems, while there are fewer male births. Mercury contamination also reduces fertility and fecundity in the mother through several routes from libido to germ cell mutations, and it's then transferred from mother to egg and/or embryo and to offspring. All of these problems combined affect reproductive success.

A couple of comprehensive publications on mercury are the USEPA's *Mercury Study Report to Congress*,[3] and the National Academy of Sciences' *Toxicological Effects of Methylmercury*.[4] Information provided by these extensive resources fully covers the issues.

Once on the landscape, heavy metals make their way to surface waters. Figure 7.2 plots the mercury entering a couple of northeastern U.S. lakes over 150 years, sampled via sediment cores in lake bottoms, reflecting the global mercury production numbers shown in Figure 3.1. The results compare a lake connected to stream with an urban stormwater system and a glacial lake (imaginatively called Glacial Lake), not connected to streams, sitting in a rather pristine watershed.[5] Groundwater seepage feeds Glacial Lake, which has a watershed that really just constitutes the lake's surface. That means that the only source of mercury is atmospheric deposition, and upstream point or nonpoint pollution sources are excluded. This scenario also means that a core sample of its sediments provides an excellent timeline of deposition, and in this plot the timeline goes back 150 years.

The watershed of Skaneateles Lake is also quite small—just four times larger than the lake's surface area—and has a typical recent history. The watershed was once a forest, cleared for both logging and agriculture, and then forests recovered as agriculture declined, and urbanization happened.

---

[1] Endocrine glands include the pineal, pituitary, hypothalamus, thyroid, pancreas, adrenal, and gonads (testes or ovaries, depending on gender). [2] Tan et al. (2009). [3] The eight-volume *Mercury Study Report to Congress* (USEPA 1997) can be found at www.epa.gov/mercury/report.htm. [4] The *Toxicological Effects of Methylmercury* NRC (2000) is available at www.nap.edu/catalog.php?record_id=9899. [5] Bookman et al. (2008) studied mercury emissions settling in New York lakes.

Mercury in lakes matches emissions.

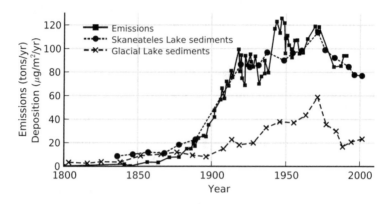

Figure 7.2: Mercury deposition in sediments of northeastern NY lakes increased in the 20th century (Bookman et al. 2008). The pattern of emissions closely matches the sediment levels both of Skaneateles Lake, connected to a stream and stormwater system, and the lower levels of the hydrologically isolated Glacial Lake.

Low sediment accumulation rates were found in both lakes, roughly 0.02 g/cm^2/year in Glacial Lake and 0.09 g/cm^2/year in Skaneateles Lake. Sediment cores were dated using ^{120}Pb, a radioactive isotope with 22.3-year half-life, and a variety of assumptions regarding sediment sources. Before industrialization, mercury accumulated at 3.0 μg/m^2/year, whereas present-day deposits from the global atmospheric pool of mercury are about 9.0 μg/m^2/year (see Figure 3.2). Given a peak flux to the glacial lake of 59 μg/m^2/year, local and regional emissions account for 80% of mercury deposition. Main sources in the area include burning medical and municipal wastes, metal smelting, coal combustion, and chlor-alkali facilities. These latter facilities produce chemicals for wood pulp production using a process involving mercury, and some of these closed plants are now USEPA superfund sites.

A fascinating change in mercury flux in Glacial Lake took place in the 1920s: a 10-fold decrease in the fraction of mercury deposition that made its way to the lake as recovering forests in the region sequestered the deposited mercury.[1] Still, higher atmospheric deposition led to higher lake fluxes.

Another estimate for mercury accumulation in lakes across the northeastern U.S. ranges over 27.1–175.4 μg/m^2/year, with the higher values occurring

---

[1] Bookman et al. (2008).

Copper and mercury recently added to lakes.

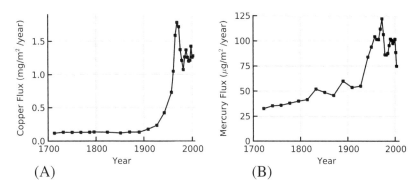

Figure 7.3: Copper (A) and mercury (B) deposition rates measured from the sediments from Otisco Lake, NY, confirm the recent high deposition rates shown in Figure 7.2 (Bookman et al. 2010).

after 1970, in line with other temporal patterns (but see Figure 3.9).[1] However, a major correction to these numbers down to a range of 10.4–66.3 μg/m2/year comes about by accounting for various features, in particular a process called focusing. This process represents within-lake movement or concentration of sediments that alters the mercury profile. Imagine, perhaps, a bucket of swirling water that picks up sediments from one area and redistributes and concentrates (focuses) them around obstructions and such located elsewhere. The data implied that the import of mercury from the watershed is several times higher than deposition directly onto one particular lake studied in detail, and that import will continue for some time.

Otisco Lake in New York sits near Glacial Lake and Skaneateles Lake, the two lakes discussed in Figure 7.2.[2] Five large streams drain the nearly 100-km^2 watershed and feed the 20-m-deep lake, which was dammed in 1869, adding another meter of depth. The residence time of water in Otisco Lake is 1.9 years, and the region gets 1 m of precipitation each year. Most of the watershed is forest, grassland, or wetlands, but significant urbanization exists along the lake's edge and transportation corridors. Figure 7.3 shows the copper and mercury fluxes in Otisco Lake measured in its sediments. Copper-based algicides had been added over a long period to prevent algal blooms, thereby improving water quality despite the increase in nutrients and organic matter

---

[1] Perry et al. (2005) examined mercury in lake sediments. [2] Bookman et al. (2010) studied heavy metals in a New York's Otisco Lake.

Urban areas contribute heavy metals to lakes.

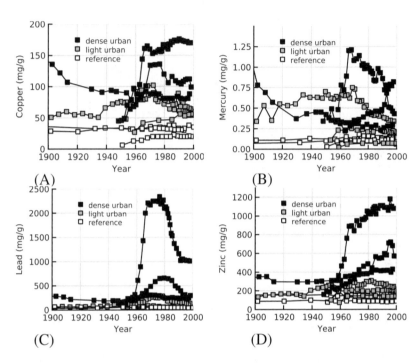

(A) (B) (C) (D)

Figure 7.4: Time course of four heavy metal pollutants in lake sediments from 35 lakes and reservoirs distributed across the U.S. (Mahler et al. 2006). Reference streams drain watersheds with less than 1.5% urban land use, and the results clearly demonstrate that dense urban land use contributes the most heavy metals.

coming in via streams. Most of the copper comes from that source, and the mercury comes from atmospheric deposition of local and regional emissions.

Going beyond mercury and copper, Figure 7.4 considers two additional heavy metals, lead and zinc, but again using lead and cesium isotopes for dating. Not shown here are additional results for cadmium, chromium, and nickel, all of which show a similar land use pattern. These results come from a USGS study examining sediment cores collected between 1996 and 2001 from 35 lakes, consisting of 23 reservoirs and 12 natural lakes, scattered across all regions of the contiguous 48 states of the U.S. and having a wide range of land uses, climates, and watershed sizes.[1] The study argues that for reference lakes

---

[1] Mahler et al. (2006) characterized pollutant profiles in lake sediment cores.

that don't have urban inputs, heavy metal concentrations show either no trend or a decreasing one. For example, lead decreased by 46%, cadmium by 29%, chromium by 34%, and nickel by 29%. Copper, zinc, and mercury levels were not showing a uniform decrease. In urban areas, lead is decreasing due to the termination of leaded gasoline, while zinc is increasing, likely due to tires, roofing, and other anthropogenic uses, as outlined in Figure 3.15. Mercury isn't changing much across this span of time, likely due to global emission sources and atmospheric deposition.

Besides the automotive-related pollutants listed in Figure 3.17, a few more show up from other sources.[1] Arsenic comes from a broad range of industrial and agricultural sources, including lawn chemicals, and as the result of being a former main component of preservative-treated lumber. Stainless steel contains nickel, released as a consequence of corrosion. In addition to automotive sources, copper comes from algicides, fungicides, and pesticides, used as an intentional ingredient taking advantage of its toxicity, which is as low as tens of micrograms per liter for snails and *Daphnia*.

Metals have a strong affinity to clays, and the concentrations in sediments are orders of magnitude (factors of ten to thousands) higher than the concentration in the water above the sediments. However, changes in streamflow, changes in pH, changes from anaerobic to aerobic conditions, and the presence of other chemical compounds can move metals from sediments to the water column and vice versa.

Sources of heavy metals and PAHs, in descending order of importance, go from highways to industrial areas to residential areas. Metal tolerance, in order of decreasing tolerance, runs from chironomids (mosquito-looking non-biting midges), trichopterans (also known as caddisflies, aquatic insect larvae that build protective cases), plecopterans (stoneflies), to ephemeropterans (the flighty, delicate mayflies).

Overall, the generally decreasing recent trends lead to the conclusion that regulations to control regional emissions are working, but under various scenarios sources from one landscape scale can overwhelm local, regional, or global trends in emissions and deposition.

As discussed in Chapter 6, nutrients are critical for growth, but too much growth in the form of algal blooms leads to problems. While algae grow and are eaten by consumer species, metal pollutants tag along, too. This uptake is shown in Figure 7.5, summarizing a detailed study of 15 Wisconsin lakes with few direct human impacts, where atmospheric deposition dominates the mercury sources with an estimated annual input around 10 μg/m^2/year.[2] Samples were taken in the early 1990s, after the mercury peak of the 1970s (see Figure 3.1).

---

[1] Beasley and Kneale (2002) reviewed heavy metals and PAH effects on stream macroinvertebrates and includes a nice table of typical sources for heavy elements. [2] Watras et al. (1998).

Mercury enters at the bottom.

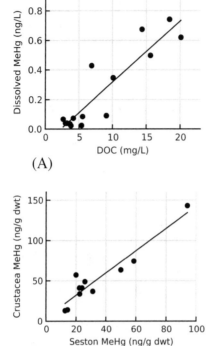

(A)

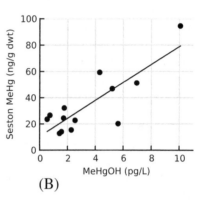

(B)

(C)

Figure 7.5: Mercury in these Wisconsin lakes makes its way to zooplankton (Watras et al. 1998). (A) Dissolved methylmercury (MeHg) increases with dissolved organic carbon (DOC). (B) MeHg bound to OH^- ions predicts seston concentrations. (C) Seston concentrations predict MeHg concentrations in filter feeders. dwt, dry weight.

In Figure 7.5A, methylmercury (MeHg)[1] increases with dissolved organic carbon (DOC), which means that it is "complexed" with organic carbon—indeed, 99% of the lake's mercury is organically complexed, or somehow attached to organic carbon. At this point, it might be useful to distinguish between methylmercury and plain old elemental mercury. Methylmercury accounts for only about 15% of the total mercury in these lakes, and elemental mercury levels were 10 times or more greater than these methylmercury concentrations. What comes next makes methylmercury more important.

At the first stage, we can think of DOC sweeping up both "metallic" mercury (Hg) and methylmercury, but other options may produce the correlation. Perhaps DOC and methylmercury are both being produced by wetlands, or DOC fuels the microbes that methylate mercury, or perhaps, once methylmer-

---

[1] Methylmercury means the Hg has a methyl group (CH_3) attached.

cury attaches to bits of carbon, the particle shades or protects the methylmercury from photodegradation.

A new term for the dictionary is "seston"—detritus floating in the water—which encompasses phytoplankton and bacterioplankton (which might object to being called "detritus"), and even bits of cells floating around in the water, all of which serve as dinner for zooplankton. For the purposes of these results, the fraction of seston considered is the smallest bits that make it through a 64-μm filter and might be referred to as "microseston." Seston-based organisms take in the DOC and, alas, the mercury complexed with it, representing its entry into the food chain. Measures of seston also include some amount of degraded cells; that fraction doesn't consume anything but it does, unfortunately, hinder the strength of some methylmercury correlations.

There is a clear preference (or some form of active uptake) for methylmercury by the lowest links of the food web. Within seston the concentrations of elemental and methylmercury averaged 170 ng/g and 33 ng/g of dry mass, respectively. Given that the total mercury concentration in the lakes was roughly 1 ng/L, or one part per billion, these concentrations reflect a one- to two-order increase in mercury concentration. However, the concentrations in DOC averaged 170 ng/g and 23 ng/g of dry mass, respectively. In other words, algae are what they eat.

To put these concentrations in context, let's consider how the mercury comes to the watershed. Mercury uptake through leaf stomata is elevated in forests due to foliage surface area, and the concomitant increase in deposition velocity (see Figure 3.5).[1] A study of mercury uptake in forests in the Adirondack Mountains in New York showed leaf tissue accumulates an estimated 0.2–0.3 ng/g/day from this deposition process, reaching 37 ± 12 ng/g at the end of the season. That latter number isn't too far off from the seston concentrations mentioned above. Researchers studied three tree species, and the American beech sequestered about 50% more than yellow birch and sugar maple. Leaves drop to the ground, and mercury in litterfall was around 17 mg/m^2/year, which exceeds the area's wet deposition of mercury of about 5.8 mg/m^2/year. Given that a hectare is 10^4 m^2, that litterfall yields 0.17 kg mercury per hectare annually— almost 0.2 kg mercury dropping from the leaves of trees collected from the air! Watershed retention of mercury was connected with the retention of organic carbon, as seen in the DOC/mercury plot of Figure 7.5 showing that the two quantities co-occur in water.

Now, the next step is to consider mercury concentrations in zooplankton in the lake. These are the beasts that eat seston, and they averaged around 53 ng/g dry mass for methylmercury but only 29 ng/g dry mass for elemental mercury. Those numbers flip the representation of the two forms of mercury, elemental and methylated, seen in the seston, and so begins the bioaccumulation

---

[1] Bushey et al. (2008).

of methylmercury. Indeed, an examination of small fish revealed concentrations of 485 versus 27 ng/g dry mass for methylated and elemental mercury, respectively.[1]

Very little methylmercury work has been performed in marine systems, despite the fact that 60% of our fish and shellfish consumption comes from marine systems, and people's greatest mercury exposure comes through seafood.[2] Asians, Native Americans, Pacific Islanders, all high consumers of fish and shellfish, have higher blood mercury than other groups. It's generally understood that microbes in coastal sediments produce methylmercury, similar to the situation with inland wetlands, and that benthic animals rooting around in these sediments either ingest it or return settled mercury to the water column. Marine bioavailability of methylmercury is also affected by the presence of other factors, such as chlorine and sulfur. From the water column the methylmercury bioaccumulates to the highest trophic levels—birds, marine mammals, and sharks, not to mention people. Given this knowledge, researchers suggest monitoring exposed populations and tracking the most commonly consumed fish and fish production areas.

As mentioned above, algae take up methylmercury four times faster than elemental mercury compared with their water concentrations, but fortunately, methylmercury constitutes just 1–35% of total mercury in water bodies.[3] As mercury continues up the food chain, its concentration increases, and as shown in Figure 7.6, the proportion of methylmercury increases to nearly 100% in loons at the very top.

Of course, a host of other problems, such as lake acidification, nutrients, and land use change, complicate the implications for mercury uptake by animals. For example, lake acidification results in mercury becoming more available as methylmercury, but higher nutrient levels increase algal blooms, which dilutes the mercury taken up (see Figure 7.10).

One study surveyed more than 150 lakes to examine the many factors associated with high mercury levels in fish and zooplankton.[4] Important watershed variables considered included lake area and the ratio of watershed area to lake area—increasing either one increased mercury concentrations in fish. Five chemical variables, four land use variables, and one ecological variable all decreased mercury levels: chemical factors were pH, ANC (acid-neutralizing capacity), SO_4, conductivity, and nutrients; land use variables were residential, agricultural, commercial/industrial, and "disturbed"; and the ecological variable was zooplankton density.

Note that these factors are correlated variables, not descriptions of causal mechanisms. However, the chemical variables generally indicate poorly

---

[1] Fun fact: the ratio of wet weight to dry weight for fish is 3.85. [2] Chen et al. (2008b) reviewed methylmercury studies in marine systems and outline a plethora of research questions. [3] Driscoll et al. (2007) cites Watras et al. (1998) for the phytoplankton data. [4] Chen et al. (2005) studied the factors associated with high mercury levels in fish in northeastern U.S. lakes.

Up the food chain, up the mercury.

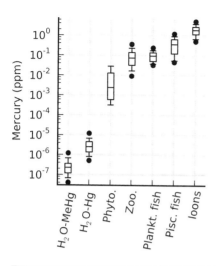

Figure 7.6: Methylmercury concentration in eastern North America increases at higher trophic levels (Driscoll et al. 2007). Mercury concentrations in water are shown as the methylmercury fraction (H_2O-MeHg) and the total concentration (H_2O-Hg).

buffered lakes (see Figure 8.7), where biological mercury uptake becomes more dominant. Land use type correlations come about because forested watersheds have high mercury transport from land to water, but human-altered landscapes increase nutrient transport, which increase primary productivity in the water, which dilutes the mercury as it is taken up. Highly productive lakes also have higher levels of DOC, which also takes up more mercury, pulling it from the biological uptake route.

Putting it all together, these results mean that relatively undisturbed watersheds, where atmospheric deposition of mercury dominates, can have the greatest mercury bioaccumulation! This paradox comes about because increasing human land use results in decreasing mercury in fish for two reasons: first, forests enhance mercury transport, and second, increased nutrients enhance productivity and dilute the mercury, diluting the bioaccumulation of higher trophic levels. Increased organic matter in sediments coming from algal blooms binds with mercury, reducing methylmercury, and the sediments produce sulfide (S^{2-}), which hinders the methylation of mercury.[1]

Yet another seemingly contradictory correlation is that a higher percentage of shoreline in wetlands increases mercury in fish because wetlands can enhance methylmercury production while those same wetlands are valuable for denitrification. It's a very strange situation that implies controlling nutrients using wetlands might enhance mercury in fish. The lesson seems to

---

[1] Chen et al. (2005).

Bottom fish, top fish, less mercury, more mercury.

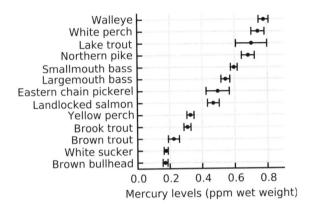

Figure 7.7: Predatory fish have much higher mercury concentrations compared with detritivores and fish feeding at lower trophic levels (Driscoll et al. 2007). Data come from the northeastern U.S. and neighboring regions of Canada.

be that interdisciplinary, watershed-scale measures need broad and careful consideration.

Downstream effects can take place way downstream, as the case of a mercury point-source incident demonstrates.[1] Mercury used in the production of synthetic fibers throughout much of the 20th century contaminated a site in Waynesboro, Virginia, upstream from the South Fork Shenandoah River. Mercury within the sediment appears limited to about 14 km downstream of the contamination site, but resident bird populations along the river had roughly 10 times higher blood mercury levels as far as 35 km downstream. Even the migrant bird populations experience mercury levels elevated five-fold over normal levels. An additional 100 km farther downstream, the blood mercury levels of local birds remain elevated above the reference sites upstream of the contamination site. Continued methylation of this mercury takes place in the stream far from the contamination site, with implications for the downstream fish.

Figure 7.7 summarizes mercury contamination in several northeastern U.S. fish,[2] particularly the game fish I recall catching in Minnesota. Summarized are the results of 24 studies on 40 fish species with 15,305 records,

---

[1] Jackson et al. (2011) examined mercury uptake as far as 137 km downstream of a historical mercury contamination site. [2] Driscoll et al. (2007) cite the project also described in detail by Kamman et al. (2005) as the source of this fish mercury data.

but the data here include only 13 highly sampled species with fish collected after 1980 from the northeastern U.S. and neighboring portions of Canada. Although the original plot separates results from reservoirs, lakes, and rivers, which show important differences, this plot shows only an overall average concentration for each species.

Disturbingly, the proportion of water bodies producing fish fillets exceeding the USEPA's methylmercury level, set at 0.3 $\mu g/g$, ranged from a low of 14% for brook trout to a high of 42% for yellow perch. The piscivorous fish coming from reservoirs, as opposed to those from lakes and streams, had the highest mercury levels.[1] Muskellunge (a top predator not plotted in Figure 7.7) had the highest mercury concentrations of 0.98 $\mu g/g$, more than three times the USEPA's level. Important features that correlated with high concentrations included water body pH and DOC, watershed size, and, for predators, fish length.

The overall conclusion is to not eat the old, big fish that come from reservoirs—if it looks like a keeper, don't do it (we'll see this warning again in Figure 7.8). Only eat fish that feed low in the food web, where bioaccumulation is just beginning (see Figure 7.6), making suckers and bullheads—yum, yum—great fish to eat.

The above-mentioned USEPA methylmercury level of 0.3 $\mu g/g$ equals 0.3 mg/kg. Toxic concentrations of methylmercury sit around 5 mg/kg (the same as 5 $\mu g/g$) wet weight in brain, while 20 mg/kg in liver has been reported in some animals.[2] Various organisms, including some seabirds and marine mammals, have a limited ability to "demethylate" methylmercury, which may detoxify the mercury to some extent. The process combines the mercury with selenium, followed by lysosomic degradation, and the results collect in the liver.

Mercury harms organisms as they take it up. Fathead minnows with 0.7–0.8 mg/kg mercury had 39% reduced spawning success, and these concentrations reduced reproductive features in white perch, walleye, and northern pike. Mink and otter, again top predators, are best-known mammalian wildlife species for mercury contamination. For these animals, diets with methylmercury concentrations exceeding 1 mg/kg caused neurotoxicity and death, although free-living mink and otter have been found with levels at 3 mg/kg. In a variety of smaller mammals, 3–11 mg/kg brain levels caused adverse behaviors, including reduced predator escape ability, and 12–20 mg/kg caused blindness and seizures.

The high mercury levels in loons are shown in Figure 7.6, and the effects on loons are pretty well understood. To compare the concentrations associated with a litany of problems with the levels reported there, 1 ppm equals 1 mg/kg. Diets with >0.3 mg/kg resulted in "severely reduced reproductive success" arising from reduced egg laying and an inability to maintain a terri-

---

[1] Kamman et al. (2005). [2] Scheuhammer et al. (2007).

tory.[1] Breeding adults with blood levels exceeding 3 mg/kg had 40% fewer fledged offspring than those with blood levels <1 mg/kg. Concentrations exceeding 15 mg/kg caused death. Eggs with >1 mg/kg have poor hatchability, and in Maine 15% of loon eggs had concentrations exceeding 1.3 mg/kg. In Florida herons, there's evidence of increased infections with increasing mercury concentration. Again, these are top predators: low-trophic-level seed- or insect-eating birds shouldn't have as great a risk of contamination.

Relatively little information on reptiles and mercury is out there, but one study showed that snapping turtles collected from a river in central Virginia contaminated with mercury showed 30- to 80-fold elevated levels of mercury in their muscle tissue and their eggs, and infertility increased by 49–174%.[2]

Silver, another heavy (or, technically, transition) metal, is yet another environmental neurotoxin, sometimes used in antimicrobial consumer products.[3] One major study organism is the zebrafish, treated much like the fruit fly or mouse of aquatic systems. A study exposing developing embryos to 0.03–0.3 μM silver (roughly 0.003–0.03 mg/L) immediately after fertilization produced adults that more rapidly learned to avoid aversive situations, like having the wall of an aquarium closing in on it. Specifically, fish faced a choice of swimming into one of two chambers, one of which would be subsequently narrowed. Fish in the control group chose the nonsqueezed chamber about 65% of the time, but those exposed to the highest silver concentrations chose it about 80% of the time. One might ask, so what? This choice represents a higher level of anxiety on the individual's part, and that anxiety has implications in the trade-off between predator avoidance and feeding rates. Anxious fish may feed less and subsequently reproduce less.

We've seen that top predatory fish bioaccumulate mercury. Within a species, small fish have less mercury than big fish, as Figure 7.8 demonstrates for bass and walleye in Lake Simcoe in southern Ontario.[4] Big fish collect mercury through their diet, storing it in their tissues over their long lives. Two advisories set up by the USEPA for mercury concentrations in fish suggest fish meal frequencies versus mercury concentrations. For fish at the lower level, with concentrations between 0.22 μg/g and 0.95 μg/g, we should limit ourselves to one fish meal per month. At the higher concentrations, above 0.95 μg/g, simply "do not eat." This study also showed that larger fish have higher concentrations of PCBs and flame retardants, making trophy fish something to catch, admire, and release. The μg/g unit of the advisory levels equals the part-per-million unit of Figure 7.7, and the "do not eat" level matches highest concentration in Figure 7.8, meaning a person in the southern Ontario and the northeastern U.S. shouldn't have a steady diet of large, locally caught predatory fish.

---

[1] Scheuhammer et al. (2007). [2] Hopkins et al. (2013). [3] Powers et al. (2011) studied silver exposure in zebrafish. [4] Gewurtz et al. (2011) show contaminant levels in Lake Simcoe in southern Ontario.

Don't eat trophy fish.

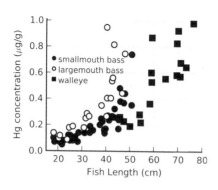

Figure 7.8: Bigger predatory fish in Lake Simcoe in southern Ontario contain higher mercury concentrations due to longer lives and consuming larger prey with higher contamination (Gewurtz et al. 2011). Concentrations measured with respect to wet weight.

Individuals of other species also increase their mercury concentration with size, but those shown in Figure 7.8 attracted my attention because I caught them in my younger fishing days. Northern pike, another species common to my native state of Minnesota, grow to nearly 1 m long, with mercury levels reaching the upper levels shown here. PBDEs, the flame retardants discussed in Chapter 8, show a similar increase in concentration with fish length; however, the fish concentrations were all below the advisory concentrations for the various isomers.

On a more positive note, fish mercury levels have trended downward over the last three decades in response to emission reductions (similar to that of moose shown in Figure 7.1), but whitefish are an interesting exception. Whitefish have switched their diet to the zebra and quagga mussels that invaded the Great Lakes region in the 1980s and Lake Simcoe in the early 1990s. These filter feeders efficiently collect mercury, which they pass on to whitefish.[1]

Figure 7.9 shows a clear connection between high mercury levels found in fish and low total phosphorus.[2] This inverse correlation seems unexpected, and indeed, disturbed watersheds should have both nutrients and heavy metals exported via streams and deposited in lakes. Experimental studies show that this correlation is mediated by lower methylmercury concentrations in the phytoplankton due to phosphorus-induced blooms.[3] Effectively, there's one methylmercury pie, as it were, and the more algal cells there are, the smaller the slice each cell gets.

The pattern repeats itself quite strongly across a nutrient gradient extending away from a wastewater discharge site in Baiyangdian Lake in the North

---

[1] Gewurtz et al. (2011). [2] Driscoll et al. (2007) review mercury contamination in northeastern U.S. forests and freshwater systems. [3] Pickhardt et al. (2002).

High mercury connects to low phosphorus.

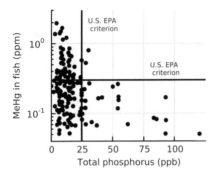

Figure 7.9: High mercury levels in yellow perch from lakes in the northeastern U.S. correlate with low phosphorus levels (Driscoll et al. 2007). As discussed previously, algal blooms that dilute mercury are the likely cause of the correlation.

China Plain.[1] Near the wastewater site, nutrients were highest while metals in fish were lowest, but several kilometers away—yet still in the same complex network of connected lakes—nutrient levels were low while the metal concentration in fish was high.

Plots in Figure 7.10 depict the situation quite clearly as mercury moves up the food chain. Data were collected in 1995 and 1996 from 20 lakes in the northeastern U.S.[2] We've seen that nutrients produce algal blooms (see Figure 6.18). I don't show it here, but the algal mercury concentration decreases with increasing algal density. Next, the algal blooms increase the densities of *Daphnia* and other small, herbivorous zooplankton, and they also reduce the mercury concentration in each individual organism (Figure 7.10A).[3] Mercury dilution continues from zooplankton to fish species of several trophic levels (Figure 7.10B).

Still, mercury levels in fish remain a multifactor problem. As just one example of the complicated correlations, lakes with yellow perch having methylmercury concentrations exceeding 0.3 mg/kg had higher levels of DOC (> 4 mg/L), lower pH (<6), lower acid-neutralizing capacity (<100 μeq/L), and lower total phosphorus.[4] The availability of methylmercury to the phytoplankton depends on many factors we won't delve into here.

There's less clarity on the consequences of other pollutants we've discussed. Mayflies feed on sediments, and sediments are where PAHs and PCBs sit; thus, mayflies tend to have high concentrations.[5] Each PAH can be characterized by its partition coefficient, K_{OW}, which describes its relative affinity for oil versus water, or fat versus nonfatty tissues. Overall, for

---

[1] Chen et al. (2008a).   [2] Chen and Folt (2005).   [3] Pickhardt et al. (2002).   [4] Driscoll et al. (2007).
[5] Gewurtz et al. (2000) studied PCBs and PAHs in Lake Erie invertebrates.

Mercury dilutes when nutrients pollute.

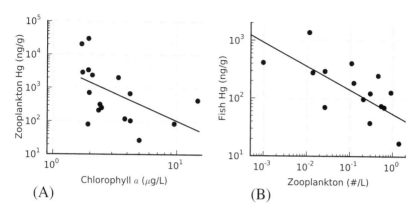

Figure 7.10: Algal blooms coming from high nutrient levels in northeastern U.S. lakes lead to lower fish mercury levels (Chen and Folt 2005). (A) Small zooplankton have lower mercury levels when algal abundance is higher. (B) Fish have lower mercury levels when small zooplankton abundance is higher. Combined, these results lead to fish with lower mercury burdens when algal blooms occur.

the large variety of PAHs, a larger K_{OW} predicts less bioaccumulation by mayflies, mussels, amphipods, and crayfish, while for PCBs an increasing K_{OW} predicts greater bioaccumulation. What this result means is that water-soluble PAHs are quickly metabolized, especially by crayfish and amphipods, whereas PCBs not so much.

Fortunately, conditions in the Great Lakes are generally improving, according to a 2007 summary.[1] Organochlorines (which contain carbon, hydrogen, and chlorine) such as pesticides are decreasing, but fish consumption advisories are still caused by PCBs, dioxin, mercury, mirex (an insecticide, along with a degradation by-product, photomirex), and chlordane (an insecticide used with crops and against termites, presently legal only for use in electrical transformers against fire ants). For PCBs, all of the lakes were at or above the level of recommending only one Chinook salmon meal per week, and Lake Superior and Lake Michigan had sites exceeding the 1 meal/month level. Again, don't eat the big fish: PCB levels were at or above the "do not eat" Ontario Sport Fish Consumption Guidelines for 60-cm lake trout for four

---

[1] USEPA (2007) reports on the contaminants in Great Lakes sport fish on a lake-by-lake basis.

Vegetarian birds have less mercury.

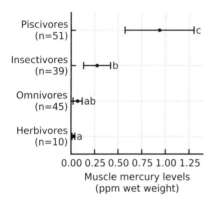

Figure 7.11: Mercury levels increase in birds that feed at the higher trophic levels in northeastern North America (Evers et al. 2005). Due to biomagnification, higher prey trophic levels have higher mercury concentrations, meaning that birds have different exposures to mercury depending on the trophic level of their diet. Letters denote significance (see text).

of the Great Lakes. Mercury poses a problem for Lake Superior and Lake Michigan, and all lakes but Lake Erie exceed the 1 fish meal/week guidelines.

Some birds eat fish, and some birds eat grass. Figure 7.11 summarizes a lot of data on birds (nearly 5,000 samples) from locations in New England, New York, eastern Canada, but mostly Maine, spanning the 1990s through the mid-2000s. Some of the data extend back as far as the 1980s and earlier.[1] Data from sampling loon populations show that mercury is present in eggs, blood, muscle, feathers, and liver, in order of increasing concentration. Astonishingly, about 83% of the methylmercury ingested by loons ended up in their bloodstream, migrating to the liver, with a concentration of 25 μg/g. However, the liver is an end point for mercury, and once there it doesn't move out. Next highest were the levels in the feathers at about 10 μg/g, and everything else was below 5 μg/g. Mercury levels in the feathers are a snapshot of the blood levels when new feathers emerge at molting.

Loons examined from the different study areas didn't show much difference in qualitative patterns, though across several species the adults typically had higher levels. Male loons have higher levels than females, and both comparisons reflect larger individuals eating larger prey items that have higher mercury levels due to biomagnification.

Figure 7.11 shows the typical biomagnification of a pollutant in bird species as one progresses up the food chain.[2] Herbivores are represented by the Canada goose; omnivores include mallard, American black duck, green-winged teal, and ring-necked duck; the insectivore is the common goldeneye;

---

[1] Evers et al. (2005) provide much of the information discussed here on mercury in birds. [2] Evers et al. (2005).

Freshwater baby bald eagles have more mercury.

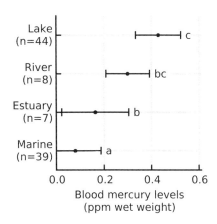

Figure 7.12: Mercury in juvenile bald eagles of different ecosystems in Maine (Evers et al. 2005). Freshwater systems have higher mercury methylation rates and fewer dilution possibilities, resulting in higher MeHg exposure rates in rivers and lakes.

and piscivores include the hooded merganser and common merganser. Letters denote statistically different values for the geometric means of the samples. The geometric mean is an averaging procedure that accounts for extreme values, be they zero or huge. I can't say I like the geometric mean,[1] but it deals with a recognized problem in reporting averages for samples with wildly varying values. This situation crops up when sampling pollution concentrations: many samples have concentrations below the level of detection, while a few have high concentrations. In such a situation, an arithmetic average obscures the importance of the magnitude and frequency of nonzero samples. Back to the data, herbivores and omnivores both have an a, meaning their mercury levels aren't statistically different, and both omnivores and insectivores both have a b, meaning their mercury levels aren't statistically different, but herbivores and insectivores don't share a letter, meaning their mercury levels are statistically different! Thus, we have a situation with statistics where, unlike math, $X = Y$ and $Y = Z$ but $X \neq Z$!

Figure 7.12 shows mercury levels for juvenile bald eagles raised in various habitats, but the pattern reflects that seen for juvenile and adult belted kingfishers.[2] The conclusion is that two processes are responsible: first, freshwater systems have greater methylation rates, and second, marine systems have better mechanisms for diluting methylmercury. The researchers noted that deposition rates across the habitats were equal, and we might hope for a reduction over time similar to that seen in moose (see Figure 7.1).

---

[1] See, for example, the discussion on the geometric mean provided by Joe Costa at www .buzzardsbay.org/geomean.htm   [2] Evers et al. (2005).

Baby egrets collect mercury.

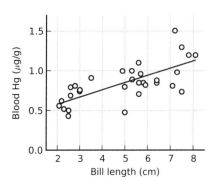

Figure 7.13: Egret nestlings in the Florida Everglades sequester mercury as they grow larger on the fish meals provided by their parents (Sepúlveda et al. 1999).

One pathway for mercury demonstrates exceptionally well the complications of ecology. Methylmercury in streams made its way to birds that ate spiders that ate aquatic invertebrates that ate algae.[1] While consuming food in the stream as aquatic invertebrate larvae, the adult insects carried that mercury as they flew right into spider webs. Isn't ecology fascinating?

Figure 7.13 shows the results from a mid-1990s study that took samples of blood and feathers from 252 great egret nestlings across southern Florida, just north of the Everglades. A clear and strong correlation ($r^2 = 0.72$) appears between mercury levels in blood and bill length (shown here) and between mercury concentrations in blood against feathers (not shown here) of 0.07–3.9 μg/g and 4.5–40 μg/g, respectively, in the higher of two sampled years.[2] Feather concentrations ranged about 10 times higher than blood levels, which were about six times higher than levels reported in other egret nestling studies from other places. Nestlings of fish-eating birds pick up and retain mercury from their parents' provisioning.

A large collaboration of 71 scientists studying many species found mercury hot spots in the Northeast.[3] Yellow perch had concentrations of 0.38–0.78 μg/g, while loons had 1.1–5.5 μg/g, so it appears the egret chicks might just be what they eat.

Let's summarize the process. Canopies capture mercury, rains take it to lakes, algae take it up, fish eat them up, and loons eat the fish. Trees take mercury out of the atmosphere in multiple ways, described by the deposition velocity process of Figure 3.5, and can be responsible for 60–75% of the flux. The USEPA estimates that present-day annual deposition rates hover

---

[1] Cristol et al. (2008) found stream methylmercury making its way to terrestrial birds. [2] Sepúlveda et al. (1999). [3] Evers et al. (2007).

around 20 μg/m^2, about four- to sixfold more than rates from 1900. Acid rain, another problem resulting from urbanization, also enhances the methylation of mercury. Various landscape features like land cover and nutrient loading also affect the methylation of mercury, which affects mercury's availability for biological uptake.

Chapter Readings

Beasley, G., and P. Kneale. 2002. Reviewing the impact of metals and PAHs on macroinvertebrates in urban watercourses. *Prog. Phys. Geogr.* 26: 236–270.

Bookman, R., C.T. Driscoll, D.R. Engstrom, and S.W. Effler. 2008. Local to regional emission sources affecting mercury fluxes to New York lakes. *Atmos. Environ.* 42: 6088–6097.

Driscoll, C.T., Y-J. Han, C.Y. Chen, D.C. Evers, K.F. Lambert, T.M. Holsen, N.C. Kamman, and R.K. Munson. 2007. Mercury contamination in forest and freshwater ecosystems in the northeastern United States. *Bioscience* 57: 17–28.

Evers, D.C., N.M. Burgess, L. Champoux, B. Hoskins, A. Major, W.M. Goodale, R.J. Taylor, R. Poppenga, and T. Daigle. 2005. Patterns and interpretation of mercury exposure in freshwater avian communities in northeastern North America. *Ecotoxicology* 14: 193–221.

Sabin, L.D., J.H. Lim, K.D. Stolzenbach, and K.C. Schiff. 2005. Contribution of trace metals from atmospheric deposition to stormwater runoff in a small impervious urban catchment. *Water Res.* 39: 3929–3937.

Weathers, K.C., M.L. Cadenasso, and S.T.A. Pickett. 2001. Forest edges as nutrient and pollutant concentrators: Potential synergisms between fragmentation, forest canopies, and the atmosphere. *Conserv. Biol.* 15: 1506–1514.

Chapter 8

Emerging Pollutants

Beyond nutrients and metals, a broad range of pollutants affect aquatic organisms, many of which have cropped up quite recently. Chemicals found in groundwater and surface water include many human-synthesized chemicals, such as agricultural chemicals, flame retardants, pharmaceuticals, swimming pool chemicals, and various consumer products. Locations that tend to be highly polluted also tend to have many different pollutants. Wastewater treatment plants eliminate some pharmaceuticals, but others pass through relatively undiminished.

These pollutants can enter ecosystems. Filter-feeding oysters take up PAHs (polycyclic aromatic hydrocarbons), which destabilize their lysosomal membranes. Mussels, seabirds, and marine mammals all take up flame retardants, and amphibians take up dry cleaning solvents. Birds also take up these pollutants, including flame retardants, with measurable effects on their health. However, human and animal pharmaceuticals, though widespread environmentally, generally appear to not be at high enough concentrations to cause extreme ecological problems.

Yet another aspect of pollution is soil and water acidity, which produces a range of effects in surface waters. Lakes in northeastern U.S. have greatly acidified over the last 150 years. On the one hand, increasing acidity mobilizes aluminum in soils, which leaches to surface waters, where it's toxic to organisms. Directly, acidity, as well as the loss of buffering against acidity changes, harms many organisms. Reductions in acid deposition also increase dissolved organic carbon, with a variety of ecosystem responses.

Anthropogenic chemicals in streams and groundwater.

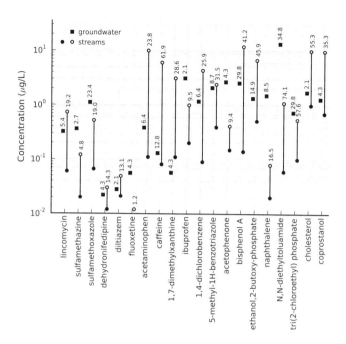

Figure 8.1: A short list of chemicals present in U.S. streams (Kolpin et al. 2002) and groundwater (Barnes et al. 2008). For the stream data, the filled circles give median concentration and the open circles give maximum observed concentration. The groundwater values (squares) show maximum concentration observed. Numbers are percentage of samples in which each chemical was present.

Up to this point, we've examined several major pollutants, but Figure 8.1 summarizes a great many more that reside in streams and groundwater. There are a lot of different pollutants out there, and in a major study one or more pollutants were found in 80% of 139 streams sampled across the entire U.S., with a median of seven pollutants each.[1] Streams were sampled just once during 1999 and 2000, but some of these streams were chosen because they were highly likely to be polluted. Hence, it was a biased sample. Many contaminants

---

[1] Kolpin et al. (2002) surveyed organic contaminants in U.S. streams. Details of the five different analytical methods can be found in the paper.

were found at such very low concentrations that they are assumed unlikely to have any environmental effects.[1]

It's also a bit fortunate, for the sake of cleanup, that many of these high concentrations are connected with adsorption onto particulates, which means that the pollutants could presumably be removed through filtration or sedimentation. As for the concentrations of the most frequently found compounds, only 5% of pollutant concentrations individually exceed a median of 1 μg/L, and most of those samples involve cholesterol derivatives and detergents. Another point for comparison is nutrients. The various chemicals listed here come in concentrations of micrograms per liter, compared with those for nutrients (Figure 6.15, e.g.), where we speak of several milligrams per liter.

The same collaboration of scientists that produced the stream results in Figure 8.1 also looked at 47 groundwater sites in 18 states and tested for 65 organic wastewater contaminants.[2] Water sources included three springs, two earthen basins collecting livestock wastewater, and 42 wells. Again, the sites were chosen specifically on the suspicion of contamination by wastewaters of either animal or human origin. Fortunately, the frequency of detection in these groundwater sources was lower than in the streams: only 10 of 38 groundwater sites had detectable contaminant concentrations greater than 1 μg/L, and half of those concentrations were between 1 and 2 μg/L. For comparison, 67 of 111 streams had a total contamination of all pollutants greater than 1 μg/L, and 23 had total concentrations exceeding 10 μg/L. The most common contaminants were an insect repellant, bisphenol A (BPA), a flame retardant, an antibiotic, and a detergent by-product. Overall, 81% of the sites had at least one contaminant present, and 35 of the 65 contaminants considered were found in at least one site. Further, the number of contaminants found decreased with well depth, with the greatest frequency of contamination being for those wells less than 10 m deep.

When the contaminants in Figure 8.1 were found in streams, the total concentration increased much more than linearly with the number of chemicals as shown in Figure 8.2.[3] The implication of this greater-than-linear correlation is that contaminated streams are really quite contaminated from either one or more major sources, and the various pollutant concentrations are somehow correlated. There was a wide variety of uses for the contaminants found most abundant, including industrial, agricultural, and residential-blame is well shared across all anthropogenic activities.

Still, even low levels of antibiotics (triclosan was one of the above contaminants) allow evolution of resistance by pathogenic bacteria. DNA (deoxyribonucleic acid) testing of microbes in animal waste lagoons and surrounding

---

[1] Every 2 years an article on emerging water contaminants updates the broad concerns, a recent one being Richardson and Ternes (2011). [2] Barnes et al. (2008) studied contaminants in groundwater.
[3] Kolpin et al. (2002).

Greater stream pollution means more chemical diversity.

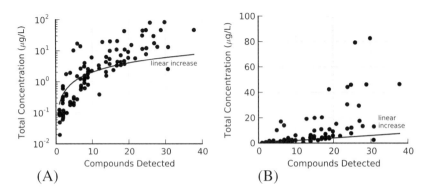

(A)   (B)

Figure 8.2: When polluted U.S. stream water contains more chemical species, the total concentration increases much faster than linearly with chemical diversity (Kolpin et al. 2002). The two panels show identical data, and only the scaling of the vertical concentration axis differs to provide different perspectives. The line indicates a linear increase with slope 0.2.

---

groundwater showed extensive presence of antibiotic resistance to tetracycline, used to treat systemic and local infections.[1] These groundwater systems are the same ones used for untreated drinking water, which represents a possible route for microbial antibiotic resistance relevant to humans. The entire blame does not rest with just our agricultural practices. Antibiotic resistance genes for tetracycline were found at frequencies 20 times greater than normal in wastewater from a high-quality treatment facility in Duluth, Minnesota, compared with levels in surrounding waters.[2] These elevated frequencies in open water and downstream went away at distances of about 1 mile.

Some of these compounds are naturally occurring. Women of childbearing age excrete, on a daily basis, 10–100 μg of several estrogen-related compounds, whereas pregnant women excrete a thousand times more, up to 30 mg daily.[3] Much of the biological processing of these excreted estrogens takes place in sewers by *E. coli* before arrival to wastewater treatment plants, and evidence suggests that wastewater treatment plants effectively, but variably, remove these estrogen-related compounds. Despite this high effectiveness,

---

[1] Chee Sanford et al. (2001). [2] LaPara et al. (2011) looked at antibiotic resistance genes at a wastewater facility in Duluth, Minnesota. [3] Baronti et al. (2000) monitored natural and synthetic estrogen moving through Rome's wastewater treatment plants.

Sewage treatment plants remove some pharmaceuticals.

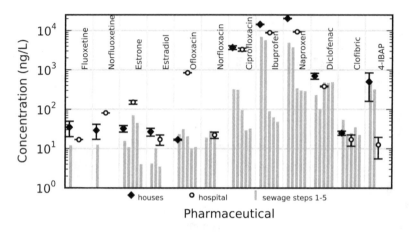

Figure 8.3: As different pharmaceuticals coming from houses and hospitals pass through the various stages of a wastewater treatment plant, they experience different changes in their concentrations (Zorita et al. 2009). Data come from a three-stage wastewater treatment plant in Sweden.

these excretions imply nanogram per liter concentrations of estrogen compounds downstream of high population centers and their wastewater treatment plants.

On the subject of wastewater treatment plants, the results of Figure 8.3 arise from a detailed study of pharmaceuticals making their way through a sewage treatment plant in Sweden.[1] Before discussing the results, it's useful to have a brief description of the plant and the five locations where measurements were taken. The first measurement taken was from the influent, and the researchers considered both hospital and residential waste streams. Influent enters a complicated set of stages, including passage through a sieve of just 3 mm (a tenth of an inch), and then enters a tank where the sewage is stirred with grit to break it down further. It then enters a second tank for settling, and during this entire stage the primary sludge is removed. The second measurement was taken upon leaving this mechanical treatment stage.

The sewage then enters a stage that removes nutrients and reduces oxygen demand through biological degradation. Here the waste first enters an anaerobic tank, and then enters an aerobic tank, and then enters another settling

---

[1] Zorita et al. (2009) studied the degradation of pharmaceuticals transiting a Swedish sewage treatment plant.

tank. Since so many wonderful microbes live in this material, pumps move 90% of the sludge, now called "activated sludge," from the sedimentation tank back into the anoxic tank to enhance the biological breakdown. The third measurement was taken post-biological treatment.

A fourth measurement was taken after a chemical treatment stage that removes more nutrients, especially phosphorus. This stage also adds 70 g/m^3 ferric chloride ($FeCl_3$), a coagulant, which mixes with the sewage to make big balls of sludge that settle out in yet another sedimentation tank. This tank also includes "lamellar settling plates," which are a set of angled, parallel plates with sewage moving up from the bottom. As the fluid moves up the plates, particles settle down onto the top of the next plate. We know from lamellar (straight-line) flow on a plate that the velocity must be zero at the surface, and the relatively heavy particles slide down the plate and drop to the bottom of the tank for easy collection. A fifth and final measurement was taken in the effluent from a sand filter, after which the treated sewage enters surface water.

Researchers concluded that the wastewater treatment plant of the type studied here was a sufficient approach to removing most pharmaceuticals. For example, nonsteroidal anti-inflammatory drugs (NSAIDs) had the highest influx because of people like me and my ibuprofen use, but 99% of the influx was removed by this sewage treatment plant.

One term for these chemicals is "pharmaceutically active compounds" (PhACs), which enter municipal wastewater systems from both excretion and waste disposal.[1] They exit treatment plants within either treated sewage or the sludge that's often spread with manure (containing its own animal medicine wastes) on fields. Both applications can result in PhACs entering surface waters, which may lead to drinking water contamination. Indeed, downstream stormwater control measures (SCMs) that recharge groundwater take certain PhACs with the water on into the aquifer. A situation of particular interest is when drinking water wells are placed near rivers, drawing stream water through the ground and up the well in a process called "bank filtration." Of particular concern is the situation in Berlin, where low streamflow couples with treated sewage coming from about 4 million people, and a bank filtration water source. Effluent water from sewage treatment plants account for up to 84% of the streamflow in one river during dry periods, a river that has important drinking water demands. Using aquifers to treat pathogens in wastewater and stormwater is a widespread practice, with examples spanning Australia, South Africa, Belgium, and Mexico.[2] Effectiveness depends quite a bit on the residence time within the aquifer. These are particularly complicated steps for PhACs that depend on consumption, processing in the body, sewage treatment plants, and environmental processes. Researchers also mention that ozonation

---

[1] Heberer (2002a, 2002b) overview PhACs in Germany's wastewater and provide nice summaries of the various types of PhACs. [2] Page et al. (2010).

(treatment with ozone) or membrane filtration at the drinking water treatment plan can remove PhACs, so all is not lost for clean water.

Removal of nutrients can conflict with the removal of pharmaceuticals. In a lab study seeking to understand degradation through wastewater treatment plants, the chemicals within pharmaceuticals and personal care products were removed faster when nitrifying (aerobic) conditions were maintained, as opposed to denitrifying (anoxic) conditions.[1] For example, removal of naproxen, ethinylestradiol, roxithromycin, and erythromycin required nitrifying conditions. Compounds insensitive to those two conditions yet removed well enough included fluoxetine, natural estrogens, and "musk" fragrances. Natural musk, a complicated PAH, can be found in quite overpowering quantities in deer glands and, for whatever reasons, is synthesized as a foundation of many perfumes. Carbamazepine, diazepam, sulfamethoxazole, and trimethoprim didn't degrade substantially. The activated sludge processes described above also helped degradation efficiency.

Notable PhACs in the streams and lakes around Berlin, Germany, included clofibric acid (from cholesterol-lowering blood lipid regulators, such as clofibrate), diclofenac, ibuprofen, propyphenazone, primidone, and carbamazepine, found at concentrations of up to several micrograms per liter in groundwater, and at the nanogram per liter level in drinking water.[2] Here is a short list of the efficiency of wastewater treatment plants in removing various PhACs, listed respectively as the average influent concentration and average effluent concentration (μg/L): carbamazepine, 1.78, 1.63; clofibric acid, 0.46, 0.48 (no removal); diclofenac, 3.02, 2.51; and caffeine, 230, 0.18. Pharmaceuticals likely represent less of a threat to aquatic organisms than do pesticides and heavy metals. Empirical examination roughly estimates that the contributions to toxicity are less than 5% for pharmaceuticals compared with more than 95% for pesticides and heavy metals combined.[3]

Now let's consider the pharmaceuticals that show up in drinking water sources.[4] In a search for 100 different compounds in 74 surface and groundwater sources across the U.S. and Puerto Rico, 63 compounds were found at least once, with an average of four compounds found in each drinking water source. In decreasing order of occurrence re the herbicide metolachlor; a nicotine breakdown product, cotinine; and a caffeine breakdown product, 1,7-dimethylxanthine. In groundwater sources the list was very different, headed by the solvent perchloroethylene (PCE), the pharmaceutical *carbamazepine*, and the plastics component bisphenol A (BPA). Concentrations were generally lower than 1 μg/L, relatively consistent with Figure 8.1. Note that these

---

[1] Suarez et al. (2010). [2] Heberer (2002a, 2002b). [3] Damásio et al. (2011) examined chemical threats to aquatic organisms in rivers near Barcelona, Spain. [4] Focazio et al. (2008) examined drinking water sources for pharmaceutical contamination. Collaboration included many of the Barnes et al. (2008) authors.

Don't drink dirty water.

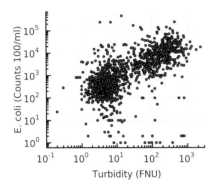

Figure 8.4: The correlation between *E. coli* and turbidity in Atlanta streams (Peters 2009). Particles just smaller than a $1\,\mu$m affect turbidity, while *E. coli* have a size of a couple micrometers. Both turbidity and pathogens come with urbanization.

samples came from untreated sources, and that the drinking water itself was subject to water treatment before going out to consumers. Posttreatment samples were not examined.

A number of detailed issues affect the environmental aspects of pharmaceuticals. Endocrine-disrupting chemicals are now understood to have complicated dose–response curves.[1] We typically think that if the concentration of some chemical increases 10-fold, then we should expect a response that's 10 times bigger, but sometimes a nonlinear response can produce much greater effects (see Figure 8.13). To make drugs more effective, fluorines sometimes replace hydrogen or hydroxyls in many drugs to reduce their metabolism, called a bioisosteric replacement, but the consequence is a larger presence of pharmaceuticals in the environment. For example, the fluorinated anti-inflammatory Celebrex is stable for days in rats, but replacing the fluorine with a methyl group results in a 4-hour half-life.

Figure 8.4 summarizes the results of a study that sought to understand how urbanization affects overall water quality. This plot shows the correlation between turbidity and *E. coli* concentrations.[2] First, a bit about turbidity. Imagine pouring a little bit of milk into a glass of water, just enough so the water loses some of its transparency. The water has become turbid. Shining light through the water causes some of the light to scatter backward and side-ways and the rest to pass through. Two basic measures of turbidity are nephelometric turbidity unit (NTU), which uses light of wavelengths 400–680 nm, and formazin nephelometric unit (FNU), which uses near infrared

---

[1] Khetan and Collins (2007) provides a lengthy and extensive review of endocrine-disrupting chemicals in the environment and provides descriptions of drug classes. [2] Peters (2009) studied water quality in Atlanta's streams.

light of wavelengths 780–900 nm.[1] Given a sample of dirty, cloudy water with an unknown set of many dissolved contaminants, one seeks to characterize the sample simply by shining light through it, rather than more expensive approaches of identifying the bacteria, chemicals, suspended solids, and so on. Certainly the results depend on many details, and those details aren't relevant here.

The results of Figure 8.4 show that turbidity measures provide an important indication of the presence of the pathogen *E. coli*, which correlates with fecal coliform and total coliform measures of bacterial contamination.[2] Data come from 21 sampling stations covering Atlanta, Georgia, and the surrounding areas, covering an area of 343 km². These drainage areas vary in size from around 4 km² up to nearly 250 km² and, remarkably, have 69–96% urban land use. Samples were taken from 2004 to 2007, along with reference stream data taken from 2001 to 2002. Most stations were sampled manually more than 30 times and by automatic samplers more than 150 times. The researchers note that, despite the best intentions, manual sampling was biased toward base-flow conditions rather than storm-flow conditions. This bias can be problematic since they also observed a "first-flush" phenomenon where the initial storm flow had higher pollutant concentrations. Runoff fraction of urban stations were 0.32–0.51, a bit lower than the runoff expected from Figure 5.11, whereas the reference forested stations were 0.25–0.4. More than 90% of their samples exceeded allowable coliform concentrations of 200 counts per 100 ml. Beyond these coliform data, data on other pollutants and sediments clearly show that urban streams are polluted in many ways, and if it's dirty, don't drink it.

Atlanta's not alone. Pathogens represent the largest fraction of reported stream impairments (see Figure 1.6).[3] Microbial pathogens in stormwater, including downstream from wastewater discharges, present an argument for promoting a well-functioning ecosystem with bacteria-consuming nematodes, ciliates, and rotifers that can control bacterial densities. Predator removal of bacteria can be quite effective, with individual microbial predators able to remove a dozen cells per minute! Two important factors for controlling pathogens include high temperatures (though well within trout tolerance temperatures of Figure 9.9) and sunlight, both of which help reduce pathogens. Water at 20°C quickly reduces *E. coli* populations, and sunlight can reduce populations by 90% in 1 hour in 20°C water. Since particulates impair light and pathogens also bind with small particles laden with heavy elements (see Figure 3.22), sediment transport truly is an important control concern. It's also useful to remember that pathogens we're familiar with, such as *E. coli* and total coliform counts, refer to indicator species, not the broad spectrum

---

[1] The USGS provides many details on turbidity measures at or.water.usgs.gov/grapher/fnu.html and water.usgs.gov/owq/FieldManual/Chapter6/6.7_contents.html. [2] Berman et al. (1988) describes coliform and turbidity. [3] Struck et al. (2006).

of stormwater and wastewater organisms that cause human health problems. They're bad, yes, but other things in polluted waters cause cholera and hepatitis, and the presence of these microbes and viruses correlates with indicator species.

Already we see some signs of conflicts between SCMs. Constructed wetland water temperatures are cooler than retention ponds, shaded by vegetation, as is the intention to prevent temperatures harmful to aquatic organisms (see Chapter 9), but these features are also less detrimental to microbes. If microbes are the main problem, perhaps the focus should be to increase temperatures, enhance sunlight, and reduce turbidity.

There's a complicated interplay among terrestrial microbial processes, atmospheric deposition, and surface water quality that goes well beyond the presence or absence of fecal coliform. Dissolved organic carbon (DOC) represents the smallest bits of carbon suspended in water[1] that can pass through a filter of around 0.25–0.5 μm pore size, at least one-tenth the size of, say, algal cells. DOC influences the carbon cycle, and biodegradable DOC contributes to bacterial growth, which can harm water quality. A long-term rise in DOC has been observed, for example, in northern England.[2] DOC might be increasing because of carbon emissions, temperature increases, nitrogen deposition, or land use changes, among other possibilities. A rise in DOC, if brought on by a climate-change-related mechanism, might portend a broader instability of soil carbon, as well as an enhanced incorporation of mercury into algae. One study that examined the causes of this DOC increase used data from monitoring programs spanning 1990–2004 and involved 522 lakes across North America and northern Europe (Figure 8.5).[3] It concluded that high sulfur emissions and, in a few places, stormy conditions in the early 1990s that deposited marine salt combined to cause low pH soil levels. Those acidic conditions reduced the solubility of organic matter in the soil, which reduced DOC in surface waters. Reduced DOC in lakes can allow greater light penetration deeper into lakes, promoting algal growth, as well as increase mortality rates for prey by enhancing predators' visual acuity.[4] Acidified soils can also leach aluminum, along with nitrate, mercury, and others (see discussion of Figure 6.3), which is toxic to fish but can be sequestered away by DOC. During the study period, reduced sulfur emissions (both sulfur dioxide and sulfate) brought on by stronger air quality regulations and reduced sea salt inputs due to less intense storms let the soil pH levels recover, and DOC levels rose to preindustrial levels. Of these two mechanisms, marine effects that bring chlorine inland dominate in the U.K. and Atlantic Canada, but everywhere else sulfur emission reductions account for more than 85% of the DOC change.

---

[1] Volk et al. (2002). [2] Worrall et al. (2003). [3] Monteith et al. (2007) connected DOC and sulfur deposition. [4] Lovett et al. (2009).

Acid rain lowered dissolved organic carbon.

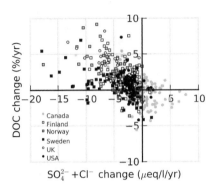

Figure 8.5: Dissolved organic carbon (DOC) in headwater lakes and lower-order streams increases as a result of decreasing sulfur and chlorine atmospheric deposition (Monteith et al. 2007), implying that reducing acid rain enhances DOC. The unit "μeq(uivalent)" places different acids on the same chemical footing.

Flushing of acids associated with soil-derived DOC, as well as deposition-related acids, can create problems in lakes.[1] The ability of a water body to neutralize or buffer the addition of strong acids is described by its acid-neutralizing capacity (ANC).[2] Higher values are better values. Lakes having negative (ANC) values are considered chronically acidic, with 0–100 μeq/L, are called acid-sensitive lakes, and lakes with higher values are resistant to acidification. Acid anions like sulfate, $SO_4{}^{2-}$, come predominantly from atmospheric deposition of power plant emissions, and the ANC represents the susceptibility of lake acidity to these inputs.

The capacity to buffer these acids affects lake ecology. ANC isn't really a condition that organisms can feel, but there are correlations between ANC, lake acidity, and levels of biologically available inorganic aluminum, which can be harmful.[3] Summarizing how acidity ranges affect biodiversity, a pH from 6.5 to 6.0 epitomizes a wonderful lake. A slight reduction to 6.0–5.5 results in the loss of minnows, dace, zooplankton/phytoplankton species, some clams, snails, and invertebrates and an increase in filamentous algae. With a further reduction to 5.5–5.0, lakes lose trout, walleye, and bass and have more filamentous algae. Once a lake reaches a pH reduction to 5.0–4.5, most fish and clams are gone, and amphibians experience reproductive failure. An important complicating point, however, is that some non-acid-related ecological changes can mimic the changes due to further acidification. For example, the removal of fish from acid-neutral lakes greatly changes the ecology of a lake, which looks similar to some of the changes of acidification.

Figure 8.6 shows the response of fish and zooplankton to ANC. Zooplankton data come from surveys of 97 lakes visited a total of 111 times through

---

[1] Wellington and Driscoll (2004).   [2] Lovett et al. (2009).   [3] Lovett et al. (2009).

Fish and zooplankton need acid buffering.

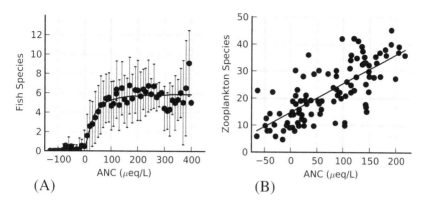

(A)   (B)

Figure 8.6: Organisms are sensitive to the rapid variations in pH that acid buffering prevents (Sullivan et al. 2006). Both fish (A) and zooplankton (B) species disappear with the loss of acid buffering in northeastern lakes.

Lakes have lost their acid buffering.

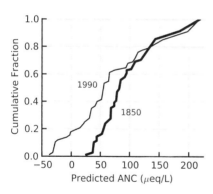

Figure 8.7: Distribution of acid buffering in Adirondack Lakes of the northeastern U.S. has changed greatly with the urbanization of the 20th century (Sullivan et al. 2006). Results come from a model of deposition, soils, and water chemistry relevant to each period.

---

several sampling efforts,[1] while the fish data come from 22 shallow lakes less than 3 m deep. ANC is indeed important for the health of surface water ecosystems.

Results of Figure 8.7 arise from a model of deposition, soils, and water chemistry. The shift to lower ANC values from 1850 to 1990 means that

---

[1] Sullivan et al. (2006) is an extensive paper estimating acid future of Adirondack lakes under various emissions scenarios.

Flame retardants in peregrine eggs.

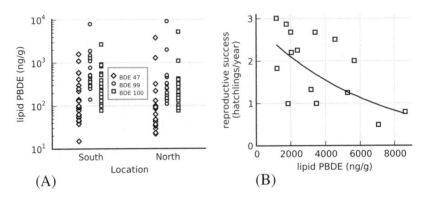

(A)   (B)

Figure 8.8: Flame retardants (PBDEs) in peregrine eggs in northern and southern Sweden (Johansson et al. 2009). (A) PBDE concentrations in eggs. (B) PBDEs at the higher concentrations lower reproductive success. Hatching success was averaged over 2–7 years.

lakes have dramatically acidified. Researchers estimated ANC for 1850s lakes, which led to an estimate that 20% of lakes supported 4.1 or fewer fish species. By 1990, emissions had caused the ANC distribution to shift so much that 20% of lakes supported 2.0 or fewer fish species.

The good news is that the models show recovery can take place by 2100 if we take aggressive emissions control steps that reduce NO_x emissions by 50% and SO_2 emissions by 70%. Emissions of sulfur dioxide (SO_2) become deposition of sulfate SO_4^{2-} after chemical reactions take place in the atmosphere, and both have been decreasing by about 5% per year over the last decade.[1] ANC recovery can be so dramatic with aggressive emissions controls that the number of chronically acidic lakes drops to one-tenth, and the number of sensitive lakes with ANC < 50 μeq/L drops by nearly half compared with present emissions controls. The bad news was that, under the baseline emissions regulations expected as of 2004, the number of chronically acidic lakes will likely remain unchanged through 2100. The main message is that if we want clean lakes, we need more emissions controls for recovery of lakes, but it will work.

Chapter 10 examines the effects of flame retardants, but here we'll briefly cover a study in Figure 8.8 that compared wild falcons in Sweden with captive falcons eating domestic chickens and quail,[2] which are flame-retardant-free

---

[1] Hand et al. (2012). [2] Johansson et al. (2009) examined flame retardants in falcon populations.

meals. Measurements showed that the eggs of captive birds had concentrations two orders of magnitude lower than those from wild populations. Just like we saw for mercury, wild birds are picking up flame retardants, and Figure 8.8A shows the distributions of various flame retardants from the various eggs—a large amount of variation.

The study authors mention that 56,000 tons of these flame retardants was produced in 2003. Among the acronyms, PBDEs are polybrominated diphenyl ethers, and HBCDs are hexabromocyclododecanes. Several PBDEs have been banned in Europe. Cross-country transport takes place, as revealed by the levels of HBCD in these falcons from Sweden, which are 40 times higher than in falcons that come from Greenland. The difference is where the birds spend their winters: Sweden's peregrines spend winters in western and southern Europe; Greenland's falcons spend winters in Central and South America. Europe uses more HBCD than Central and South America.

Are flame retardants affecting wild populations? Fecundity data in Figure 8.8B come from the southern peregrine population, where individual females were followed for 2–7 years, and reflect the number of fledged offspring. It does seem suspicious to me that the trend line, reproduced from the original plot, likely comes from the last two points; take them off and there's likely no trend. However, it's these high concentrations where the problems would appear, and the data come from wild populations where there's no possibility for experimental control on the exposures.

Further evidence of problems comes from a study of loon eggs and carcasses collected between 1968 and 1980 across Ontario, Canada. The study involved 98 eggs and 215 found carcasses of loons that were drowned in nets, shot, or strangled or died from swallowing hooked fishing lines or of disease or unknown causes.[1] Thirty of these loons were emaciated, while the others, though dead, were otherwise considered "healthy." Concentrations in the emaciated birds compared with the healthy birds showed that organochlorines (DDTs, dieldrin, and PCBs) were 100 times higher: brain DDTs, 25–49 vs. 0.2–0.9 μg/g; dieldrin, 0.5–1.2 vs. 0.01–0.05 μg/g; and PCBs, 39–63 vs. 0.6–2.0 μg/g. Other fatty tissues had much higher concentrations. Mercury levels weren't as drastically different, with emaciated birds having around 1 μg/g, twice that of healthy birds. Eggs collected in 1969 and 1970 had high DDT concentrations and shell thicknesses of 0.56 ± 0.02 mm^2, compared with museum specimens having thicknesses of 0.64 ± 0.02 mm. That's why we banned DDT—to prevent crushed eggs while birds incubated their clutches.

Cellular function depends, in part, on stable membranes separating all of the different cell parts. Pollutants (including PAHs, PCBs, and metals) can end up in these membranes and cause problems with their function and

---

[1] Frank et al. (1983). [2] I averaged their two results in two different places in the two different years.

PAHs destabilize oyster membranes.

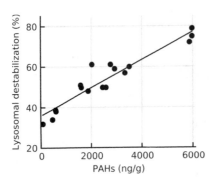

Figure 8.9: High PAH concentrations in oysters cause their cells' lysosomes to break down, spilling their enzymatic contents (Hwang et al. 2008). Oysters can recover when their water clears. These are dry weight measurements.

stability. Lysosomes are enzyme-filled small vesicles, generated by the Golgi apparatus, that fuse with other cellular bodies, ejecting enzymes that degrade rogue biomolecules and toxins and then excrete the resulting products from the cells. Lysosomes also collect pollutants and sequester them from the rest of the cell. These functions are protective to the organism at low levels, but some pollutants destabilize the lysosomal membranes, which then spill their contents into the cell, degrading vital organelles, which then leads to cell death.

Destabilization of the lysosomal membrane takes place because of PAHs and copper, but also because of a range of nonchemical stresses, such as heat and oxygen depletion.[1] Lysosomal destabilization serves as a general measure of organismal health in oysters, as oysters and other mollusks have high lysosome density. You might imagine lysosomes being a necessary part of the life of a rather nonselective filter feeder.

The experiment yielding data in Figure 8.9 placed oysters into water with a mixture of 24 PAHs maintained at 1,000 ng/L for 25 days.[2] A so-called critical body residue for oysters was measured as 2,100 ng/g dry weight, a level at which half the cells displayed leaking lysosomes, spilling their contents into the cell volume, releasing the enzymes to damage the cells instead of protect them. Fortunately for the oysters, resuming a life in clean water can reverse these destabilization levels.

Contrast these lysosomal destabilization levels for oysters with the effects on other organisms: when exposed to the PAH fluoranthene, mussels experience a 50% reduction in feeding at 5 times greater concentrations, copepods experience 50% reduction in reproduction at 50 times greater concen-

---

[1] Moore (1990) discusses lysosome issues.   [2] Hwang et al. (2008) used dyes to indicate lysosomal destabilization in contaminated oysters.

trations, and amphipods experience 50% death at concentrations about 500 times greater concentrations.[1]

Indeed, experiments that exposed mussels to two specific PAHs, fluoranthene and benzo(a)pyrene, for up to 4 weeks at levels up to 6 μg/L produced tissue concentrations reaching as high as 7,300 ng/g and 16,000 ng/g dry weight, respectively. Feeding rates were reduced by one-half at tissue concentrations of 9,500 and 16,000 ng/g dry weight, respectively, for the two contaminants.[2] An experiment that exposed mussels to "harbor" sediments[3] contaminated at a level of 100 mg/L concentration yielded an exposure for the mussels of a PAH concentration of 57 μg/L. The study found the mussels terminated feeding after 2 days of exposure.

These levels of PAH concentrations are out there. Limpets collected from the Sicilian coast of Italy had PAH concentrations ranging from 34 to 750 ng/g dry weight.[4] Interestingly, seaweed was collected from 4 of the 16 limpet collection sites but its PAH concentrations showed no correlations with the limpet PAH concentrations, which means that one organism does not represent the entirety of an ecosystem.

PAH concentrations in amphipods at the head of the Mississippi basin in the Gulf of Mexico measured 5 μg/g dry weight in adult-dominated June populations and 0.34 μg/g dry weight in juvenile-dominated August populations.[5] These were some amazingly dense concentrations of amphipods, exceeding 10,000 individuals per square meter. That's more than one amphipod per square centimeter! Samples of the surrounding sediments showed that the PAHs correlated with barium concentration (as well as clay and sand fractions), and the presence of barium indicates that drilling mud coming from oil well platforms was the likely source of PAHs. Concentrations in amphipods were roughly four times greater than in the surrounding sediments, posing a biomagnification scenario similar to that of mercury (see Figure 7.6), but the ratio certainly depends on the specific PAH and its partition coefficient, K_{ow} (see discussion of Figure 7.10). Details from the amphipod PAH data indicated that they're getting the pollutants indirectly via the water column rather than directly from the sediments.

On an encouraging note, bacterial degradation of PAHs with four or more rings, called high-molecular-weight (HMW) PAHs, was discovered in 1988.[6] These naturally occurring bacteria consume four-ring PAHs, and the rings themselves get split apart during the degradation process. Further, bacteria were found in creosote-contaminated soils that can use four-ring PAHs as a sole source of carbon and energy. Sediment collections from the Elizabeth River at Norfolk, Virginia, near oil tanks and refueling stations revealed 101

---

[1] Hwang et al. (2008). [2] Eertman et al. (1995) studied mussels exposed to PAHs. [3] Eertman et al. (1995) did not specify "which" harbor, but it was likely one in the Netherlands. [4] Gianguzza and Orecchio (2006). [5] Soliman and Wade (2008). [6] Kanaly and Harayama (2000).

Flame retardants in French mussels.

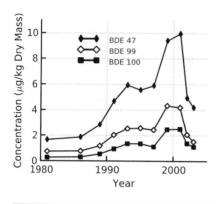

Figure 8.10: BDE flame retardants from consumer products make their way into mussels from Villerville, located at the mouth of the Seine estuary, France (Law et al. 2006).

PAH-degrading "isolates" (unique bacterial colonies).[1] Of these, seven were novel discoveries, never seen before in genetic databases. There may be hope for the bioremediation of PAH-contaminated sites.

Flame retardants, discussed in Chapter 3, are another class of molecules making their way into organisms. First off, environmental levels of total BDEs in sludge from European sewage plants were 130 ± 60 μg/kg dry weight (that unit is the same as ng/g or ppb).[2] This sludge gets applied to agricultural fields, where it accumulates in the soils, and worms can biomagnify it by a factor of 5. After that, it can work its way up the food chain. In 2004, worldwide production of BDEs totaled 204,000 tonnes, or 204 million kg. Following the calculation at Figure 3.2 for mercury, these BDEs spread out uniformly over Earth's surface gives a layer of 400 μg/m^2. BDEs were banned in the European Union in 2004[3], and in the U.S. they are no longer used in furniture foams (driven by California) but have been replaced by another flame retardant, TDCPP (tris(1,3-dichloroisopropyl) phosphate).[4] In the U.S., deca-BDEs remain in use.

Bioaccumulation of flame retardants results in levels of a few micrograms per kilogram dry mass for mussels collected at the mouth of the Seine River in France, as shown in Figure 8.10, though the sudden drop-off from disuse is remarkable. Concentrations in the flatfish dab and flounder livers from the Baltic Sea also sit around a few micrograms per kilogram live weight. Marine mammals, mostly fish-eating pollutant biomagnifiers, collect BDEs and reach total concentrations on the order of 290 μg/kg in seal blubber, reaching into the thousands for some samples.[5]

---

[1] Hilyard et al. (2008). [2] Law et al. (2006). [3] Johansson et al. (2009). [4] Stapleton et al. (2012).
[5] Kalantzi et al. (2005).

Flame retardants in humans.

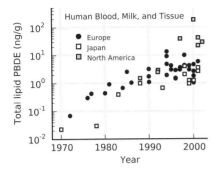

Figure 8.11: Flame retardants make their way into human blood, milk, and tissues (Hites 2004). Notice that human levels in later decades match those of the harbor-water filter feeding mussels in Figure 8.10.

Flame retardants in sea birds and marine mammals.

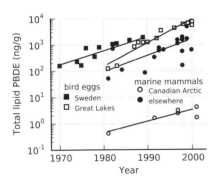

Figure 8.12: Flame retardants make their way into sea birds and marine mammals (Hites 2004). Except for marine mammals of the Canadian Arctic, these concentrations are orders of magnitude greater than those found in humans.

The plots in Figures 8.11 and 8.12 summarize much research (up until August 2003) on the environmental presence of the five PBDE congeners mentioned above, and in a variety of organisms, including humans.[1] PBDE-47 makes up the greatest fraction in these samples. Variation in human samples at any particular location spans an order of magnitude or more, but overall, there's a doubling time of about 3 or 4 years. Presumably with the recent bans on some PBDEs, these levels may level off or decrease.

A classic and tragic situation took place with another class of flame retardants, polybrominated biphenyls (PBBs), which are just like PBDEs except the phenyls are connected directly without an intermediate oxygen atom. The

[1] Hites (2004, 2005) provides background.

exposure took place in Michigan in 1973 and resulted from a mix-up between the PBB flame retardants and a nutritional supplement for cattle, magnesium oxide, used as a buffer for the acidity of chopped corn plants stored in silos that's fed to cattle.[1] One company produced both chemicals, both were packaged in plain paper bags, and some PBB (about a dozen 50-pound bags) was included in a feed supplement shipment to a feed mill producing animal feeds containing 0.4% magnesium oxide. Severe health problems in dairy herds arose in the fall of 1973, though the contamination wasn't identified until April of 1974.

Interestingly, a gas chromatograph accidentally left on too long uncovered the peaks associated with PBB. The event resulted in lots of contaminated livestock: ultimately, 9,400 cattle, 2,000 hogs, 400 sheep, and 2,000,000 chickens were killed due to PBB tissue concentrations exceeding 1 ppm. Twenty thousand more cattle were killed when the threshold was lowered to 0.3 ppm, and later lowered to 0.02 ppm in 1977.

In the meantime there was a significant consumption of contaminated milk across much of Michigan, particularly by farm families. More than 20 herds yielded milk fat concentrations exceeding 10 ppm. Aspects that contributed to its severity included a general lack of understanding, at the time, of the health effects of PBBs and a hesitance to act in the face of "uncertainty."

The event's consequences continue.[2] Four thousand people were enrolled in long-term studies of the effects. One study found that girls exposed to high levels of flame retardants *in utero* and through breastfeeding reached menarche about one year early.[3] High exposure levels were considered to be more than 7 ppb, but individual cases reached as high as 500 ppb. Clearance of PBBs from the body were sped up by smoking and breastfeeding and slowed down by high body mass index and age.[4] Still, for a study population with PBB concentrations ranging over several orders of magnitude, individual PBB concentrations decreased only by an order of magnitude over a couple of decades.

It's not just filter feeders and humans that collect flame retardants. One study looked at uptake of PBDEs in laboratory frogs that were provided a tank with a polyurethane foam "island" that was 32% PBDE: the frogs were fed crickets, and the crickets fed on the foam.[5] PBDEs were not detected in crickets unexposed to the foam island, and only crickets that managed to elude the frogs upon entering the tank had the chance to eat foam. Their later ingestion by the frogs represented the frogs' source of PBDEs and a mini-biomagnification experiment. Crickets reached a concentration of 14.4 μg/g wet weight (257 μg/g lipid), while the frogs reached just 10.1 μg/g wet weight (209 μg/g lipid).

---

[1] Fries (1985) describes the Michigan PBB incident and cites additional similar mix-ups. [2] Hites (2005). [3] Blanck et al. (2000). [4] Terrell et al. (2008) update the PBB results. [5] Hale et al. (2002).

Figure 8.12 shows that animals in the wild, birds and marine mammals at least, accumulate PBDEs.[1] Eggs collected from herring gulls in the Great Lakes of North America and guillemot eggs collected in Sweden demonstrate large and increasing concentrations, as do marine mammals sampled from across the world. Indeed, these levels are orders of magnitude higher than the concentrations found in humans in Figure 8.11. PBDEs also make it into the wild and clean Canadian Arctic, but at quite low concentrations, relatively speaking, comparable to human levels.

Humans are great at making chemicals that get out into the environment. Another class includes perfluoroalkyl and polyfluoroalkyl acids (with many variants going by such acronyms as PFAS, PFSA, PFCA, and PFAA) that can be applied to paper, carpets, clothes, and so on, to repel water and oils. Polar bears in Greenland show bioaccumulated PFAAs that reach wet weight concentrations exceeding 50 ng/g in various parts of the brain, and several thousand ng/g in the liver.[2] Harbor seals in the North Sea off the coast of Germany had brain concentrations that reached 100 ng/g. However, these levels are a couple orders of magnitude lower than lab studies showing toxicological effects, and little is known about the physiological effects at these residual levels.

Two of the many PAHs are PCE (perchloroethylene or tetrachloroethylene), also called "perc," and TCE (trichloroethylene). Their improper handling during dry cleaning (as well as other solvent-driven uses) led to environmental contamination. TCE and PCE are carcinogens that affect reproduction and development separately from the endocrine system (see Chapter 7).[3] These chemicals are also not to be confused with perchlorate, a component of rocket fuel and an environmental contaminant that also affects reproduction. TCE and PCE can both be taken up by developing amphibians (Figure 8.13 shows data for PCE).[4] These chemicals can contaminate groundwater drinking water systems; a famous incident occurred at a Marine Corps base in North Carolina, where drinking water concentrations of PCE reached 1,400 ppb in some wells in the early 1980s.[5] Concentrations of PCE up to 75 mg/L in groundwater sources in Canada have been reported, and up to 150 mg/L at U.S. industrial site. For reference, the USEPA maximum concentration is 4 ppb (1 μg/L is 1 ppb).[6] That means some of these concentrations are several hundred times higher than the guidelines. These chemicals evaporate quite quickly when exposed to the atmosphere, meaning they fall under the description of a volatile organic compound (VOC), but they have long residence times in groundwater, and groundwater-fed streams can have chronically high PCE levels.

The experimental results in Figure 8.13 came from 4-day-long exposures at the developing embryo stage at concentrations just a few times higher than the

---

[1] Hites (2004). [2] Greaves et al. (2013). [3] Berger and Horner (2003). [4] McDaniel et al. (2004) studied deformities in amphibians due to chlorinated solvents. [5] ATSDR, www.atsdr.cdc.gov/ sites/lejeune/index.html [6] NRC (2010).

Dry-cleaned amphibians.

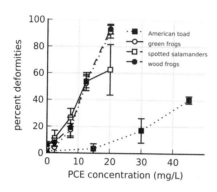

Figure 8.13: PCE (perchloroethylene), a dry cleaning chemical, makes its way into amphibians, causing deformities in the developing offspring (McDaniel et al. 2004). Concentrations used in these lab experiments reflect environmental concentrations observed in North America.

USEPA maximum. At these concentrations, mortality rates were under 10%, despite the high incidence of deformities. Of course, out in the real ecological world, these deformities might well be a death sentence, considering predation risks and foraging demands. Similar results were found for TCE, which arises from the degradation of PCE, as well as from pollutant sources.

Many human-created chemicals have environmental implications. A study of sources of endocrine-disrupting chemicals examined 21 sites around San Francisco, California, of which 16 sites were residential, commercial, and/or industrial locations, and 5 were from the wastewater treatment plant for those 16 sites (2 influent, 3 effluent).[1] For these various locations, the frequencies of detection for several compounds were reported. Phthalates are fragrances added to everything—pill coatings, adhesives, inks, flexible plastics, and many other products—but also help make flexible plastics; these were widely detected at 19 of 21 sites. Bisphenol A (BPA), a constituent of plastics and dental sealants, was found at 5 of 21 sites. Triclosan, an antibacterial compound in consumer products, was detected at 9 of 21 sites. TCEP (tris [2-chloroethyl] phosphate), a flame retardant, was detected at 4 of 21 sites. All of these chemicals affect estrogen and androgen systems and the thyroid and show up as various problems in fish and amphibians; when found, they had concentrations on the order of 0.1–100 μg/L.

In terms of biological consequences, BPA exposure, for example, at 16 weeks' gestation has been correlated to "externalizing" behaviors (hyperactivity and aggression) in 2-year-old girls, but not boys.[2] In this study, BPA

---

[1] Jackson and Sutton (2008) studied sources of endocrine-disrupting chemicals. their report has a nice table describing various chemicals' sources, concerns, and ecological impacts. [2] Braun et al. (2009) examined BPA exposure and childhood behavior.

concentrations in the mothers' urine averaged around 2 μg/L, and these exposures brought the girls' "externalizing" scores up to match the boys' scores.

Some environmental chemicals can make changes to offspring DNA.[1] Essentially, one of the bases in DNA strands can have an attached methyl group that, in effect, can turn off a gene. During fertilization and cell division, these methyl groups get detached and reattached, with especially complicated consequences during reproduction. The inheritance of these methylation patterns—and changes therein—spans generations, all the while making no changes in the DNA base patterns themselves, giving the methylation patterns the moniker "epigenetic," or heritable activities in genes that depend on features beyond the gene patterns. A study exposing gestating rats to a mixture of BPA, DEHP (bis[2-ethylhexyl]phthalate), and DBP (dibutyl phthalate) examined the health effects in the offspring and the offsprings' offsprings' offspring ("great grandrats") at 1 year of age. The last generation is particularly interesting. The situation involves telescoped generations. A gestating female has a female fetus. That fetus develops the germ cells of its eggs while she herself is developing. Chemical exposures of the adult thus can directly affect the following two generations. Testicular, prostate, kidney, ovarian diseases,[2] obesity, and early or delayed puberty were found in one or both generations. However, the offspring of the females exposed as germ cells in the developing fetuses also demonstrated inherited ovarian diseases. Granted, daily exposures via injections at 50–750 mg/kg of body weight are very high concentrations indeed, but the environmental exposures of the rats affected their great-grandrats.

Many insecticides, herbicides, and fungicides have clear effects on animals at low doses[3] and, indeed, call into question the concept of "low dose." For example, 40 of 100 studies examining effects of BPA below the 50 μg/kg/day "lowest observed adverse effect level" (LOAEL) found negative effects. The problem arises because the endocrine system is fine-tuned to respond to very low concentrations of hormones produced by organisms. For example, natural levels of estradiol in the body are around 10–900 ng/L, and testosterone levels are less than 10 μg/L. For comparison, dioxin at a concentration of 1 μg/kg kills 50% of exposed guinea pigs. BPA has estrogenic effects at 0.05–30 ppb (ppb- μg/L).[4]

Beyond the concerns of effects at low doses is a growing awareness of dose–response curves shaped like a U, or having a hump shape or switch-like off–on effects. In these situations—nonlinear dose–response curves—responses to chemicals are terribly difficult to predict without detailed research. For example, low doses can trigger cell proliferation while high doses

---

[1] Manikkam et al. (2013) examined endocrine disruption exposure in rats. [2] Nilsson et al. (2012). [3] Vandenberg et al. (2012) provide an extensive survey of effects from endocrine-disrupting chemicals. [4] Welshons et al. (2003).

can kill cells, while intermediate concentrations can result in chemical binding with different types of receptors with different responses.

To summarize, then, four features make the low doses of endocrine- disrupting chemicals in the environment problematic: (1) the fine-tuning of natural hormonal systems; (2) the present empirical testing approach that usually relies on the assumption of a linear extrapolation downward from high-dose effects; (3) nonmonotonic responses with dosage; and (4) the endocrine system already functioning in living organisms—any external dosage piles onto those levels.[1] These features mean that toxicological studies must examine a broad range of dosages, from higher lethal concentrations all the way down to environmentally observed concentrations.

Other nonphysical reproductive difficulties can also occur, such as finding and identifying mates of the right species. For example, exposure of two shiner species (*Cyprinella venusta* and *C. lutrensis*) to BPA at a concentration of 1.28 mg/L resulted in both males and females being unable to distinguish between males of the two species.[2] It's hard to think like a fish, contaminated or not, but part of the problem was a significant loss of male color, including much of their bright orange coloring in the fins and belly.

By design, pharmaceuticals aren't lethal at prescribed dosages, let alone environmental concentrations.[3] While the understanding of the environmental effects isn't clear, these chemicals are becoming widespread in the environment with increased human and animal use; stream and groundwater concentrations of a few are depicted in Figure 8.1. Generally speaking, concentrations of human pharmaceuticals are well below the concentrations at which ecological effects are seen,[4] which might be roughly described as an effect in a species not prescribed the drug.

The drugs in Figure 8.14 show that effect levels are typically 100× or more than the maximum concentrations found at wastewater treatment plant outfalls. Because of dilution, surface water concentrations are even lower than the wastewater outfall concentrations plotted here. The drugs summarized include common ones: acetylsalicylic acid, which is just aspirinan (NSAID: nonsteroidal antiinflammatory drug), and its metabolite, salicylic acid; diclofenac (NSAID); propranolol (a beta-blocker, which targets heart (and other) cells to reduce the effects of stress); clofibric acid (metabolite of clofibrate, a cholesterol-lowering drug); carbamazepine (antiepileptic and bipolar drug); and fluoxetine (antidepressant, brand name Prozac).

Other than obvious things like spills, ecological effects are rare and the result of unintended consequences. A commonly cited incident is one high mortality event (>95%) in which vultures in India fed on dead livestock that

---

[1] Welshons et al. (2003). [2] Ward and Blum (2012). [3] Ankley et al. (2007) summarize the ecological risks of endocrine-disrupting pharmaceuticals, providing an excellent list of references and detailed background. [4] Fent et al. (2006) summarize the ecotoxicological effects of pharmaceuticals.

Chronic toxicity of pharmaceuticals.

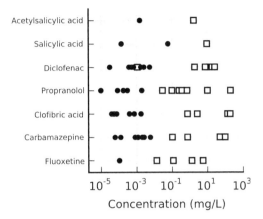

Figure 8.14: Chronic toxicity of pharmaceuticals are higher than sewage outfall concentrations (Fent et al. 2006). Circles are the maximum concentrations observed at the outfalls of sewage treatment plants, and squares are the lowest observed effect concentrations for a variety of aquatic organisms ranging from phytoplankton to fish.

had been treated with the NSAID diclofenac, used in humans and animals. Just like loons collect mercury as a top predator (see Figure 7.6), the vultures suffered from the biomagnification of diclofenac, resulting in mortality-inducing kidney problems.

As a general statement, chronic effects of pharmaceuticals at high enough concentrations in water include lesions on organs, delayed hatching, reproductive reductions, and the feminization of male fish. In the laboratory, ecotoxicological effects are usually measured at high, pulsed durations, much like the experiments performed with bull trout and warm water in Figure 9.10. Chronic toxicity levels are very different from lethal concentrations, so making predictions from these lab studies is difficult. For example, aspirin has chronic toxicity in *Daphnia* at a concentration of 1.4 mg/L, but its acute toxicity is a thousand times higher, 1.3 g/L. A factor of 100–1,000 is common for concentrations causing of these two classes of effects.

So the arguments with respect to pharmaceuticals work both ways. It certainly seems a little mistaken to worry about *human* health effects of environmental concentrations of human pharmaceuticals when the levels don't get anywhere near the concentrations delivered by ingesting the drugs themselves.

However, we can expect that human pharmaceuticals have chronic effects in animals similar enough to humans, which is a very broad class because evolution has conserved many important physiological pathways. Even effects in plants may be important, with phytoplankton being affected by ibuprofen and fluoxetine at the milligram per liter level. One also shouldn't be surprised if effects in microorganisms are quite different and unexpected, simply because evolution conscripts similar molecules in different species for different tasks. We should be doing more studies on other organisms for screening purposes,[1] and a fine argument can be made that, given high uncertainty in each chemical's ecotoxicity, simple production volume and environmental persistence provide good baseline regulatory prioritization.[2]

Chapter Readings

Damásio, J., D. Barceló, R. Brix, C. Postigo, M. Gros, M. Petrovic, S. Sabater, H. Guasch, M.L. de Alda, and C. Barata. 2011. Are pharmaceuticals more harmful than other pollutants to aquatic invertebrate species: A hypothesis tested using multi-biomarker and multi-species responses in field collected and transplanted organisms. *Chemosphere* 85: 1548–1554.

Fent, K., A.A. Weston, and D. Caminada. 2006. Ecotoxicology of human pharmaceuticals. *Aquat. Toxicol.* 76: 122–159.

Focazio, M.J., D.W. Kolpin, K.K. Barnes, E.T. Furlong, M.T. Meyer, S.D. Zaugg, L.B. Barber, and M.E. Thurman. 2008. A national reconnaissance for pharmaceuticals and other organic wastewater contaminants in the United States: II) Untreated drinking water sources. *Sci. Total Environ.* 402: 201–216.

Gewurtz, S.B., S.P. Bhavsar, E. Awad, J.G. Winter, E.J. Reiner, R. Moody, and R. Fletcher. 2011. Trends of legacy and emerging-issue contaminants in Lake Simcoe fishes. *J. Great Lakes Res.* 37: 148–159.

Hites, R.A. 2004. Polybrominated diphenyl ethers in the environment and in people: A meta-analysis of concentrations. *Environ. Sci. Technol.* 38: 945–956.

Hwang, H.-M., T.L. Wade, and J.L. Sericano. 2008. Residue-response relationship between PAH body burdens and lysosomal membrane destabilization in eastern oysters (*Crassostrea virginica*) and toxicokinetics of PAHs. *J. Environ. Sci. Health* A43: 1373–1380.

McDaniel, T.V., P.A. Martin, N. Ross, S. Brown, S. Lesage, and B.D. Pauli. 2004. Effects of chlorinated solvents on four species of North American amphibians. *Arch. Environ. Contam. Toxicol.* 47: 101–109.

---

[1] Fent et al. (2006). [2] Ankley et al. (2007).

Richardson, S.D., and T.A. Ternes. 2011. Water analysis: Emerging contaminants and current issues. *Anal. Chem.* 83: 4614–4648.

Zorita, S., L. Martensson, and L. Matthiasson. 2009. Occurrence and removal of pharmaceuticals in a municipal sewage treatment system in the south of Sweden. *Sci. Total Environ.* 407: 2760–2770.

Chapter 9

Thermal Pollution

Humans affect stream temperatures both directly and indirectly. Thermal pollution comes as heat contained within wastewater, precipitation warmed through the urban heat island phenomenon, and runoff from hot pavements. Other routes include riparian deforestation, warmed groundwater, stream base flow reductions that limit thermal dilution, and climate change.

Runoff temperatures and subsequent stream temperature pulses depend on features as varied as dew point, watershed imperviousness, riparian forest loss, reservoir discharges, and distance from a stream's source. Warm stormwater runoff also affects groundwater through water-quality measures, such as conductivity and dissolved oxygen.

The size of a stream is important for thermal buffering, with small, headwater streams generally being close to groundwater temperatures, except for sudden heat pulses from urban storms, whereas large streams have higher temperatures with little daily variation. Temperature also plays an important role in the ecological competition between various phytoplankton species, with high temperatures leading to the dominance of green and blue-green algae associated with algal blooms. As discussed in Chapter 6, these blooms lead to other water quality problems. Fish, namely, trout, as we'll see below, prefer lower stream temperatures. Typically, urbanization results in heat-tolerant invasive trout species displacing native trout with lower-temperature requirements. This temperature effect means native trout suffer four simultaneous and direct problems: nutrients, heavy metals, heat, and interspecific competition. Perhaps even more directly, a fifth problem comes about because fish eggs get either washed away by flashy flows or dried out by reduced base flows.

Warm streams start with urban heat islands.

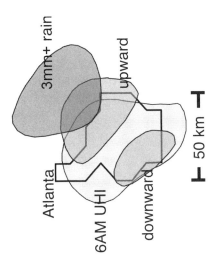

Figure 9.1: An urban heat island developed over Atlanta, GA, on July 30, 1996 at 6:00 a.m. Fifteen minutes later, the atmospheric instability created downward and upward air flows, and generated a 15-min thunderstorm at 6:30 a.m., with the indicated >3 mm rainfall pattern (after Bornstein and Lin 2000).

We now consider a completely different type of pollution: higher water temperatures. Urban areas have sacrificed vegetation and produced much higher temperatures than surrounding rural areas, creating so-called urban heat islands (UHIs) that poke out of the cooler rural background.[1] UHI refers to the phenomenon that urban areas have temperatures roughly 5–10°C higher than the surrounding rural and suburban areas. The effect is most profound at night, when the treed rural areas cool off quickly, while the buildings and paved surfaces of the urban areas retain their heat.

UHIs can change weather patterns at local, regional, and continental scales, from shifting the time of day that precipitation occurs to inducing localized thunderstorms. Studies from Atlanta, Georgia, for example, even demonstrate that more lightning strikes take place in urbanized areas compared with outlying regions, and lightning strike data suggest a similar pattern for urban areas of Durham County.[2]

At the scale of individual cities, urban areas affect the weather. The contour plots shown in Figure 9.1, traced from the originals, caricatures a thunderstorm that took place during the 1996 Atlanta Summer Olympics.[3] A UHI persisted over nearly the entire urban core in the early morning of July 30, around 6 a.m., having a central temperature roughly 2.5–4°C higher than surrounding areas. This UHI produced a thermal instability, leading to two regions of high and low pressure, where air moves upward in one spot and downward in

---

[1] Oke (1973). [2] Wilson (2011) [3] Bornstein and Lin (2000).

Rain cools the city and warms streams.

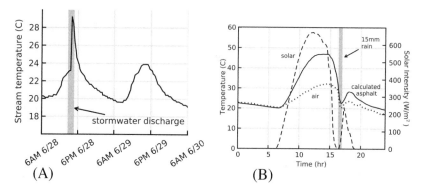

Figure 9.2: (A) Temperature surges in Maryland Piedmont streams due to urban rains reach up to 7°C but average 3.5°C (Nelson and Palmer 2007). (B) Measurements and calculations show how warm summer rains cool hot urban pavement in Albertville, MN (Herb et al. 2008).

---

another.[1] Think about a rolling boil taking place in a shallow pan of heated water: heated water in one spot becomes less dense than nearby spots—maybe due to the vagaries of heating or random shapes and pits in the pan's bottom—and as that water rises to the surface, slightly cooler and denser water from the surface sinks to take its place. Similarly, air over the city heated by UHI processes becomes less dense in various other spots. The warm air mass starts to rise, reducing the pressure at that ground location, and the surrounding high-pressure air mass pushes more air into that location of rising air. The rising warm air quickly cools in the upper atmosphere, and the water vapor that it holds condenses like the drops on a glass of ice water. Suddenly you've got a thunderstorm downwind of the UHI.

Now imagine these sun-baked city surfaces being washed with a warm afternoon rain shower.[2] The cooling rains absorb heat from the air and the impervious surfaces, while the precipitation temperature rises, bringing heat to urban streams. Study of the phenomenon of warmer urban streams during the summer and colder streams during the winter dates back to at least the 1960s.[3]

Figure 9.2A depicts an example of the thermal effects on urban streams. Researchers took stream temperature measurements 10 cm below the stream surface every 30 min in 16 streams in five watersheds in Maryland's Pied-

---

[1] Dixon and Mote (2003). [2] Herb et al. (2008) developed the models predicting stream runoff temperatures. [3] Leopold (1968) discussed urban stream temperatures.

Runoff temperature increases with dew point and surface temperatures.

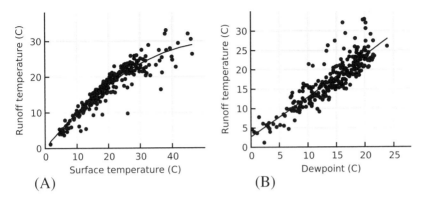

Figure 9.3: Data-based model results for stormwater runoff temperatures increase with surface temperature (A) and dew point (B) (Herb et al. 2008).

mont region north of Washington, DC.[1] Over nearly 1,500 monitored summer days in 2002 and 2004, researchers observed 37 stream temperature surges exceeding 2°C, with a maximum rise of 7.4°C. These surges were really surges: there was an average temperature rise of 3.5°C, and except for one case, stream temperatures came back to normal within 3 hours.

The details behind the surges are shown in Figure 9.2B. This study concerns the Minnesota cities of Albertville and Minneapolis and relates solar energy input, air temperature, and precipitation. Pavement temperature was calculated, not measured, and the project's goal was a predictive one that incorporates a multitude of factors, not all of which could be measured simultaneously in one study. Note the rapid rebound of pavement temperature after the rainfall: heat quickly transfers up to the surface from the subsurface pavement.

The numbers from Figure 9.2B generated stormwater runoff temperatures using a series of models that predict thermal effects of future developments and, again, aren't measured directly, but the results are shown in Figure 9.3. The study used several detailed climate data sets, but in the end, dew point, air temperature, and the amount of solar radiation (sunshine) before the storm explained 91% of the variance in runoff temperature. Highest runoff temperatures came from late-afternoon, light summer rainfalls.[2] Just like the storm development associated with the Atlanta UHI depicted in Figure 9.1 connected

---

[1] Nelson and Palmer (2007) studied stream temperature surges. [2] Herb et al. (2008).

high dew points with UHI storms, higher dew points indicate higher heat export by stormwater runoff.

Related mathematical models showed a 1°C increase in air temperature and dew point preceding a storm increased runoff temperature by 0.5–0.7°C.[1] That dependence means that a commonly occurring 10°C UHI increases runoff temperatures by 5–7°C. Solar radiation had an important effect, with a 20% increase in light leading to a 0.3°C increase in temperature. Parking lot slope had no effect. Specific heat and thermal conductance of impervious materials were also important parameters, as might be expected.

After predicting runoff temperature, the next step is predicting the increase in stream temperature, which is given by

$$\Delta T = (T_{\mathrm{ro}} - T_{\mathrm{s}}) \frac{Q_{\mathrm{ro}}}{Q_{\mathrm{s}} + Q_{\mathrm{ro}}}, \qquad (9.1)$$

where the variables are the temperatures of the runoff and stream, T_{ro} and T_{s}, respectively, and the flow rates of the runoff and stream, Q_{ro} and Q_{s}, respectively. Note that the expression works whether the runoff temperature is less than or greater than the stream temperature. Streams with a low base flow—in other words, urban streams (see Figure 5.2)—bear the full impact of high runoff temperatures. Short-duration storms with high flows were the worst, leading to 1°C or more increases in stream temperatures.

Put in context, these differentials can seem small. Researchers have observed higher summer urban stream temperatures of 10–15°F (see Figure 9.7A), and 5–10°F lower winter temperatures.[2] The summertime causes include the loss of shade from the loss of riparian vegetation, and stormwater temperature changes during transit through control measures. Both seasonal differentials are affected by comparisons with natural streams that derive base flow from groundwater rather than surface runoff, particularly during the winter when groundwater temperatures are higher than surface water temperatures. In this case stormwater runoff cools streams.

It's important to realize that most storms have little thermal effect, but a few each year have big effects. This fact likely corresponds to the rather limited times of the year—the summer months—when the UHI phenomenon is most apparent.

Parking lots aren't the only problem, and heat export from roofs can be significant.[3] Based on data coming from a set of Minnesota-based sites, heat export from residential shingled roofs was just one-third that from the standard gravel roof of commercial buildings, while the heat export from concrete

---

[1] Herb et al. (2009) presents heat transfer models for stormwater runoff, and used these models in the Herb et al. (2008) work. [2] Galli (1990). [3] Janke et al. (2011) studied heat export from various surfaces, including roofs and driveways.

Imperviousness warms streams.

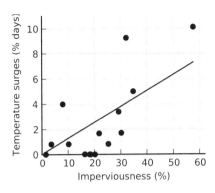

Figure 9.4: Stream temperature surges increase with impervious surface fraction in the Piedmont north of Washington, DC (Nelson and Palmer 2007). Impervious surfaces, surface temperatures, and dew point all heat up together to affect runoff temperatures.

driveways was twice that of commercial buildings. Though commercial roof temperatures can exceed that of driveways, the high thermal mass of driveways stores more heat. Roof temperatures also drop about 5°C as the clouds gather in the hour before a rainfall, shading the roof and letting radiative cooling do its thing. In conclusion, residential roofs usually contribute little heat to stormwater, but commercial roofs can be significant.

Putting together the above thermal features, the frequency of stream temperature surges increases with watershed impervious surface fraction, as shown in Figure 9.4.[1] That connection just makes sense, and the statistics show a correlation of $r^2 = 0.48$, meaning that imperviousness explains nearly half of the variation in stream surge frequency. Furthermore, the study authors found that climate change has further doubled the number of temperature surges.

One report documents that Tokyo's winter stream temperatures increased by 3.2–4.2°C spanning the years 1978–1999 and attributes that increase to the heat export of wastewater treatment discharges.[2] During the winter, the stream's "natural" flow rate drops dramatically, and wastewater represents more than 60% of the flow. The report also shows strong connections between air temperatures and stream temperatures, but in such a way that it is clear that stream temperature increases are not driven by air temperature. Essentially, air temperature increases were small, wastewater effluent temperature increases were large, and stream temperature increases were large; therefore, wastewater heat is responsible. The study authors suggested that the solution might involve cooling wastewater effluent before it enters streams.

---

[1] Nelson and Palmer (2007). [2] Kinouchi et al. (2007) connected anthropogenic heat with urban stream temperature increases.

Riparian forest loss warms streams.

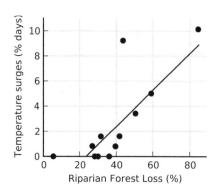

Figure 9.5: Stream temperature surges also increase with riparian deforestation in 50-m buffers in the Piedmont north of Washington, DC (Nelson and Palmer 2007).

Just as imperviousness produced stream temperature surges, so does riparian buffer deforestation. In Figure 9.5, riparian forest loss explains 58% of the variation in temperature surges. Certainly a correlation exists between riparian forest loss and imperviousness, but the experiment wasn't designed to separate these two factors.

Beyond the more specific issue of temperature surges, riparian forest loss affects several additional factors, including average temperature. In natural headwater streams, deciduous vegetation blocks light from reaching streams, and the amount of photoactive radiation reaching the stream explains about 70% of gross primary productivity over the year, but it's even more important in spring, explaining about 84%.[1]

In forested streams, solar radiation is the dominant energy flux, and then longwave radiation, or the blackbody radiation that, for example, gets trapped in greenhouses.[2] This longwave radiation is most critical for small streams with little thermal buffering. After these energy fluxes comes evaporation. Channel width affects surface area; narrow and deep versus wide and shallow have different heat transfer effects on these various fluxes. Riparian vegetation both shades and affects wind speed over surface water, the latter affecting evaporation rates. Once disturbed by deforestation, stream temperatures take about a decade to recover.

Of heat input to streams from another South West England study, radiative fluxes (that means "sunshine") accounted for 70%, and water friction with the stream bed and bank gave 15.8%. In winter, friction accounted for about 75%.[3] Just like the brakes on a car heat up as it stops, stream water warms as

---

[1] Roberts et al. (2007) studied carbon cycling in headwater streams in eastern Tennessee. [2] Caissie (2006) reviews the thermal aspect of rivers. [3] Webb et al. (2008).

Streams get warmer farther from their source.

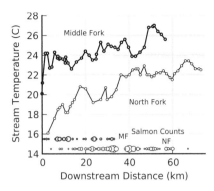

Figure 9.6: Eastern Oregon tributaries of the John Day River warm up as they flow downstream, and salmon seek the upstream cool waters (Torgersen et al. 1999). The study was performed in summers of 1993 and 1994. Circles size reflects salmon count numbers.

it hits rocks and pebbles and sticks. Any falling mass, water included, converts gravitational potential energy to the kinetic energy of motion, and the sudden "thud" of the contact with the ground dissipates that energy as heat and sound. In comparison, heat transfer from air was just 10%.

Releases from dams drastically cool streams in summer but warm them in winter (see Figure 9.13), causing an overall warming averaged over a year. This warming can be responsible for advancing brown trout hatching times by over 50 days.[1]

An interesting feature of riparian areas with streams requires a historical view. Imagine some stream that meanders through some bottomlands. We all know that the stream twists and turns, and slow erosion and fast floods can change these stream meanders, but below the surface of the riparian area exist old stream beds (called paleochannels) covered with organic matter and sediments.[2] Water also flows through these subsurface channels, and that pathway has important influences on stream water temperatures by storing water and buffering temperature swings. When these flows are disturbed by humans, they, too, can produce stream temperatures changes.

Figure 9.6 demonstrates one of the consequences of warming stream temperatures. In these eastern Oregon streams, temperatures increase rapidly moving downstream in late July and early August.[3] Chinook salmon that migrate upstream from the ocean to spawn have an upper thermal tolerance of 25°C, and it's clear from their distribution along these two streams that they seek the lower temperatures of headwater streams in search of higher growth rates for their offspring (see Figure 9.10). In the Middle Fork, temperature was the most important habitat variable explaining salmon distribution, followed by

---

[1] Caissie (2006).   [2] Poole and Berman (2001).   [3] Torgersen et al. (1999).

pool density (number of pools per kilometer), but in the cooler North Fork, pool volume and density were most important. The North Fork sits in a pre-served wilderness area, but at the time the Middle Fork experienced intensive land use from logging and grazing. Stream temperature differences reflect these activities. Along these rivers, smaller streams entered larger streams with either colder or warmer water, producing various temperature deviations. Water temperature was highest in wide, shallow, and shaded sections. Over-all, stream temperatures increase downstream with increasing stream order, changing as much as 14°F from first-order to third-order streams.[1]

A mandate by the state of Georgia to reduce riparian buffers from 100 feet to 50 feet (from about 30 m to 15 m) motivated the study that connected watershed and riparian cover (see Figure 11.4A). An after-the-fact study in the northeast corner of Georgia determined that increasing the newly mandated buffers from 50 feet back to the original 100 feet would decrease summer peak stream temperatures by 2.0 ± 0.3°C and fine sediments by 25%.[2] Fur-thermore, the act of decreasing the buffers reduced young trout biomass by about 87%. Researchers also estimated that with 15-m (50-feet) buffers, only 9% of second-through fifth-order streams could support trout, but with 30-m (100-feet) buffers, the fraction of productive streams increases to 63%. In summary, they expected 18.8% of streams to have high thermal habitat qual-ity with 30-m buffers, but just 0.7% with 15-m buffers. However, it was an after-the-fact study.

Beyond the riparian and watershed scales, climate change also affects stream temperatures.[3] In the northwestern U.S., summer stream temperatures increase at around 0.2°C per decade, with increasing air temperatures—about 0.13°C per decade—deemed responsible for 82–94% of the rise. A survey of historical data from 40 U.S. streams and rivers with data covering from 20 to 100 years showed that half of streams and rivers have warmed 0.09–0.77°C per decade.[4] Only 2 of the 40 streams had a significant cooling trend, and, showing the continued importance of human disturbance, the largest stream temperature increases took place near urban areas.

Figure 9.7 contrasts urban and rural stream temperatures and comes from an easily readable and extensive document prepared for the Maryland De-partment of the Environment.[5] The data are now more than 25 years old, but the science hasn't changed. Both streams sit within the basin of the Anacos-tia River that flows through Washington, DC. The urban stream is the White Oak tributary, with a 225-acre watershed, which at the time was 60% im-pervious with a 0.35 cubic feet per second (cfs) base flow, and the natural stream is the Lakemont tributary, with a 400-acre watershed, at the time an

---

[1] Galli (1990). [2] Jones et al. (2006). [3] Isaak et al. (2012) studied thirty years of stream tempera-tures in the northwestern U.S. [4] Kaushal et al. (2010) compared rising stream and air temperatures across the U.S. [5] Galli (1990) discusses the broad issues of temperature and stormwater, including temperatures found in SCMs like constructed wetlands and ponds.

Urban streams are warmer than rural ones.

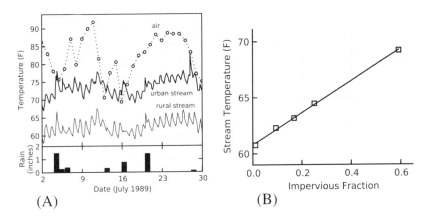

(A)   (B)

Figure 9.7: (A) Urban streams are warmer than rural streams, while stream temperatures in both environments respond to air temperatures (Galli 1990). (B) Temperatures in these Maryland streams also increase with a watershed's imperviousness.

undeveloped area with 1% impervious and 0.86 cfs base flow. No stormwater control measures (SCMs) were in place for the urban watershed. Although only a few miles separate the two streams, the urban stream was typically 4–15°F warmer, as shown in Figure 9.7A. In fact, during the study period the urban stream's maximum temperature was 82.6°F, whereas the rural stream's maximum was just 67.8°F.

Imperviousness increases stream temperature. As Figure 9.7B shows, stream temperature increases by 1.4°F for every 10% increase in imperviousness.[1] These headwater streams were located in Montgomery County, in the Piedmont of Maryland, and stream temperatures were averaged over the late spring through summer, but there's only a single stream for each imperviousness value. This dependence would imply an increase of 14°F for a 100% impervious area, a number that closely matches the scale of UHI temperatures.[2] One conclusion of this early paper was that local air temperature had the biggest effect on stream temperature most of the time, at least as far as these small, free-flowing headwater streams are concerned.

Thermal pollution also helps produce algal blooms, and Figure 9.8A summarizes the outcome of algal competition along temperature gradients.[3] This summary comes from the same study of the Piedmont area of the Anacostia

---

[1] Galli (1990). [2] Wilson (2011). [3] Galli (1990).

Ecosystems change with water temperature.

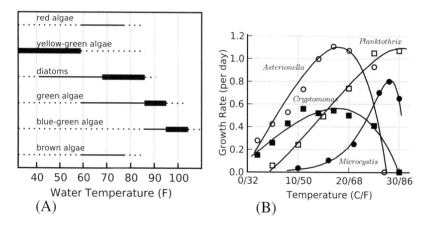

Figure 9.8: The community of primary producers changes with water temperature. (A) Thick bars indicate the relatively dominant species, whereas the solid line reflects high species abundance (Galli 1990). (B) Curves reflect individual species growth rates that underlie competitive outcomes (Paerl and Huisman 2009). *Asterionella formosa* is a diatom; *Microcystis aeruginosa* and *Planktothrix agardhii* are cyanobacteria; *Cryptomonas marssonii* is a cryptophyte.

River that flows into Washington, DC, and whose results are depicted in Figure 9.7. The study mentions that storm-flow temperatures are most often lower than base-flow temperatures, even in the highly developed Lower White Oak tributary, contrasting with the results of Figure 9.4. For the streams in this basin the base-flow temperatures usually exceed storm-flow temperatures, but 1–5% of the time storm-flow temperatures are higher.

As mentioned before, the loss of riparian vegetation can increase summer stream temperatures by 11–20°F and decrease winter temperatures by 5–7°F.[1] Remarkably, on some streams where the temperature difference is measured between a 150-foot section of stream where the canopy has been opened, there's a temperature increase of 8.5°F. Smaller base flow, wider channel, and longer travel time all contribute to higher stream temperatures when the sun shines. Once the water has warmed, cooling it back down is quite slow. The water warms up due to solar radiation, but the cooling takes place through

---

[1] Galli (1990).

Trout prefer cool temperatures.

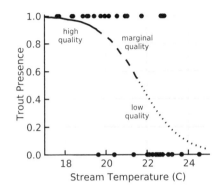

Figure 9.9: Rainbow and brown trout need cool stream temperatures (Jones et al. 2006). Each dot reflects the presence or absence of young-of-the-year trout in a particular stream in Georgia, and the line reflects an occurrence likelihood given a stream's temperature.

lower-intensity channels of conduction, evaporation, blackbody radiation, and groundwater exchange.

In the context of this thermal environment, then, we see the ecological competition of Figure 9.8A. Really, all it says is that different species of algae grow better under different environmental conditions, and the plot shows the environmental factor of temperature. Another factor not shown involves light intensity: red algae prefer low light; yellow-green algae, brown algae, and diatoms prefer medium light; and green and blue-green algae survive at high light.

Figure 9.8B provides a bit more detail on four primary producers, two of which are cyanobacteria, also known as blue–green algae, that create algal blooms when nutrients and temperatures are high.[1] At least one of these species, *Planktothrix agardhii*, produces toxins. What this all means is that aquatic insect growth depends on stream temperatures directly and indirectly through degree-days of primary productivity, physiological temperature tolerances, and algal resource species composition.[2]

Trout temperature preference data of Figure 9.9 come from the same stream buffer study as that shown later in Figure 11.4A, using the Georgia streams associated with the described primary sites in 2001 and 2002.[3] While surveying the streams, if a young trout was found in the stream (sampled via electric shocking), there's a dot shown at one—which means "present"—for the stream's temperature, but if no trout was found, there's a dot shown at zero, meaning "not found." The line comes from a parameterized statistical model

---

[1] Paerl and Huisman (2009). [2] Caissie (2006) reviews stream temperature consequences for stream life. [3] Jones et al. (2006) provide the trout presence data and summarize the trout temperature tolerance data.

Trout survival depends on water temperature.

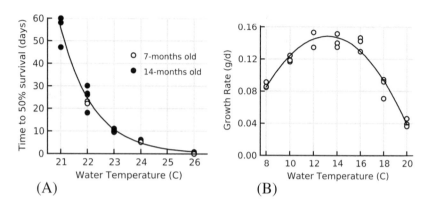

Figure 9.10: (A) Bull trout perish when stream temperatures are quickly raised from 8°C to a temperature above 21°C, (B) and prefer a nice, cool 13°C where they have maximal growth rate (Selong et al. 2001).

that estimates the probability of finding a trout given the stream's temperature; solid means a high-quality environment for trout, whereas the dashed and dotted lines reflect marginal and low-habitat quality streams, respectively. Only rainbow and brown trout are represented in this plot because the researchers found no native brook trout—they persist in streams that were smaller than the primary sites.

Figure 9.10 depicts the underlying physiological basis for these stream occurrence probabilities. These survival (Figure 9.10A) and growth rate (Figure 9.10B) data are for bull trout (*Salvelinus confluentus*), which range over the northwestern U.S., where it is a native, threatened species.[1] Results of the experiments shown here demonstrate their heat tolerance limits. Fish were first held at a cool 8°C and then brought to experimental temperatures with increases pegged at 1°C/day. Experiments then ran for 60 days or until the fish died. Survival was more than 98% at 18°C and below, but no fish survived for 60 days above 22°C. At 22°C half of the fish perished after about 25 days, and there was 100% mortality within a day at 26°C (79°F). Consider these physiological facts in light of the storm-induced heat spikes in Figure 9.2A or the higher-temperature urban streams depicted in Figure 9.7A, though, granted, these latter results come from the southeastern U.S. Nevertheless, heat is a deadly pollutant for trout in our urban streams.

---

[1] Selong et al. (2001) examined thermal issues with trout and salmon.

Cutthroat trout dominate urban streams.

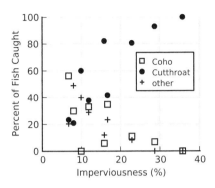

Figure 9.11: Cutthroat trout outcompete coho salmon in the competition for urban streams in western Washington State (Booth and Reinelt 1993). Important competitive factors include temperature and the frequency of pools and specific substrate in the streams.

Given the evolved physiological limits for native trout, in come the nonnatives as local thermal environments change. Chinook salmon (salmon and trout are closely related) possess the highest optimal growth temp of 18.9–20.5°C. Compare that with the optimum growth rate of 13.2°C for bull trout, shown in Figure 9.10B. Between these values sit brook trout at around 14.4–16°C, and brown trout at 13.9°C.

An important measure for these fish species is its "critical thermal maximum," which measures its tolerance to rapid temperature change, like the tolerance depicted in Figure 9.10A. Bull trout is the least tolerant species by far, followed more closely together cutthroat trout, and then rainbow and brown trout. Both bull and cutthroat trout are presently limited to high-elevation streams where the water temperatures are low (but see Figure 9.11). The presence/absence data of Figure 9.9 match quite well the upper temperature tolerances for trout, which are around 20°C for native brook trout, 21°C for the rainbow trout (introduced to eastern North America from its native western North America), and 24°C for the introduced European brown trout.[1]

The damage from pollution spikes isn't limited to temperature. Fluctuating toxin concentrations, like those found in stormwater discharges, tend to be more damaging than constant toxin levels, in a way similar to temperature variation.[2] Higher toxin concentrations can be tolerated for shorter pulse durations, but even brief pulses of some pollutants, such as cadmium and zinc, lasting just 15 min led to physical effects lasting up to 7 days on standard test organisms.[3] Phenol, a simple six-carbon ring with an OH attached, also called carbolic acid, is used in products ranging from construction materials

---

[1] Tabulated by Jones et al. (2006).   [2] Burton et al. (2000) describe toxicity definitions and issues with variability.   [3] Brent and Herricks (1998) examined the effects of pulsed toxicants.

to medicines. It, too, causes harm with short-term spikes. Some organisms can acclimate to pulsed exposures or recover from sublethal exposures of toxins that can be cleared, but not all organisms can handle all pollutants.

A major concern involves the working definitions of "toxicity." Acute toxicity implies lethality in <96 hours, whereas chronic toxicity means lethal in 7–9 days.[1] Real systems have longer and more variable exposures, with changes and durations extending over seasons. Other difficulties range from using grab samples from streams to making broad conclusions after testing on just one or two laboratory organisms, usually fathead minnows or *Daphnia*. That said, laboratory testing plays an important role for controlled tests and has important predictive uses.

We've just discussed the temperature tolerance differences between the various species of trout in Figure 9.10, and we'll see related problems with finding fish eggs in urban streams in Figure 11.18. The broader-scale consequences of those life history and physiological issues show up in the population density data of Figure 9.11, collected in King County in the western part of Washington State.[2]

What details of physiology or life history promote this switch in species dominance? We've seen the temperature issues, but these researchers mention that cutthroat trout tolerate small, homogeneous streams, whereas coho salmon require the pools and substrates of a variable, natural stream. This plot demonstrates that urbanization tips the ecological balance between these two species.

Chapter 10 considers the connections between streams and groundwater, and Figure 9.12 foreshadows this connection via thermal consequences. We know of groundwater's importance because many municipalities use it for drinking water. Groundwater temperature is important because temperature affects many aspects of water quality, in particular, arsenic concentration, but also including dissolved organic carbon, pH, and various other chemicals at higher temperature.[3] This study compared groundwater temperatures below infiltration basins with other areas having direct infiltration by rain.[4]

Overall, the higher temperatures resulting from infiltrating stormwater in retention basins drove higher microbial activity, which drove down oxygen levels. These higher temperatures were mostly found in the higher layers of groundwater, prior to great amounts of mixing within the aquifer.

In both panels in Figure 9.12, the horizontal axis measures the amount of potential heat released into the groundwater by the runoff; in figure 9.12A the vertical axis simply reflects the temperature increase of the groundwater. Stormwater runoff temperature correlated to rainfall temperature, air temper-

---

[1] Burton et al. (2000). [2] Booth and Reinelt (1993) report and cite fish catch data from a 1993 conference presentation by Luchetti and Fuerstenberg. [3] Bonte et al. (2013). [4] Foulquier et al. (2009) examined the thermal influences of stormwater on groundwater.

Warm runoff warms groundwater.

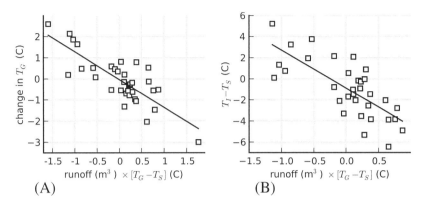

Figure 9.12: (A) Groundwater temperatures rise with the infiltration of warm water in Lyon, France; however, cold winter temperatures led to cooling (Foulquier et al. 2009). Groundwater temperature, T_G, was measured at 1 m below water table; infiltration water temperature, T_I was measured just above the water table; and the water temperature measured in stormwater basins is T_S.

ature, and solar radiation before the storm, energy transfers shown in Figure 9.2B. Yes, in the warm months all the things that make the UHI also increase stormwater temperatures and, subsequently, increase groundwater temperatures: warmer urban air temperatures, sun-drenched asphalt, and heat from everything powered by electricity or fuels.

However, between September and February the stormwater heated up as it traveled through the vadose zone (the soil above the groundwater table), while between March and August the water cooled down. That means hot summer rains heated up the vadose zone while the stormwater cooled off, and winter rains cooled down the vadose zone while the stormwater heated up. These processes determine the sign on the vertical axis of Figure 9.12B. T_I is the temperature of the infiltrating stormwater as it passes through the groundwater table, and T_S is the temperature in the stormwater pond.

It's not shown here, but the researchers also measured the water's specific conductance, which served as a fingerprint for stormwater since groundwater's specific conductance is about three times higher than stormwater. The data showed clear plumes of lowered specific conductance below stormwater ponds. Ultimate temperature change depended on vadose zone thickness, and

Reservoir discharges complicate stream temperatures.

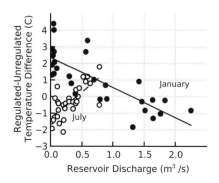

Figure 9.13: Two nearby streams in South West England, a regulated one with reservoir discharge water and the other without, show seasonal temperature changes due to the reservoir (Webb and Walling 1996; B. Webb, pers. comm.). The two streams show seasonal differences absent discharges, but the reservoir cools the stream in January, and warms the stream in July.

the "half-length" of the temperature profile was 5 m. Time delay of recharge ranged from 33 to 337 hours, and the temperature plume traveled "downstream" at about 2 m/hour.

Things other than stormwater also transfer heat to groundwater, including geothermal heat pumps and hot water flowing down sewer pipes. This study estimated that groundwater was recharged with 60 m³/hour of 14.4°C stormwater draining a 180-hectare basin but cited the presence of at least one groundwater-based heat pump per 80 hectares discharging 50 m³/hour of water exceeding 20°C! An interesting comparison is that the heat transferred to lower groundwater temperature by 2°C throughout a 20-m-thick layer via heat pumps represents two-and-a-half times the total residential demand for heating! Clearly, stormwater-associated groundwater warming might not be the most serious factor.

Some evidence links UHIs and warmed groundwater. For example, groundwater temperatures outside Dublin, Ireland, average 10°C, but sites within the city get to 14°C.[1] Estimates for the UHI-related increased temperature of subsurface groundwater in Cologne, Germany, and Winnipeg, Canada, sit around 5°C.[2] Other groundwater warming examples demonstrate an increase of around 3–4°C, and cost–benefit calculations concerning geothermal applications show a economic payback time of 5 years or less for combined heating–cooling systems, an increased efficiency that's in part due to raised groundwater temperatures from urbanization. Indeed, comparing the efficiencies of

---

[1] Allen et al. (2003) and references therein. [2] Zhu et al. (2010).

the rural and urban areas for geothermal energy, the UHI-raised temperatures yield a 50% increase in heat recovery. Complicated effects may also
occur with climate change since the atmosphere is the source of precipitation, and precipitation is the source of groundwater.[1] Changes in precipitation patterns will lead to potential changes in streams and alter groundwater
availability.

Data in Figure 9.13 show temperature effects of reservoirs on streams from
the River Pulham, Exmoor, in South West England.[2] In this plot, temperatures
were measured in two streams that come together, one from a reservoir where
the stream temperatures are drastically affected by the discharged reservoir
water, and another unaffected stream.[3] The numbers indicate that the temperature difference between the two streams depends, perhaps as expected,
on flow rate from the reservoir, but they also show the reservoir's effects on
groundwater. During winter, the reservoir warms the groundwater, but in summer it cools the groundwater. In contrast, water discharge cools the stream
in winter but warms the stream in summer. Groundwater effects control the
vertical intercepts of the plotted temperature differences, whereas reservoir
discharge determines the slopes. One measurement series from 1995 (not
shown here) demonstrated a 4°C decrease in stream temperature over a short,
5-hour time period when reservoir withdrawal came from below the reservoir's
thermocline.[4]

Chapter Readings

Foulquier, A., F. Malard, S. Barraud, and J. Gibert. 2009. Thermal influence of
urban groundwater recharge from stormwater infiltration basins. *Hydrol.
Proc.* 23: 1701–1713.

Galli, J. 1990. *Thermal impacts associated with urbanization and stormwater management best management practices.* Metropolitan Washington
Council of Governments (Washington, DC).

Herb, W.R., B.D. Janke, O. Mohseni, and H.G. Stefan. 2008. Thermal pollution
of streams by runoff from paved surfaces. *Hydrol. Proc.* 22: 987–999.

Janke, B.D., O. Mohseni, W.R. Herb, and H.G. Stefan. 2011. Heat release from
rooftops during rainstorms in the Minneapolis/St. Paul metropolitan area,
USA. *Hydrol. Proc.* 25: 2018–2031.

---

[1] Green et al. (2011) link groundwater changes with climate change. [2] Webb and Walling (1996),
and Bruce Webb, pers. comm. [3] Webb and Walling (1996). [4] Webb and Walling (1997).

Kaushal, S.S., G.E. Likens, N.A. Jaworski, M.L. Pace, A.M. Sides, D. Seekell, K.T. Belt, D.H. Secor, and R.L. Wingate. 2010. Rising stream and river temperatures in the United States. *Front. Ecol. Environ.* 8: 461–466.

Nelson, K.C., and M.A. Palmer. 2007. Stream temperature surges under urbanization and climate change: Data, models, and responses. *J. Am. Water Res. Assoc.* 43: 440–452.

Chapter 10

Responses in Streams and Groundwater

Streams represent the uppermost, exposed portion of groundwater. In fact, streams, groundwater, wetlands, and riparian systems are so interconnected that the effects of transpiration by riparian vegetation show up in daily stream water-level variation.

Recharge of groundwater takes place across an entire watershed as some fraction of the falling precipitation but depends on many factors, including rainfall amounts, geology, and soil types. As this infiltrated water reappears in the low spots, streams emerge from the soils.

Biology takes place throughout streams and groundwater, with microbial action dependent on factors such as oxygen levels, the amount of organic matter in the water and soils, and nutrient availability. Along with these natural components, groundwater carries chemicals unique to human manufacturing, such as gasoline additives, that deposit across the watersheds and infiltrate with the recharging precipitation. These chemicals trace groundwater flow paths and emerge in streams, allowing scientists to age groundwater recharge times and identifying when stream contaminants originated as surface applications.

High flows also lead to increased erosion in steeper streams and increased sedimentation in low-slope areas. Another, more natural, process influencing groundwater and stream interactions are the sticks, branches, and stumps that come from dead trees alongside rivers. These inputs structure rivers, slow down the erosive forces, and provide a variety of biological habitats for many types of organisms. With land use changes like logging and urbanization, this large woody debris is lost, and streams lose a stabilizing force.

Many small streams, few large streams.

Stream Order	Number of Streams	Total Stream Miles	Average Drainage Area (mi^2)
1	1,570,000	1,570,000	1
2	350,000	810,000	4.7
3	80,000	420,000	23
4	18,000	220,000	109
5	4,200	116,000	518
6	950	61,000	2,460
7	200	30,000	11,700
8	41	14,000	55,600
9	8	6,200	264,000
10	1	1,800	1,250,000

Figure 10.1: Most U.S. stream miles consist of low-order, headwater streams, with most being the small first-order streams that drain 1-square-mile basins. As these streams join, stream order increases, and basins combine to form large watersheds (Leopold et al. 1964; USEPA 2005c).

The table in Figure 10.1 relates stream order, the number and length of streams, and watershed size.[1] First-order streams are headwater streams, the ones that first begin to drain the upper reaches of watersheds, and can be either intermittent or permanent.[2] When two first-order streams come together, they make a second-order stream, but when additional first-order streams run into a second-order stream, it remains a second-order stream. Two second-order streams join together, however, to make a third-order stream. This process provides a classification scheme for streams, and various properties correlate with stream order.

Watershed terminology is more confusing. Under one classification scheme, catchment is the smallest-scale drainage basin, essentially the area drained by a single stormwater outfall that intersects with a stream.[3] A sub-watershed is the area drained by a second-order stream and marked by the boundaries defined at the point where two second-order streams meet and create a third-order stream. Sometimes a watershed is even defined as the largest area under the control of a local planning authority, the seemingly unpleasant situation of putting the definition at the mercy of some political boundary as

[1] These numbers originate with Leopold et al. (1964) but have lived on in USEPA (2005c).
[2] Strahler (1957). [3] USEPA (2005c).

opposed to a natural structuring. More generally, a watershed simply defines the drainage area to a particular stream segment. I use the various watershed terms rather loosely.

Note that the numbers in Figure 10.1 don't preserve order-to-order drainage areas obtained by multiplying the number of streams by the average drainage area. All this problem means is that these numbers really are just estimates with uncertainties. Still, the estimation process used data from quite a few watersheds across the U.S., and those data show some very robust scaling of the variables shown in Figure 10.1 with stream order, lending pretty good support to the results.

How useful is stream order? One large study examined stream order and various attributes using more than 2,000 first- to eighth-order streams across the U.S., in part to test the ability of stream order to predict complicated stream variables.[1] It would be wonderful if that simple classification scheme correlated with information available through more complicated GIS analyses. Results showed that stream order provides a good approximation to catchment size, discharge, stream width at various stages, and so on, but these characters can have large variations, of several orders of magnitude. Indeed, stream order, catchment size, discharge, and distance to source headwaters were all highly and positively correlated. Overall, the best agreement was in mountains, and the worst agreement was in the plains. Despite the correlations, the authors recommend using quantitative measures—things like discharge or stream width—for most purposes, not qualitative stream order, which has a very limited mechanistic utility.

While we're at it, we can define some terms associated with streams.[2] A *pool* is a deep (perhaps a couple feet or deeper) portion of a river with leaves, sticks, and debris at the bottom in some places, and sediments covering small stones in others. The low flow speeds allow heavy things to drop out. A *riffle* is a shallow area with high flow speeds over stones with areas of fine gravel and rocks with algae growing on them. A stream *reach* is a series of pools and riffles bounded by *boulder cascades* and *debris dams*. Think of a terraced hillside as an analogy for a terrestrial stream: each level area bounded by terraces is a reach, and the terraces are the dams and cascades. River *segments* are bounded by locations where streams enter the river.

Streams also have other parts with various descriptors.[3] *Eddy drop zones* are where water flows backward and allows sediments to drop, *glides* make up stretches that have shallow and smooth flow with gravelly bottoms, and *runs* are deeper and slightly more turbulent glides. Gradients for riffles vary, with 1–2% slopes being shallow with lots of surface turbulence and 2–4% slopes grading into white water rapids.

---

[1] Hughes et al. (2011). [2] Allan et al. (1997) describe stream terminology. [3] Marcus (2002) provides some stream terminology.

Urbanization destroys headwater streams.

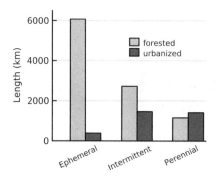

Figure 10.2: An extensive stream mapping and modeling study from Hamilton County, OH, demonstrates that urban land use destroys ephemeral and intermittent headwater streams (Roy et al. 2009). Development preferentially takes place in drier parts of a watershed, leading to the loss of these small streams.

First-order streams are the most endangered by urbanization. People construct houses where it doesn't flood, and that means in the higher-elevation areas, not in the bottomlands. Those higher-elevation areas are precisely the locations where headwater streams form at the uppermost parts of the watersheds, making up the catchments where water first collects into the smallest channels. As more people come together, the surrounding surfaces become impervious, and various structures direct the precipitation over and around constructed features, often via stormwater pipes that are dug into the remnants of the naturally formed channels. The headwater streams are then lost, transformed into conveyance structures, as shown by the Durham, North Carolina, stormwater pipe system in Figure 1.5.

From a more practical standpoint, one study defined headwater streams as those that would be first-order when viewed on a 1:100,000-scale map, and can be perennial, intermittent, or ephemeral.[1] Ephemeral streams have water for only short periods after a rainfall, while surface waters collect in their small channels. Intermittent streams flow during seasonally wet periods and are barren during the dry season. Perennial streams are those that carry water all year long. Under the 1:100,000 definition, 53% of all continental U.S. streams are headwater streams. Other definitions use a map with a different scale, or designate first- and second-order streams as headwater streams.[2] No matter the definition, these headwater streams are important (see Figure 12.7). Their connections to the general hydrological cycle range from direct contributions to the streamflow of higher-order streams and water bodies to completely indirect connections through infiltration to groundwater.

Sadly, science doesn't have great bearing on the legal system. The Supreme Court ruled in 2001 that nonnavigable waters needed a "significant nexus" to

---

[1] Nadeau and Rains (2007).   [2] USEPA (2005c).

navigable waters for protection under the Clean Water Act (CWA), but that legal concept has no scientific definition. Human navigation is not a physical or hydrological principle that guides the formation of headwater streams.[1] This lack of protection holds despite the CWA's objective to "restore and maintain the chemical, physical, and biological integrity of the Nation's waters."[2]

An excellent examination on the effects of urbanization on headwater streams comes from Hamilton County, Ohio, home of Cincinnati.[3] The 1,074-km^2 area is 50% developed, 36.1% forest, and 11% agricultural. The study randomly selected 150 of 6,686 headwater stream channel origins and then "ground-truthed" them to see what was there in both the wet and dry seasons. First of all, study authors noted that standard maps of streams with scales of 1:100,000 and 1:24,000 drastically underestimated channel lengths by 85% and 78% respectively. Our maps undercount headwater streams. Beyond that, the county has lost 93% (5,680 km) and 46% (1,276 km) of its ephemeral and intermittent channel length.

Catchment areas for ephemeral streams were 0.66 hectares in forested areas but a much larger 5.13 hectares in urban areas; for intermittent streams the numbers were 3.6 and 6.79 hectares, respectively. Interestingly, the reverse is true for perennial streams: catchment size was 48.1 hectares in forests and 31.2 hectares in urban areas. Furthermore, urban areas gained 22% (260 km) in perennial channel length. Those numbers suggest that reduced urban forest cover resulted in reduced evapotranspiration (see Figure 10.4), thereby pushing streams from intermittent toward perennial.

Astonishingly, 40% of perennial streams originated from stormwater pipes! Since rain isn't the permanent weather condition in Cincinnati, this permanent flow means there are permanent contributions from some sources that enter stormwater pipes. Thus, the increase in perennial streams comes about from human water discharges due to irrigation of lawns, leakage from sewer pipes, and/or stormwater pipes acting like groundwater drainage ditches.

Of the nitrogen and water volume found in second-order streams, the watersheds associated with the headwater streams contribute about 65% of the nitrogen and about 70% of the water volume.[4] Of this nitrogen load in the headwater basins, about 70% comes from atmospheric deposition. Thus, relatively little of the nutrients and streamflow comes from groundwater or surface flows that enter the downstream higher-order portions. Even for fourth-order and higher streams, these fractions from headwater streams stay quite high, at roughly 40% and 55%, respectively. Now, this factoid makes it sound like headwater streams are terrible beasts, but it doesn't express the amount of nutrient processing and water storage volume contained in these catchments while in their natural state. Land use changes in these basins are magnified

[1] Nadeau and Rains (2007). [2] 33 U.S. Code 1251. [3] Roy et al. (2009). [4] Alexander et al. (2007) look at the importance of headwater streams.

Streams connect to groundwater.

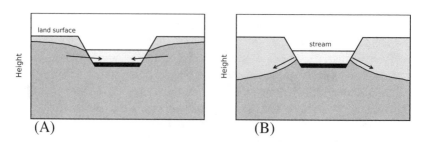

Figure 10.3: Streams are deeply involved with the water contained in the surrounding soils. (A) Rains replenish aquifers feeding receiving streams. (B) Losing streams sit above aquifer levels, replenishing groundwater storage (after Peterson and Wilson 1988). Dark gray represents the fully saturated groundwater, light gray is the unsaturated vadose zone, and the dark stream bottom represents the clogging layer (see Figure 10.6).

in their downstream consequences, because their problems are compounded and their benefits are reduced.

Figure 10.3 depicts how streams interact with the surrounding groundwater.[1] Diagrams represent steady-state conditions; Figure 10.3A typifies a gaining stream, while figure 10.3B represents a losing one. "Gaining" and "losing" refer to the stream gaining or losing water from or to the groundwater.

A further extreme is the case when the groundwater level is below the stream entirely, a situation in which the stream and groundwater are considered "disconnected," and without replenishment, the stream ultimately goes dry somewhere downstream as the water completely infiltrates through the stream bed. The dark area at the bottom of these diagrams is called the clogging layer, caused by colmation (see Figure 10.6). That layer can allow the stream to retain water even when disconnected.

As the water table drops further below the stream, the saturated soil pinches off and hangs above the water table, remaining attached to the stream bottom, slowly seeping water from the saturated "bulb" down to the water table. The region where surface water meets and mixes with groundwater is called the "hyporheic zone." Denitrification can take place at this interface (see Figure 12.9), making it a really important region.

Some early work that calculated flow patterns in the *vadose zone* (the soils above the saturated groundwater table) used approaches originally designed to predict the movement of radioactive contaminants in groundwater.[2] A study

---

[1] Peterson and Wilson (1988). [2] Reisenauer (1963).

in the early 1960s produced diagrams like those in Figure 10.3 and arose from the flow paths calculated using the underlying differential equations. At the time, the calculations used a program written for a multimillion-dollar IBM 7090 with 128 kilobytes of memory! At a glacial 400 grid point calculations per second, it required 30 min or more for a single analysis. Of course, despite technological advances, the mathematics and physics haven't lost any value in the last few decades, so the results are, of course, still valid.

Effects of urbanization can be considered within the framework of these diagrams. A deeply incised stream, like those typical of the urban stream syndrome (see Figure 5.14), act much like a drainage ditch or, equivalently, the tile drains in a midwestern farm field, lowering the water table around the stream. Likewise, the occasional flashy flows prevent or remove the buildup of a clogging layer, increasing infiltration.

Like many Piedmont cities, Durham, North Carolina, relies on reservoirs for drinking water: less seepage into groundwater might mean reservoirs fill more quickly, but the lack of base flow and limited groundwater recharge deplete groundwater storage. Groundwater represents additional storage to recharge reservoirs over long time scales. Impervious surfaces reduce infiltration, decreasing the recharge of groundwater, and prevent sustained base flows.

Natural streamflow depletion usually arises from a combination of either a water table sitting well below the stream bed due to geologic structure, or evapotranspiration from the perimeter of surface water bodies (see Figure 10.4).[1] Water going into the aquifer may reappear elsewhere in the river as the watershed topography changes, with the groundwater draining to rivers in flatter areas or near coasts.

Of course, diversions and withdrawals occur from human activity as well as vegetation and streams. Aquifers in some areas of Oklahoma have dropped by 7–15 m, 30 m near irrigation wells, and these drops have caused perennial streams to become intermittent.[2]

Overall, global groundwater depletion since 1960, when I was born, totals somewhere around 4,000 km^3, enough water to raise sea level by 1 cm.[3] Groundwater wells in the U.S. pumped out around 115 km^3 during the year 2000 alone, giving a net depletion of 17.7 km^3. Since a cubic kilometer equals a billion cubic meters, 17.7 of them divided between the 300 million Americans equals about 600 m^3 each. Each cubic meter equals a thousand liters, so that groundwater withdrawal means about 1,600 liters (about 420 gallons) per day per American.

Trees also pull out groundwater. Figure 10.4 shows the intimate connection between streams and groundwater described in the previous few pages, and

---

[1] Winter (1999, 2007) also discusses flow paths. [2] Winter (2007). [3] Church et al. (2011) summarizes groundwater depletion in an examination of sea-level rise.

Trees drink from streams.

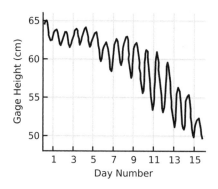

Figure 10.4: Daily variations due to transpiration of trees are visible in stream height measurements of the San Pedro River in Charleston, AZ (Winter 2007). Zero streamflow occurs at 50 cm height. Nightly groundwater recharge replenishes the water drawn out during the day.

it's revealed because of the tremendous amount of water transpired by trees in the riparian zone.[1] Chapter 12 covers riparian systems more fully.

In Figure 10.4, the vertical axis plots gage height: think of the stream's surface measured against, essentially, a meter stick stuck into the stream bed. We see from this plot that as trees transpire during the day, the stream height is drawn down by several centimeters, and then, when transpiration shuts down during the night, groundwater seeps back in and recharges the river, causing the rise in stream height. Over the course of a couple weeks, the water drains into the riparian areas and the stream height plummets until its flow stops, in this case at a gage height of 50 cm.

Rain fills streams and recharges groundwater. Figure 10.5 shows groundwater recharge rates from South Africa, Botswana, and Zimbabwe, plotted against rainfall.[2] The researchers warn that such averaging removes important temporal information, such as the effect of extended wet or dry periods, as well as ignoring such issues as soils and vegetation. However, the data clearly make the general point that recharge saturates with precipitation.

A bit of groundwater terminology is useful.[3] *Infiltration* covers all of the water going from the surface into the ground. *Net infiltration* refers to the infiltrated water that makes its way below the root zone, which, of course, is relevant only to situations where the root-zone soil is not saturated with water. If the water doesn't make it through the root zone, it was pulled up and out, making its way back into the atmosphere. This net infiltration fraction may also

---

[1] Winter (2007) studied the groundwater–stream connections. [2] De Vries and Simmers (2002) discuss groundwater recharge, its variability, and how hard it is to measure. Data for Figure 10.5 come from a 1998 Ph.D. thesis by E.T. Selalo, cited within, of Vrije University in Amsterdam. [3] Scanlon et al. (2002) introduce groundwater terminology.

Groundwater recharge rate increases with increasing rainfall.

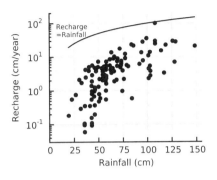

Figure 10.5: Recharge versus precipitation for South Africa, Botswana, and Zimbabwe (de Vries and Simmers 2002). The curve marks where the line of recharge equals all of the precipitation. Low rainfall leads to even lower recharge rates, endangering groundwater supplies.

be called *drainage* or *percolation*. *Recharge* refers to the water that makes its way to the saturated zone, also called the *aquifer*. *Diffuse recharge* comes from broad-scale water sources like precipitation, but *localized recharge* comes from collected water sources like streams, lakes, and ponds. Diffuse sources also yield *direct recharge* in which the recharging water infiltrates right where it fell, whereas localized sources produce *indirect recharge* in which the water has collected from all over the watershed to the place where it infiltrates.

Various approaches might be used to estimate recharge. Surface water approaches include a water budget analysis that sets precipitation equal to various partitions that are relatively easy to measure, and groundwater recharge is left over. Another approach measures the loss or gain in streams: if there's more or less water flowing in a stream at two different points with no intervening streams, the water had to have come from or escaped into the ground (perhaps after correcting for evapotranspiration), and that difference estimates recharge or discharge from a watershed. Another fraction can be estimated by driving a pipe into a lake bed and measuring seepage at the lake bottom. The water level in the pipe would drop while the lake water infiltrates since the pipe water is disconnected from the lake's recharging sources. That drop provides an estimate for localized groundwater recharge under the lake. Finally, one can spike water with a tracer isotope, or even heat (see Figure 9.12), and watch how it progresses through the groundwater using measurements at test wells.

Other techniques measure recharge in the vadose zone. One approach uses a lysimeter. Imagine a huge tub (on the scale of meters in diameter) filled with soil and vegetation but specially decked out to measure water coming in from precipitation and water going out through the bottom on its way to the aquifer. The lysimeter acts like a miniature vadose zone. Some lysimeters are also built on a scale to continuously measure the weight of the isolated soil. Knowing the water that enters and the water that drains out provides the recharge estimate.

Another approach makes use of the "zero flux plane," the depth at which there's a hydrological balance between roots pulling water up and infiltrating water that meanders down through the soil pores. Measuring the variation in the depth of this plane gives an estimate of the change in water storage, and that change determines recharge rates. Another set of approaches uses applied tracers, which can be chemical or isotopic, or even unfortunate historical spills of toxins, to follow the hydrological course of water through the soils (see Figure 10.9).[1]

In urban areas, groundwater recharge includes pathways such as water main leakage, sewer pipe leakage, downspout runoff, and various stormwater control measures (SCMs) designed for groundwater recharge. These leakages are so great that they can fully offset the increased runoff from impervious surfaces. In those situations, urban areas have greater discharges to groundwater than do rural areas, as evidenced through examples spanning Australia, Thailand, the U.K., Mexico, and Venezuela.[2]

Figure 10.3 addressed the connection between streams and groundwater, and how losing streams recharge groundwater via infiltration through the stream bed. An important aspect of the stream–groundwater connection is a process called *colmation*, or the clogging of stream beds by siltation and algal mats, promoted by disturbance and excess nutrients.[3] It's really just like particles getting trapped in filter paper: the filter clogs, and liquid doesn't drain. Fine sediments in streams get into the small pores that infiltrate water through the stream bed, clogging them and reducing the exchange with groundwater sources. It sounds simple, but colmation is an incredibly detailed process that reduces groundwater recharge, dries out riparian buffers, turns perennial streams into intermittent ones, and results in bank erosion during peak flows due to a loss of stream-bank stability. It's a global problem, getting worse with more human disturbances, directly harming the habitat needed for some aquatic invertebrates and fish larvae.

Colmation involves all three of physical, chemical, and biological processes.[4] Physical processes include flow conditions, sediment load, particle size distribution, and the hydraulic gradient (strength and direction) of water flows. Elevated iron and magnesium concentrations produce coagulates that can plug up pores, and microbes with adhesive-producing features round out the dependence of colmation on chemical and biological processes.

Figure 10.6 plots the measured hydraulic conductivity of a stream bed, essentially a measure of how easily water flows through the soil, against the

---

[1] These and other approaches are discussed in Scanlon et al. (2002). [2] Lerner (2002) discusses groundwater recharge in urban areas. [3] Sophocleous (2002) explains the connection between surface water and groundwater. [4] Veličković (2005).

Fine sediments clog stream beds.

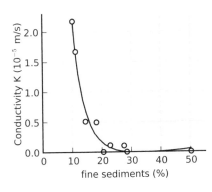

Figure 10.6: Fine sediments clog streams and reduce groundwater infiltration, measured as conductivity, in a study of streams in southeastern France (Descloux et al. 2010). Here the amount of fine sediments was measured 20–40 cm below the stream bed.

fine sediments contained in the stream bed.[1] The data come from three streams, one dammed and two others with different geological situations. These streams span the silt fraction values shown. Essentially, streams clog up from the smaller particles from urbanization (see Figure 4.8). Dams and "flood harvesting" make the problems worse because these things change hydrology and reduce the export of sediments to riparian zones during floods. Added nutrients also clog stream beds by promoting the growth of algal mats and biofilms.

Soils are more than just dirt that water flows through—they're a place where biology happens. Figure 10.7 shows the dependence of two biologically important quantities with depth, measured at twenty-one sites around the eastern edge of Lyon, France. Measurements were done on groundwater collected just below the water table with vadose zone thickness varying from 1.7 to 28.2 meters. The study area is mountainous with highly permeable soils from glacial processes, and examination well pipes were drilled around recharging stormwater ponds.[2]

Microbial activity, or the breakdown of organic matter, is limited by dissolved oxygen (DO). Unlike the situation for the water-holding ability of the atmosphere—the warmer the air, the more water vapor it can hold—increases in water temperature decrease the amount of oxygen it can hold. Urbanization adds organics and biologically available nutrients and increases runoff temperatures, all of which drive down oxygen levels in stormwater. Decreases in oxygen harm the life forms that need it. The oxygen is depleted in stormwater ponds, so microbial respiration of dissolved organic carbon (DOC) doesn't

---

[1] Descloux et al. (2010). [2] Foulquier et al. (2010) examined DO and DOC below stormwater ponds.

Biology happens in groundwater.

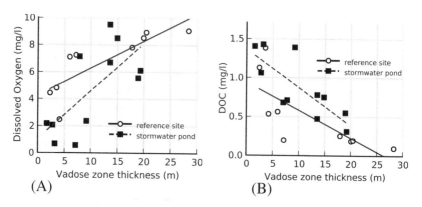

Figure 10.7: (A) Measurements below an infiltration pond in Lyon, France, show that as stormwater descends through the vadose zone it replenishes its dissolved oxygen content. (B) With that oxygen, microbial respiration increases, which removes dissolved organic carbon (DOC) from the descending plume of infiltrating stormwater (Foulquier et al. 2010).

take place in the groundwater directly below these ponds. However, as the DOC-rich, DO-poor water moves downward, oxygen in air-filled pores of the soil dissolves into the stormwater, increasing DO and allowing respiration to take place. That respiration then decreases DOC. Essentially, the soil replenishes the stormwater's oxygen. Once the percolating water hits the water table, both DOC and DO take on the aquifer values as the stormwater gets diluted with groundwater.[1]

Biological oxygen demand (also called biochemical oxygen demand, abbreviated BOD) represents the amount of oxygen that would be consumed in the biological transformation of all of the organic matter and ammonium into nitrate. Chemical oxygen demand (COD) represents a measure similar to BOD but using chemicals instead of organisms to fully consume a sample. Since chemicals can always do what biology can do, but not vice versa, COD always exceeds BOD.

Urbanization does not always lead to decreases in DO. For example, DO and COD conditions improved in a watershed near Portland, Oregon, between the mid-1990s and mid-2000s even while residential land increased 10%.[2] Many aspects contributed to this improvement, and different factors

---

[1] Pierre Marmonier, pers. comm. [2] Boeder and Chang (2008) examined oxygen trends in an urbanizing watershed of Portland, Oregon.

were important for the different scales ranging from stream buffer up to sub-basins. At the full watershed scale, forests helped reduce COD at several scales. Riparian buffers mitigated DO concerns, likely through temperature control, but in some situations the buffers correlated with increased nitrogen. At the subbasin scale (about 1 km), agriculture also correlated with increased nitrogen during the dry season. However, stormwater management correlated with increased COD, and there are worries that development pressure might overwhelm natural and constructed processes that mitigate oxygen demand issues.

Keep in mind that streams don't remove a great deal of carbon from watersheds. One estimate for stream export of carbon through DOC is around 1–5 g/m^2/year, whereas primary productivity averaged across the U.S. is around 500 g/m^2/year.[1]

As stormwater infiltrates, groundwater collects many chemicals, and the following contaminant history demonstrates an unintended consequence of solving air pollution problems but trading them for water quality concerns. To clean up our air, the 1990 Clean Air Act required that a certain amount of oxygen be incorporated within gasoline for more complete burning, resulting in fewer pollution emissions. Methyl *tert*-butyl ether (MTBE) was a primary fuel additive up until 2005, and looks like an oxygen atom guarded by a whole bunch of methane (CH_4) molecules. The additive represented about 11–15% of gasoline by volume, making it quite the substantial additive. In 2001, the U.S. produced about 12 billion liters,[2] but it has been mostly phased out and replaced by ethanol. In groundwater, MTBE is highly mobile and persistent. Unfortunately, it mixes both in gasoline and in water and also easily vaporizes. Ethanol also has its problems, including a potentially carcinogenic metabolite that arises from photodegradation in the atmosphere.[3] On the other hand, the addition of MTBE reduced emissions of benzene, a known carcinogen, and other by-products by up to 30%, thereby reducing gasoline fume cancer risks.

A bit comforting, only 0.3% of samples from a broad study of USGS data of nearly 4,000 wells[4] demonstrated levels of MTBE exceeding the USEPA's lowest level of concern at the time, 20 μg/L. Of the urban wells examined (Figure 10.8), only 3% exceeded 20 μg/L. Not surprisingly, the frequency of finding MTBE (which averaged 7.6%) was greatest in urban land use types and increased with human population size, appearing in more than 1,500 public water supply systems between 1996 and 2002.

The data in Figure 10.8 come from shallow wells of unspecified depths, not deep municipal drinking water wells, that were collected as part of a USGS national water quality assessment program. Of 210 urban wells, MTBE contaminated 27%, but just 1.3% of 549 rural wells. These data come from 20 years ago, but at the time, MTBE was the second most commonly found

---

[1] Schlesinger (1997). [2] Moran et al. (2005). [3] Ahmed (2001). [4] Moran et al. (2005).

Gasoline additives in urban groundwater.

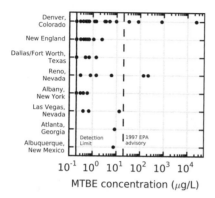

MTBE concentration (μg/L)

Figure 10.8: Although it's now phased out, the gasoline additive methyl *tert*-butyl ether (MTBE) was present in U.S. urban groundwater in 1993–1994 (Squillace et al. 1996). Chemicals like MTBE are unique to human manufacturing and can be used to age and trace groundwater flows, and occasionally are a cause for concern.

chemical in shallow groundwater sources, although no MTBE was found in the five shallow drinking water wells of this studied set.

MTBE levels reached 2.36 μg/L in urban German rivers and 8.7 μg/L in U.S. stormwater, but these levels are much, much lower than the 10^6 μg/L (or 1 g/L!) for some contaminated groundwater sites.[1]

Human health effects include some reports of dizziness, headaches, and irritation when exposed through air, yet many results conflict, giving a very uncertain picture.[2] MTBE caused several problems in rats and mice at high concentrations, but many orders of magnitude higher than the concentrations seen in groundwater.

Once something gets into the groundwater, it just goes with the flow, and that can be a lazy pace. Results in Figure 10.9 demonstrate that it can take many decades for a compound to travel through an aquifer and show up in streams. That delay, or time lag, can lead to faulty conclusions of cause and effect, unwarranted accusations, and inappropriate regulations. Past agricultural practices may not have been the best ones for keeping pollutants out of our groundwater and streams, but we shouldn't necessarily blame today's farmers for present stream pollution.[3] Finer details from this study show that the groundwater flowing into streams leads to an interesting pattern of older groundwater coming up in the middle of the stream and younger groundwater coming up at the stream's edge. Those patterns arise from the complicated flow paths occurring in groundwater.

Even more bizarre, unintended processes can take place. One such interesting process was observed whereby a layer of sulfide minerals below the

---

[1] Rosell et al. (2006). [2] Ahmed (2001). [3] Böhlke (2002) studied agricultural pollutants and groundwater recharge.

Old groundwater sits deep.

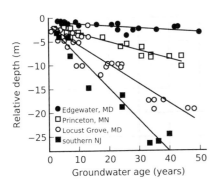

Figure 10.9: Variation of groundwater age and its depth below the water table for various locations across the U.S. (Böhlke 2002). Using various tracer chemicals, groundwater can be aged according to its infiltration date. Important factors include aquifer depth, soil porosity, and groundwater recharge rate.

water table connected with nitrate contaminants. The resulting processes denitrified nitrate and created SO_4, and in the process precipitated arsenic farther down the groundwater flow. Agricultural fertilizers end up creating higher arsenic levels, despite having no arsenic components.[1]

Other contaminants serve a useful purpose. Scientists use chlorofluorocarbons—an entirely human-made chemical—to date groundwater age, but other techniques use by-products of atmospheric nuclear weapons testing from the 1950s and 1960s. We invent unique chemicals and start using them on a large scale (or have a widespread deposition pattern), and then they start percolating into the groundwater. It's certainly a fortuitous happenstance, but sad to think that we can use our pollutants as scientific measurement tools.

For example, perfluorochemicals (PFCs) undergo long-range transport, with deposition of nonvolatile PFCs and subsequent uptake by organisms. These chemicals are used in firefighting foams, insecticides, and water and stain repellants.[2] They have no natural occurrence, so they conveniently mark flow patterns, much like rubber duckies spilled by the shipload can track ocean currents. Wastewater is an important source of PFCs, and they've been found everywhere—in fish, birds, and mammals nearly everywhere on Earth. Compounds coming from the degradation of these chemicals are found in urban surface waters at around 2–50 ng/L and in remote lakes at up to 1.2 ng/L. Concentrations were found as high as 700 ng/L in Korea and Osaka, Japan, and up to 140 ng/L downstream of Tennessee manufacturing plants. Even Arctic lakes have seen sediment concentrations of up to 85 ng/g dry weight,[3] although some possibility exists of local contamination from a nearby airport.

---

[1] Böhlke (2002). [2] Simcik and Dorweiler (2005) study transport of perfluorooctane sulfonate (PFOS) and perfluorooctanoic acid (PFOA). [3] Stock et al. (2007).

Old fertilizers move slowly through groundwater.

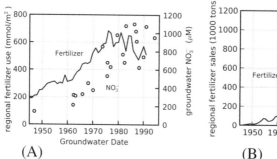 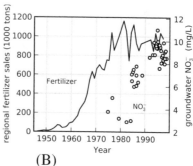

(A) (B)

Figure 10.10: (A) The release of nitrogen carried with groundwater into NC streams takes place years after the nitrogen first infiltrated agricultural soils (Kennedy et al. 2009). (B) Water drawn from Iowa wells 145–275 feet deep has nitrate concentrations that reflect earlier fertilizer use (Schaap 1999).

Using such dating approaches, Figure 10.10 shows directly how nitrogen applied as fertilizer takes time to come back out into streams, in this case estimated as a 30-year time lag.[1] These results come from the coastal plain of North Carolina and northeastern Iowa, both locations where agriculture dominates land use.

For the North Carolina results of Figure 10.10A, fertilizer application data come from records held by the USDA, and the groundwater age seeping into the examined stream was estimated by the techniques described above. This time lag has important implications for attributing stream nitrogen to current agricultural practices versus outdated practices of the past. We can't afford to charge cleanup costs to present-day farmers for the perhaps careless— with hindsight—approaches of the past. An analogy is the infamous chemical DDT that's still cycling through our ecosystems: is its continued presence a sufficient reason to penalize today's farmers for the use of that pesticide decades ago? Likewise, fertilizers generously applied to fields when it was quite cheap and when stream buffer preservation was less regulated, which was certainly an unfortunate period, but those approaches have changed. Still, the data provide good evidence for further updating fertilizer, herbicide, and insecticide application practices.

The well water data shown in Figure 10.10B look remarkably similar to the stream data of Figure 10.10A. This study comes from Cedar Rapids, Iowa,

---

[1] Kennedy et al. (2009) and Schaap (1999).

using wells ranging in depth from 145 to 275 feet. Soils in the area were built upon 100 feet or so of glacial deposits, making them quite porous. Results clearly show that the recharge date was the mid-1970s for water drawn out in 1998.[1] Nitrate levels nicely match the shape of the fertilizer curve, and another plot (not shown) demonstrates a similar connection between fertilizer applications and recharge date for entrained nitrates from Princeton, Minnesota, spanning the entire period from 1960 on.[2]

Nitrogen removal in streams is an important process ameliorating our enhanced nitrogen runoff, and which is covered in Chapter 12. One study monitored nitrate fluxes in base flows coming from agricultural streams (less than 10% urban land use) in northwestern France.[3] Somewhat obviously, surplus nitrogen applied to agricultural lands played an important role in stream concentrations, but a little less obviously, two other factors, the surface area of riparian wetlands and the length of the path nitrates had to flow, explained more than 60% of the remaining nitrate variation. Upon reaching sixth-order streams, 53% of the nitrates had been removed, equaling 21.1 kgN/ha/year. In this situation, higher-order streams were more important; specifically, the nitrate concentrations in sixth-order streams were 47% those of second- and third-order streams. Nitrate removal reached 91% during low water levels (during the dry period) due to long residence times in wetlands.

In an effort to find a quick, dirty, and cheap way to estimate stream nitrate concentrations, one research team drove sticks into streams and observed depth of color change after 2–3 weeks, and connected that to sediment, nitrate, and oxygen concentrations.[4] The plots in Figure 10.11 show the results. Color change taking place at a deeper level means that colmation is low (fewer fine sediments), and stream water can get deeper into the stream bed. With that water comes oxygen and nitrate, and their concentrations reach higher levels at deeper (15 cm) depths. With more fine sediments and greater colmation, freshwater doesn't reach very deep, and nitrates are removed in the anoxic conditions just below the stream bed.

In perhaps a tenuous segue from sticks in streams, large woody debris (LWD) structures the hydrology in streams, and the induced water flows create pools and riffles around the obstructions and promote stream health. A study examining 27 streams between 1982 and 1993 tested the connection between LWD and logging in the Olympic Peninsula of Washington State.[5] Time since logging in the surrounding area classified these streams into four groups: unlogged, old-growth forests; old second-growth sites logged between 1920 and 1942; middle-aged second-growth sites logged between 1945 and

---

[1] The Schaap (1999) plot uses data attributed to J. Sawyer via personal communication. [2] Böhlke (2002). [3] Montreuil et al. (2010) estimated nitrate removal by riparian wetlands. [4] Marmonier et al. (2004) connected visual changes in stakes driven into stream beds with nitrate concentrations. [5] McHenry et al. (1998) discusses LWD and logging in streams of the Olympic Peninsula, Washington.

Stakes in streams reveal nitrate depth.

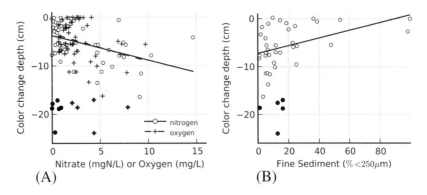

(A)   (B)

Figure 10.11: Color change depth taking place with stakes driven into French streams depends on nitrogen and oxygen (A) and sediment concentrations (B) (Marmonier et al. 2004). Filled and bold symbols come from forested streams and are not included in the regressions. Color change marks the onset of oxygen depletion with depth.

1955; and young second-growth sites logged between 1956 and 1972. During the study period, all but one old-growth site was logged, and most sites had upstream disturbances.

This study defined "key" pieces of wood as those with 80 cm diameter and more than 10 m in length. These pieces remained where they were from year to year, but their volume decreased by 25% (from 51.7 m^3/100 m to 38.2 m^3/100 m) or more in logged old-growth sites, and old-growth pieces decayed significantly over the decade-long interval. There was also a roughly 50% loss of old-growth LWD over 5 years for a 10-m-wide stream, with lower rates for smaller streams and faster rates for bigger streams. The likely mechanism is that higher flow rates are flushing out the debris. The distribution of second-growth pieces increased in volume and size, but not enough to make up the loss of first-growth LWD. Recovery of this material will take a century, and the study authors urge protection of riparian zones and restoration of LWD.

LWD was missing from western North Carolina streams of the Blue Ridge Mountains when more roads and buildings were found in their catchments.[1] A correlation wasn't found with forests, likely because the forests were logged

---

[1] Scott et al. (2002).

Woody debris and pools lost from urban streams.

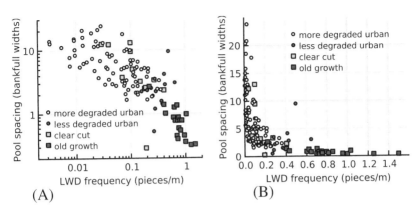

Figure 10.12: The frequency of pools in streams is correlated to the presence of LWD with >10 cm diameter and >1 m length (Larson et al. 2001). Using the identical data to serve as a visual comparison, (A) shows logarithmic axes, whereas (B) shows linear axes.

within the last century and the regrowth has yet to reach the end of its life span and full debris-contributing potential.

There are arguments for common metrics in the developing field of LWD studies, like metrics defining "large."[1] There are many important variables for these pieces of wood, such as length, diameter, flow orientation, rootwad presence, jams, and decay state. One research study suggested no less than 35 variables—23 involving in-stream LWD—though many variables concerned features regarding stream structure, and a dozen involved watershed concerns. However, good arguments justify the level of detail on such a seemingly arcane topic.

Figure 10.12 shows Washington State data concerning a simple form of stream restoration in urbanized watersheds—dropping LWD, sometimes called coarse woody debris, into streams.[2] Adding LWD to urban streams increased pools and "channel complexity," but much of the LWD was either buried or transported downstream (overwhelmed by streamflows). The study concludes that added LWD pushes a stream's physical characteristics toward those of less disturbed streams but doesn't do much for their biological condition—no biological improvements were seen after 2 years.

---

[1] Wohl et al. (2010). [2] Larson et al. (2001) examined urban stream restoration in Puget Sound, Washington.

This study qualitatively defined "key" pieces of wood as those that are so huge they're not going to move no matter what, and what's more, they trap and stabilize other LWD. Streams with key pieces kept their other LWD in high-flow conditions, but in streams without them the LWD moved significantly. Key pieces were responsible for about one-third of stream sediment storage, and sediment storage increased by 50–100% when LWD was increased. Think of LWD like trees capturing air pollutants: induced turbulence promotes deposition. Importantly, the authors conclude that, although nothing really stops us from adding LWD to urban streams, it's probably a waste of resources to do so without correcting the problems farther upstream that lead to flashy flows.

Biological improvements with added LWD are seen in some cases. One study showed that adding LWD to agricultural streams works quite well, but urban streams often fail.[1] Deforestation and riparian clearing prevent leaf litter and woody debris from entering streams, and the resulting habitat loss connected to declines in some Australian amphibians. Some LWD is actively removed to increase flow speed and decrease flooding, reflecting the motives behind adding stormwater pipes to drain urban areas. Aside from flow-speed changes, turbulent flow around LWD adds oxygen and causes vertical flows that produce a structured bottom, enhancing water retention and nutrient processing.

A survey of "natural streams" found 71 pieces of woody debris per 50 m; researchers than anchored 13 branches of 5–35 cm diameter, with similar total volume, per 10 m of stream treatment length.[2] These were placed together and singly, and treatment sites were compared with control sites without added wood. After a half-year or more, wood-added sites had abundances of 989 organisms from 21.6 taxa, whereas the control sites had just 522 organisms from 18.2 taxa, an abundance increase of 89% and diversity increase of 19%. It's a cheap and easy restoration approach and made the organisms very happy, evidenced by their leaving less preferred habitats for the safety and feeding opportunities of the woody debris. Note that woody debris also helps spawning by reducing flow speeds and trapping gravel.

One study of the restoration of woody debris in streams of the U.S. West Coast used readily available streamside alder trees.[3] After roughly doubling the woody debris by number, but increasing the volume by just 25%, via several methods, the levels remained high after 3 years. Alder rots quicker than other species but is more plentiful at many sites. This added debris was associated with aggregations, but not clearly causative, meaning they may or may not have added the key pieces. At many locations researchers simply

---

[1] Lester et al. (2007) examined adding LWD to Australian streams, while Lester and Boulton (2008) provide a literature survey of LWD projects performed across the world. [2] Lester et al. (2007). [3] Keim et al. (2000) added alder trees to west coast streams.

Eroded streams recover woody debris slowly.

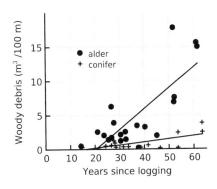

Figure 10.13: Once logging removes the source of woody debris, its presence in streams comes back slowly (Booth 1991). The large pieces of trees that stabilize streams take many years to grow, die, and then fall into nearby streams.

pulled down alder trees into the streams, leaving their stream-bank roots intact, providing a nice anchor. Root-stabilized alders acted as key pieces but decayed quickly, and the study concluded that adding woody debris helps restore natural streamflow.

A helping hand is needed with LWD because there's a naturally slow recovery, as Figure 10.13 shows for the case of the LWD-related aftermath of logging a second-growth forest.[1] Here the first data points are 10 years after logging, so much of the debris might have washed away or decayed, but even after 60 years the numbers reach nowhere near the level of 30–50 m³/100 m seen in old-growth forests of the area.[2]

Most woody debris in Washington streams is less than 50 years old, and the age distribution follows an exponential decay with a time constant of about 20 years.[3] Hardwoods decay quicker than conifers, but erosion and channel movement can uncover very old debris buried near streams. The implication is that loss of woody riparian buffers—the source of woody debris—has effects within a couple of decades. The role of LWD for stream health is another reason to preserve riparian buffers (see Chapter 12). When you walk along a stream, throw in some branches to brighten some crunchy animal's day.

Another study considered a southeastern Alaskan watershed involving mostly natural areas, except for timbering in about 5% of the area that left behind buffers 20–30 m wide.[4] The standing biomass of the forest was about 625 m³/ha, yielding LWD recruitment from dying trees falling into streams at 0.1–8.1 m³/km/year. Bank erosion rates ranged from 1 to 30 cm/year for the

---

[1] Stream channel data on woody debris in Booth (1991) comes from a 1985 thesis by Grette cited within the paper. [2] McHenry et al. (1998). [3] Hyatt and Naiman (2001) examined woody debris in Washington streams. [4] Martin and Benda (2001) studied a southeastern Alaskan watershed for LWD.

Imperviousness creates flashy, unstable streams.

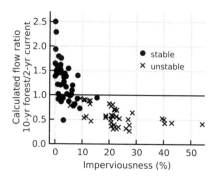

Figure 10.14: Development removes tree cover, adds imperviousness, and creates the flashy flows that incise streams, which marks a loss of stability for the groundwater–stream system (Booth and Reinelt 1993). For example, a forested watershed's 10-year peak flow is seen more than once every 2 years at just 10% imperviousness.

largest watersheds. LWD input from bank erosion was 1–16 m³/km/year, with the highest values coming from the high flows of a large 60-km² watershed. Total LWD in the streams ranging from 10 to 50 m³ per 100 m. The study found equal contributions from mortality and bank erosion in about 20-km² watersheds having 10-m-wide channels. In small streams jams were spaced about 50 m apart and LWD moved about 200 m, while in large rivers jams were spaced about 200 m apart and LWD moved about 2.5 km. Median age of LWD ranged up to 30 years, and wood decay and burial caused the loss of LWD.

Figure 10.14, although quite complicated, brings together much of the previously covered material.[1] There are three ingredients. The first is simply imperviousness. The other two ingredients include the relative flows for urban and forested watersheds and the stability of streams.

Taking the last first, a stable channel means little or no erosion of bed or banks, whereas an unstable channel means the presence of bare banks indicating downcutting and widening (e.g., the stream diagnosed in Figure 5.14 is *not* a stable channel). Data come from streams in two drainage basins in King County, Washington, home of Seattle, with areas ranging from 2 to 110 km². Researchers measured channel width along a 2-km stretch in one stream and found that channel width increased by 0.6 m (compared with 3–4 m average width) wherever the riparian buffer had been removed or altered. Both scenarios also showed a steady widening downstream.

The ratio of flows comes from a hydrological model created by the USEPA that calculates the flow rates for forested and urban land use conditions. No flows were actually measured, except in the initial model construction that led

---

[1] Booth and Reinelt (1993) connect flow rates, imperviousness, and stream stability.

to abstractions like the curve numbers of Figure 5.13. A ratio of 1.0 means that the current maximum flow seen in an urbanized watershed every 2 years is the same as the maximum flow seen every 10 years in a completely forested, zero-imperviousness watershed. We see that ratio at an imperviousness around 5%. In other words, a big storm in the city has high flows that flush and scour out urban streams, as we know from Figure 5.7, whereas the same storm in a forested area might be a real nonevent.

The stream stability consequences of urbanization start to go away when impervious cover is below 10%. Below that imperviousness value the flow rate ratio gets bigger, like it should, because in an "urban" watershed with 0% imperviousness (in other words, not an urban watershed), the ratio should simply be the ratio of the maximum flow seen every 10 years in a forest to the maximum flow seen every 2 years in a forest. That ratio is always greater than 1.

As Figure 10.14 shows, impervious surface fractions above 10% mean that the highest discharge seen every 2 years from an urban watershed is *higher* than the highest discharge seen every 10 years from a forested watershed, giving a ratio less than 1.

A rule of thumb for correlating stream health with impervious cover (IC) is the "model" that IC < 10% denotes sensitive streams, 10% < IC < 25% denotes impacted streams, 25% < IC < 60% denotes nonsupporting streams, and situations with 60% < IC < 100% really aren't streams—they're simply urban drainage systems.[1] A summary of more than five dozen studies published since 2003 examining various mechanisms of stream degradation concludes that a large proportion of research essentially reinforces the validity of the IC proxy.[2] There also remains the important issue of connected imperviousness, and the many studies discussed in Chapter 4 that clarify its harms. The IC model fails for less than 10% imperviousness because other features like road crossings and forested fraction become dominant factors, and imperviousness doesn't give sufficient insight. The study noted that base flow was not well predicted by imperviousness because contributions from irrigation and wastewater system leakage had important influences. It also recommended a slight reformulation of the imperviousness model to include high variability at low imperviousness and low variability at high imperviousness.

Putting all of the material from this chapter together, another study used information from the "Wadeable Streams Assessment," a reporting required by the Clean Water Act, to summarize 1,392 randomly selected sites all across the U.S.[3] Extensive surveys, including physical, chemical, and biological sampling, were performed between 2000 and 2004 in low-flow conditions. In the U.S., 42%, 25%, and 28% of stream length is in poor, fair, and good condition,

---

[1] Schueler (1994) [2] Schueler et al. (2009) examine how well impervious cover (IC) can be used as a proxy as an indicator for urban stream health. [3] Paulsen et al. (2008).

respectively. In the West, 45% were in good condition, whereas in the eastern highlands only 18% were in good condition. Nearly 20% of the stream length in eastern highlands had lost more than 50% of stream taxa, and nationally 13% of stream length had lost more than 50%. Nutrients and sedimentation were the biggest problems, followed closely by riparian buffer disturbances. All in all, streams that were rated as poor in these respects were two to three times more likely to have a poor biological condition.

Chapter Readings

Böhlke, J.-K. 2002. Groundwater recharge and agricultural contamination. *Hydrogeol. J.* 10: 153–179.

Booth, D.B., and L.E. Reinelt. 1993. Consequences of urbanization on aquatic systems: Measured effects, degradation thresholds, and corrective strategies. In: *Watershed 93 conference proceedings,* March 21–24. Tetra Tech (Alexandria, VA).

Foulquier, A., F. Malard, F. Mermillod-Blondin, T. Datry, L. Simon, B. Montuelle, and J. Gibert. 2010. Vertical change in dissolved organic carbon and oxygen at the water table region of an aquifer recharged with stormwater: Biological uptake or mixing? *Biogeochemistry* 99: 31–47.

Kennedy, C.D., D.P. Genereux, D.R. Corbett, and H. Mitasova. 2009. Relationships among groundwater age, denitrification, and the coupled groundwater and nitrogen fluxes through a streambed. *Water Resources Res.* 45: W09402.

Larson, M.G, D.B. Booth, and S.A. Morley. 2001. Effectiveness of large woody debris in stream rehabilitation projects in urban basins. *Ecol. Engin.* 18: 211–226.

Lerner, D.N. 2002. Identifying and quantifying urban recharge: a review. *Hydrogeol. J.* 10: 143–152.

Roy, A.H., A.L. Dybas, K.M. Fritz, and H.R. Lubbers. 2009. Urbanization affects the extent and hydrologic permanence of headwater streams in a midwestern U.S. metropolitan area. *J. N. Am. Benthol. Soc.* 28: 911–928.

Squillace, P.J., and M.J. Moran. 2007. Factors associated with sources, transport, and fate of volatile organic compounds and their mixtures in aquifers of the United States. *Environ. Sci. Technol.* 41: 2123–2130.

Winter, T.C. 2007. The role of ground water in generating streamflow in headwater areas and in maintaining base flow. *J. Am. Water Resources Assoc.* 43: 15–25.

Chapter 11

Ecosystem Responses

Human disturbances range from agricultural to urban and include simple deforestation. However, not all disturbances equally affect streams and their riparian buffers. The harms include the typical channeling of the urban stream syndrome, higher stream temperatures, and a loss of stream organisms. In return, people lose ecosystem services that include the processing of organic matter and the uptake of nitrogen.

Imperviousness most consistently correlates with stream biodiversity damage, with loss estimates of three-quarters of stream taxa in watersheds with 30% imperviousness. Across widely dispersed geographical locations, significant reductions in aquatic invertebrates occur within 10% imperviousness within a watershed; sometimes up to 40–50% of species are lost by 15–20% imperviousness. The reduction is particularly apparent in large watersheds, which tend to be the scale that unavoidably suffers from development somewhere or another. Particularly affected are a set of species classified as "sensitive," which represent indicator species for stream health. Yet, some aquatic invertebrate species thrive under urbanization, like their terrestrial vertebrate counterparts—squirrels, rats and raccoons.

While urbanization always demonstrates harm to streams and their inhabitants, agricultural land use presents a more mixed picture. In midwestern areas of the U.S., row crops can lead to significant species loss compared with pastures, and the losses can be as great as 50% moving from 0% to 100% agricultural land use. However, eastern U.S. agriculture affects stream biodiversity much less than imperviousness, and in the Southeast agriculture has a relatively insignificant effect while urban land use presents serious harm.

Disentangling the effects of land use on stream health.

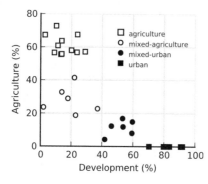

Figure 11.1: A Maryland study examined 29 sampling locations in four watersheds, each location having a different mix of agricultural and development (Moore and Palmer 2005). The four different land use categories represent relatively distinct mixtures used in following plots.

Forested watersheds provide the best stream environment for aquatic organisms. Humans transform this natural land use into agricultural and urban land uses, and figures in this chapter show how combinations of the two uses compare in preserving stream health. A study from four watersheds in Montgomery County in the Piedmont region of Maryland, just north of Washington, DC, examined this question for that region.[1]

The study used 29 sampling locations in forested riparian buffers with wide variation in the fraction of forest within 30 m of the stream channel. For the most part, urbanization in this area means residential development at low and medium density, 0.08–0.8 and 0.8–3.2 dwellings/hectare, respectively, along with some more intensive high-density residential, commercial, and industrial uses. Catchment areas of the sites ranged from 2.7 to 9.2 km^2.

Perhaps it's obvious, but the inverse relationship in Figure 11.1 shows that land can't be used for more than one thing: if it's used for agriculture, it can't be urban. The plot divides the continuum of land use allocations among the various sampling locations into four discrete classes, with the mixed classes being tilted toward one use or the other. Interestingly, the sum of land use fractions is maximum when a watershed specializes in one use or the other compared with mixed-use watersheds.

Figure 11.2 uses the sites introduced in Figure 11.1 to show an example of biodiversity decreasing with watershed imperviousness. The separation of land use types also confirms the basic message seen later in Figure 11.22, that East Coast agriculture isn't as damaging to stream taxa as urban land use.

---

[1] Moore and Palmer (2005).

Imperviousness is a bad thing for aquatic insects.

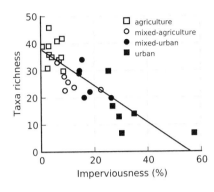

Figure 11.2: Across the land use categories of Figure 11.1, the biodiversity of aquatic insects responds negatively to imperviousness in Maryland watersheds (Moore and Palmer 2005).

The lowest five taxonomic levels for animals, from lowest to highest, are species, genus, family, order, and class. Scientists identify the organisms they collect to as low a level as possible and then report taxa richness at a level that the organisms have been confidently identified. The lowest level is not always the species level because that specificity demands deep knowledge, often requiring the identification of obscure variations in various body parts.

I've been describing results of aquatic insect surveys. Macroinvertebrates make great indicator organisms because they're everywhere in large quantities and don't move around too much and have long enough life spans to integrate over broad environmental conditions, and the variations between species provide a range of tolerances to a great many stormwater contaminants.[1] They're a natural water-quality measuring tool.

As we begin this chapter, we can define *riparian ecosystems* as "the complex assemblage of organisms and their environment existing adjacent to and near flowing water," and sometimes more simply, "ecosystems next to a river."[2] Chapter 12 examines how these ecosystems can help us deal with stormwater. The related term "buffer" implies some sense of protection of the buffered riparian area, as well as the river, from land use outside the buffer, farther away from the river. Some scientists do not include grassy buffers or, more formally, vegetated filter strips, as valid riparian systems. These strips often replace forested buffers to make buffers more palatable to neighboring residents and other political forces, giving a park-like setting adjacent to a stream, sacrificing the important ecosystem functions of real buffers.

---

[1] Beasley and Kneale (2002). [2] Wenger (1999) describes this definition, attributed to Lowrance et al. (1985).

Urban land use removes riparian forest cover and aquatic insects.

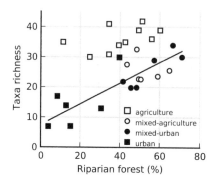

Figure 11.3: Across the land use categories of Figure 11.1, the biodiversity of aquatic insects responds positively to increasing riparian forest cover (within 30 m of streams) in urbanized Maryland watersheds (Moore and Palmer 2005). Here the line is the best fit to urban and mixed-urban sites; no significant correlation arises for the agriculture and mixed-agricultural sites.

Figure 11.3 shows a new and interesting finer point regarding the importance of riparian vegetation.[1] In an easily understood conclusion, but perhaps a bit surprising, agricultural land use preserves stream biodiversity regardless of riparian forest buffers, with no statistically significant trend between the two variables. Messing up the simplicity of this result a little bit is a difficulty in remote sensing. Agricultural lands used herbaceous riparian buffers that weren't revealed in the GIS mapping; thus, a dependence on these types of riparian vegetation may be obscured. For example, right where there's low riparian forest cover in the agriculture data points, there may be a reliance on herbaceous buffers to filter organism-harming sediments and nutrients. To the GIS analysis, these buffers simply looked like plain old fields. Indeed, a majority of the farms in the region already used various low-impact, sustainable agricultural approaches to reduce sedimentation, including contour farming, grassed waterways, and nutrient management, which means the countermeasures taken to account for riparian forest loss might just be working.

In contrast, another study in southern Pennsylvania and northern Maryland compared streams having forested riparian buffers with deforested ones having grassy buffers.[2] These were areas of no urbanization, crops, or animal intrusions into the streams draining pastures. In this comparison, deforestation caused a narrowing of stream channels, increased flow rates, a much decreased uptake of nitrogen and other pollutants, and reduced biodiversity. So, if it can be done, small streams should be restored with forested buffers rather than herbaceous cover.

---

[1] Moore and Palmer (2005). [2] Sweeney et al. (2004)

Coming back to Figure 11.3, the harms from the imperviousness of urban land use can, to an extent, be ameliorated by forest cover within 30 m (100 feet) of streams. In this case the linear fit to urban and mixed-urban points explains 74% of the variation to a highly significant degree. That phrasing takes a positive spin, but an equally valid phrasing is that high urbanization results in less riparian forest and lower taxa richness, much like riparian forest loss increases stream temperatures (see Figure 9.5). Those temperature-related negative results can be turned around to say that forested riparian buffers in urban areas can cool streams.

Confirming these overall results, another study in the Maryland region indicated that imperviousness, forest fraction, and crop fraction were the best predictors of aquatic biodiversity over the entire region.[1] In that study, imperviousness decreased biodiversity, while the other two land use types correlated with a biodiversity increase.

For water-quality purposes, an oft-used measure of invertebrate biodiversity is the sum of the three orders, Ephemeroptera, Plecoptera, and Trichoptera, abbreviated as EPT. Ephemeroptera are the flighty, delicate mayflies; Trichoptera are caddis flies—little aquatic larvae that build protective cases from sticks and stones; and Plecoptera are the stone flies. Much the same picture emerged when the study considered just the sensitive EPT taxa, though crops had a more variable effect.

The message from agriculture is always a mixed one: other studies dispute this friendliness of agriculture (see, e.g., Figure 11.20). A study that collected nearly 20,000 fish of 63 species from 50 sites in the Upper Wabash River watershed in Indiana showed that agriculture harmed aquatic biodiversity due to decreased habitat variability; it led to streams with much less woody debris than forested reaches, and erosion led to finer substrates on the stream bottom.[2]

It remains safe to conclude that wide riparian buffers in urban areas promote biodiversity even when those areas have high fractions of imperviousness. It also seems reasonable to conclude that agriculture with good agricultural practices, at least in the southeastern U.S., can be friendly to water quality concerns as indicated by biodiversity measures.

Figure 11.4A reveals the correlation between forest cover at the watershed scale and riparian buffer scale and comes as a side benefit of a study performed in Georgia that sought to understand the role these two types of forests play in determining stream temperatures. Temperature plays an important role in the health of trout, shown as a part of the study first discussed in chapter 9 (see Figure 9.9), which underpins the economics of the trout fishing tourism industry.[3] More than 4,000 data points from stream segments longer than 1km sitting in basins spanning 5–50 km^2 from 12 counties make up the plot. From

---

[1] Goetz and Fiske (2008). [2] Hrodey et al. (2009). [3] Jones et al. (2006) performed the watershed–riparian forest cover study. Thanks to both Krista Jones and Geoff Poole for providing the data.

Developed watersheds have reduced riparian cover.

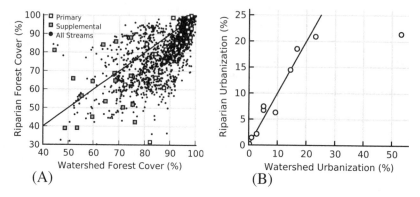

Figure 11.4: (A) In an extensive Georgia study, forest cover measured at the watershed and riparian scales is strongly correlated (Jones et al. 2006). (B) A study of Spring Creek, PA, watersheds demonstrates that watershed urbanization leads to riparian urbanization, which results in less effective riparian buffers (Chang and Carlson 2005).

---

this set, the scientists first chose and used 28 primary sites representative of the entire set to measure stream temperatures; then they realized that a correlation existed between the two forest covers within the subset. That correlation prevented an easy understanding of the separate effects on stream temperature, so they next chose 18 additional secondary sites that, when put together with the primary sites, gave a collection of sites with uncorrelated forest covers at the two scales.

I've drawn one-to-one line in Figure 11.4A, and the clustering of points below the line shows that riparian cover is lost more so than watershed forest cover. For example, at a watershed cover of 90% there appears a clustering of riparian cover points at around 80%, and at a watershed cover of 80% the clustering of riparian cover ranges from 60–80%. As we'll see in Chapter 12, riparian forest cover greatly affects stream temperatures (see Figure 12.15).

Figure 11.4B demonstrates the same correlation between watershed and riparian forest cover, but from the opposite direction: watershed development implies riparian development. The data comes from a study near Spring Creek, Pennsylvania, during the years 2000–2002.[1] Included in the study were 10 subbasins ranging in size from 3.9 to 71.7 km^2, with land use types ranging

---

[1] Chang and Carlson (2005).

from 0.3–53.2% urbanization, 29–78% agriculture, and 11–60% forest. Riparian cover was measured within 100 m (328 feet) of streams, and the cover was then represented as a percentage. The data show that urban land cover in riparian buffers increases with urbanization in the subbasin: development respects riparian buffers only to the extent forced by regulations. In Durham, North Carolina for example, regulations preserve buffers of only 50 feet, less than one-sixth of the distance considered here. Regarding the data point with the highest watershed urbanization, streams had been shortened, essentially wiping out the small intermittent streams, making them fully urban (or piped underground) and not counted in the riparian buffer GIS measure. Hence, it's far off of the trend line of the other points because the urbanization was so complete that it didn't even count as "riparian." In other words, how much of downtown Durham's headwater streams should be listed as "riparian" when, in fact, it's completely urbanized (see Figure 1.5)?

The discussion of harms to stream biodiversity should also stress the benefits of stream biodiversity, and that brings up a discussion of ecosystem services and ecosystem function. Nutrient processing is one such service that removes nutrients before they enter reservoirs and create algal blooms and all their consequences. Perhaps we don't concern ourselves much with this ecosystem function because we all focus on plant growth when we garden and when we watch the trees leaf out in the spring. But the flip side of plant growth is that all of these leaves and plant material needs to be broken down into their constituent parts. Granted, deeply buried coal seams are a testament to the fact that this biomass breakdown function has always been an incomplete one. Many leaves fall into streams or get carried in by stormwater runoff, where they provide resources for the base of stream ecosystems, growing microbes and invertebrates that feed larger organisms. Leaf breakdown that releases nutrients is just one aspect that falls under the concept of ecosystem function carried out by healthy ecological systems.

Figure 11.5A shows, interestingly, that the highest breakdown rate comes at intermediate values for total impervious area (TIA), due to a mix of biological and physical processes. An intermediate maximum is also seen in the study's species richness data (see Figure 11.8). One suggested possibility for the intermediate peak is that biologically mediated breakdown processes increase with a bit of added nutrients, like adding fertilizer to a garden, but then breakdown decreases as other pollutants take hold.

Putting together the data on leaf litter breakdown (Figure 11.5A) and biodiversity (Figure 11.8) from this Florida study shows increasing breakdown with higher biodiversity, depicted in Figure 11.5B. This dependence makes sense because both measurements come from the same bags of leaves submerged in streams: more critters, greater breakdown. It's also interesting that sweet gum trees also have one of the highest emission rates for volatile organic compounds, which are chemicals used to reduce thermal stresses and

Leaf litter breakdown rates vary.

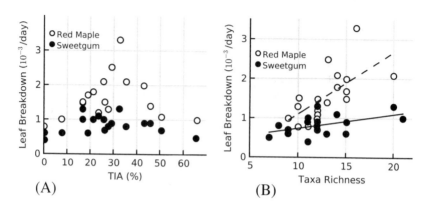

(A)   (B)

Figure 11.5: In urbanized Florida watersheds, leaf litter breakdown rate shows an intermediate maximum as a function of total impervious area (TIA; A), but a steady increase with aquatic invertebrate biodiversity (B) (Chadwick et al. 2006). Snail biomass and biodiversity are important contributors to much of these patterns.

---

herbivory.[1] Although sweet gum shows a slow breakdown rate here, another comparison showed its rate is much faster than oak, with mixed leaf bags demonstrating an inhibition of sweetgum leaf decay by the presence of oak leaves.[2] Ecology is complicated.

Figure 11.6 depicts results from a Melbourne, Australia, study examining leaf breakdown under increasing watershed urbanization, measured as the percentage of connected imperviousness.[3] These results supplement the biodiversity loss with imperviousness from the same study shown in Figure 11.7. Both results come about from bagged leaves of two tree species left submerged in six streams, ranging from 0% to 20% imperviousness for their watershed, for about 2 months. In one species, *Pittosporum undulatum* (sometimes called cheesewood or mock orange), the breakdown rate, measured as a fraction of mass lost per day, increased with imperviousness because of higher stream temperatures (a couple of degrees) and an increased phosphorus concentration, which increased microbial activity. Both factors correlated with imperviousness, and the researchers couldn't isolate which factor drove the increased activity. It certainly seems reasonable, though, that fertilizing and warming a slurry of microbes and organic matter would lead to quicker decomposition. After 21 days, much of the *P. undulatum* leaves were gone.

---

[1] Wilson (2011). [2] McArthur et al. (1994). [3] Imberger et al. (2008) examined the mechanisms of leaf breakdown.

Imperviousness increases leaf litter breakdown rates.

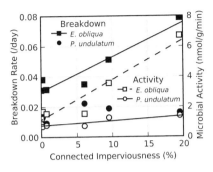

Figure 11.6: For the leaves of the *Eucalyptus obliqua* tree submerged in Melbourne streams, both leaf breakdown rate and microbial activity strongly increase with connected imperviousness. In contrast, for *Pittosporum undulatum* (mock orange) leaves, only a weak response in microbial activity is observed (Imberger et al. 2008).

Imperviousness reduces leaf shredder abundance.

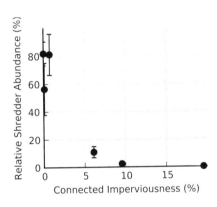

Figure 11.7: Leaf shredder biodiversity decreases sharply with increasing connected imperviousness in Australian watersheds (Imberger et al. 2008). Shredders are the macroinvertebrates that consume detritus like leaves that drop into streams, much like the role of aboveground pill bugs (a terrestrial isopod) and termites.

In a bit of a twist, researchers went into the study thinking that physical abrasion increased with urbanization, with the visual picture that higher flows and more sediments effectively sand-blasted leaves away. However, they tested this idea with blocks secured in place in streams to measure directly the abrasion rates. Surprisingly, mass loss *decreased* with increasing urbanization, which led the researchers to conclude that higher base flows with natural sediment levels were more erosive than the sporadic peak flows with more urban sediments.

Decreasing abrasion with increasing urbanization isn't universal, however. Measured fractional rates of *Acer barbatum* (southern sugar maple) leaf

Aquatic insect diversity peaks at intermediate imperviousness.

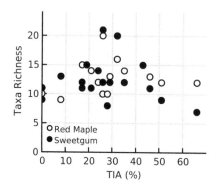

Figure 11.8: Biodiversity found in leaf bags suspended in streams peaks as a function of total impervious area (TIA) fraction in urbanized Florida watersheds (Chadwick et al. 2006). An intermediate species richness takes place when many habitat types ranging from undisturbed to disturbed support a broad range of ecological communities.

breakdown were found to be around 0.02/day at 10% high intensity urban, with three times higher breakdown rates, 0.06/day, at 30% high intensity urban.[1] Though no direct measure of physical abrasion was performed, it was concluded that the higher litter breakdown rate was from abrasion.

Part of the confusion concerning the mechanisms behind leaf breakdown rates is revealed in Figure 11.7. So-called shredders are invertebrates that chew up the leaves and such that drop into streams, and this study from Melbourne, Australia, included nine orders of aquatic insects, one of which were Plecoptera, the stone flies, representing the "P" in EPT taxa.[2]

Shredder abundance was measured from packs of leaves of two different tree species (see Figure 11.6) suspended in streams for 3 weeks, kind of like setting out a baited trap but without the need for a trap. Overall, as the data show, shredders were all but absent in urbanized streams with imperviousness >5%.

Shredder abundance drastically decreases, yes, but notice the "hump" in taxa richness with imperviousness in Figure 11.8. Shortly below I describe the general ecological phenomenon underpinning the hump. Data shown here were taken at 18 stream sites on the St. Johns River near Jacksonville, Florida, along with the data of Figure 11.5.[3]

Researchers took newly fallen leaves from red maple and sweet gum, air dried them, remoistened them, and put 5g of one or the other into 5-mm mesh bags. They then tied four bags of each type to metal stakes driven into the stream bed. After collecting the bags 2, 4, 8, and 12 weeks later, they carefully

---

[1] Meyer et al. (2005). [2] Imberger et al. (2008). [3] Chadwick et al. (2006) studied ecosystem function in north Florida streams.

surveyed the biological community in the bags, washed the contents, dried the leaf samples, and weighed them to determine the lost mass.

Though total impervious area (TIA) ranges uniformly from 0% to 66%, the urban land use fraction has two sites with 0%, one site with 12%, and the rest above 60%. One small 39-hectare site, for example, is listed as 94% urban but has just 8% TIA. That contradiction can only mean that urban land is not necessarily impervious, which could come about by a subdivision with large residential lots. Indeed, a comparison of their location sketch with satellite images shows the presumed area to be a rather fully developed residential area with abundant tree canopy and large yards.

Humps in biodiversity like this one sometimes arise from a process that goes by the name "intermediate disturbance hypothesis."[1] The classic description (and study) of the phenomenon involves boulders moving in the waves of a rocky coastline with organisms clinging to life on the rocks. Small rocks easily tumble in the powerful waves, all the while these motions crush and kill most organisms hanging on to them. Indeed, a survey of taxa on small rocks shows very few species and taxa. Very large boulders are rarely, if ever, moved even in the largest of storms. Again, a lowish diversity of taxa is seen, but a different set from those on the small rocks because a few species outcompete the rest. Now, intermediate-sized boulders aren't moved in everyday waves, but perhaps every year or so a storm with huge waves tosses them about. Such boulders have a mosaic of colonizing communities and long-lived communities with the representative diversity of both types mixing all around the surface. As a result, the highest amount of biodiversity is often found in "environments" with intermediate disturbances because a greater variety of communities are represented.

In Figure 11.8 the conceptual application may be that more organisms can tolerate intermediate values of the disturbance—here reflected in the measurement of TIA—compared with the extremes of a natural environment or a fully impervious watershed.

Of course, one shouldn't make too much of the higher shredder species richness at intermediate imperviousness. Counting the number of different species found in bags of leaves suspended in a stream isn't the only way to measure stream health. In Figure 11.9, many biological metrics measuring such things as species abundance and richness, resource use, and traits were analyzed to find the best representation of sensitivity and tolerance to environmental stresses.[2] In the end, EPT richness, filterer richness, and percent clingers made up the final biological condition index. The urban gradient represented by "relative urbanization" is a measure from least to most urbanized and includes population density, percent urban land use, and road density. Data

---

[1] A classic paper is Sousa (1979). [2] Purcell et al. (2009) characterized the effect of urbanization on biological condition using Bressler et al. (2009).

Urbanization reduces the biological condition of streams.

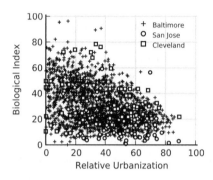

Figure 11.9: Data from Baltimore, MD, San Jose, CA, and Cleveland, OH, show similar effects from urbanization (Purcell et al. 2009). Relative urbanization is a complicated aggregate measure but includes population density, percent urban land use, road density, and housing unit density.

come from extensive state, county, and municipal surveys and include nearly 2,000 points from Baltimore, Maryland, used to design a statistical model, and around 80 points each from Cleveland, Ohio, and San Jose, California, to test its applicability elsewhere.

Note that many streams sit at the lower left of the plot, representing streams in a low-urbanized condition but having the same low biological index as much more urbanized streams. These streams at relatively low urbanization likely suffer from other forms of stream-disturbing land use practices not captured in the urbanization gradient used here, such as agriculture, riparian buffer impairments, and streamflow and water quality issues. The upper limit in the cloud of points that decreases with increasing urbanization represents a hoped-for biological condition for stream restoration projects. In other words, stream restoration to the relatively pristine conditions of biological index value 80 when the urbanization level is 60 (where the highest observed biological condition reaches a value of 40) simply represents a completely unrealistic goal. It helps to have reasonable, managed expectations.

A study that examined toxicity in Boston, Massachusetts, streams within the Aberjona watershed did, in fact, use some surrounding streams as reference sites in their examination of metals in stream sediments, vegetative buffers, and physical habitat condition.[1] A total of 10 habitat quality parameters were incorporated, each rated 0–20, and included in-stream cover, epifaunal substrate, embeddedness (habitat buried by fine sediment), channel alteration, sediment deposition, the variety of velocity–depth combinations, channel flow status, bank vegetative protection, bank stability, and riparian

---

[1] Rogers et al. (2002).

vegetative zone width. These results showed that contaminants in vegetation were three times more important than contaminants in the sediments toward the effect on biological condition. Still, physical habitat condition was slightly more important than contaminants, but the two together explained 49% of the biological condition of their streams. A very important point was that the physical habitat assessment required at each site only requires *15 minutes* on the part of a trained observer to score the 20 physical parameters, as opposed to the intensive chemical and biological sampling, and yet added a great amount of information on causes and effects.

The index of biological integrity (IBI) emphasizes the goal of the Clean Water Act (CWA) to restore the biological health of the nation's surface waters.[1] Along with the IBI are other indices, for example, the invertebrate community index (ICI), mentioned in Chapter 12, and other similar measures mentioned throughout this book. The details of these many indices require ecological discussions beyond the scope of this book, but suffice it to say that these biotic measures make sense because everything that affects living organisms is integrated into the final composition of the ecosystem. These measures also contrast with an often seen ove-emphasis on chemical and physical indicators, as well as programmatic benchmarks, that don't necessarily lead to a healthy ecological community. For example, an Ohio study using bioindicators found 30% more impaired streams than previous approaches using chemical indicators: nearly half of the streams found unimpaired via chemical indicators were found to be impaired via bioindicators.[2]

One problem with the CWA is the very basis of section 1251, which has as its stated goal "to restore and maintain the chemical, physical, and biological integrity of the Nation's waters." Yet, the law itself doesn't define biological integrity, and ecologists have taken up the slack by proposing meaningful measures depending on local and regional conditions. Various scientists are coming to a consensus on the need for a more universal measure, with one such measure, the biological condition gradient (BCG), using 10 ecological attributes that can be applied more broadly across different places with regional tweaks.[3] This unification is encouraging: posing classification problems to 33 macroinvertebrate biologists and 11 fish biologists from 23 states and one tribe found 81% agreement among the macroinvertebrate researchers and 74% agreement between the fish researchers.

Results of Figure 11.10A from streams across the northern portion of the U.S. demonstrate another problem with stormwater runoff: chloride (salt) concentrations.[4] Over a period of 20-some years the concentrations nearly doubled across all levels of urban land cover. Chloride comes primarily from the salt used to deice roads and parking lots, and Figure 11.10B shows that

---

[1] Yoder and Rankin (1998). [2] Yoder and Rankin (1998). [3] Davies and Jackson (2006). [4] Salt concentrations in streams are discussed by Corsi et al. (2015).

Lots of road salt in U.S. streams.

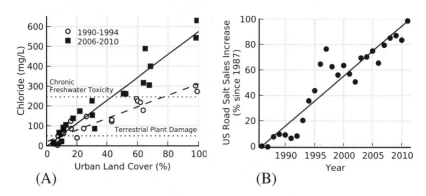

(A)   (B)

Figure 11.10: Chloride concentration in streams across the northern U.S. increases with urban land cover, surpassing toxic levels established by the USEPA (A). Deicing with road salt is responsible, and the higher 2006–2010 slope in A reflects doubled sales (and application) of road salt since 1987 (B) (after Corsi et al. 2015).

---

sales of road salt doubled over the period spanning 1987–2011. Along with the stormwater runoff, salt ends up in streams that empty into reservoirs, lakes, and aquifers.

These data reflect salt concentration studies in Baltimore area streams that increase with a watershed's impervious surface fraction, where salt concentrations ranged upward of 5,000 mg/L (5 g/L), or about one teaspoon of salt in one quart of water.[1] These concentrations exceed toxicity levels defined by Environment Canada and the USEPA for healthy streams and far exceed levels tolerated by terrestrial plants, implying bad things for streamside vegetation. All in all, these freshwater streams become toxic to aquatic animals when watersheds have about 35% of their area covered by impervious surfaces.[2]

Interestingly, in places like Minnesota, sand is bought at $2–4 per ton to throw on wintry roads and then costs $6–11 per ton to dispose in landfills after street sweeping.[3] To bring these prices down, the reuse of swept materials could be considered, though there may be a buildup of contaminants mixed in with the sand. Salt, however, just melts away with the ice with no immediate disposal costs.

We've seen earlier that nutrients are another class of pollutant that harms our streams. One major study sampled 240 wadable first- to fourth-order

---

[1] Kaushal et al. (2005). [2] Kaushal et al. (2005). [3] Curtis (2002).

Stream biological condition decreases with more nutrients.

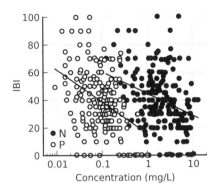

Figure 11.11: A fish index of biological integrity (IBI) decreases with median values of measured nitrogen and phosphorus concentrations (Wang et al. 2007). The fish IBI decreases at threshold levels for the Wisconsin streams, which is somewhat visually muted by the logarithmic nutrient scale.

streams across all of Wisconsin, documenting 30 habitat variables describing the stream and riparian cover, nitrogen and phosphorus concentrations, and macroinvertebrate and fish communities.[1] In addition to the broad biodiversity measure plotted in Figure 11.11, scientists also found that carnivorous and omnivorous fish species, intolerant species, and salmonid abundance were highly correlated in a negative way with nutrient concentrations. Using fish communities as a criterion, the threshold nutrient concentration range for total phosphorus is 0.06–0.09 mg/L, and for ammonia and total nitrogen, 0.02–0.04 mg/L and 0.54–1.8 mg/L, respectively. For macroinvertebrate communities, the threshold concentrations are around 0.09 mg/L for total phosphorus and 0.03–0.04 mg/L and 0.6–1.7 mg/L for ammonia and total nitrogen, respectively.[2] Compare these numbers with, for example, data in Figure 6.15. Above these values, ecological communities are seriously degraded.

To put these numbers into context for human health, the USEPA limits nitrate concentrations to no more than 10 mg/L for children less than 6 months of age.[3] High nitrate concentrations in drinking water can cause methemoglobinemia, where a variant of hemoglobin that can't carry oxygen is produced at extreme levels.

Of course, an immediately obvious point in Figure 11.11 is that there's lots and lots of scatter in the dependence of IBI on nutrient concentration. Despite that scatter, all of the variables considered in the study explained a total of 54% of the variation in the fish communities and 53% for the invertebrate communities. Of this explained variation, structural habitat variables such as

---

[1] Wang et al. (2007).   [2] These numbers agree generally with the Dodds et al. (1998) classification.
[3] Mayer et al. (2005).

Imperviousness makes streams worse places for plants and animals.

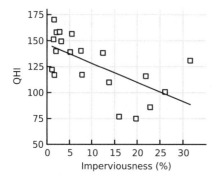

Figure 11.12: Stream characteristics favorable to organisms, as measured through the 10 components of the qualitative habitat index (QHI), decrease with increasing imperviousness in southern and central Maine watersheds (Morse et al. 2003).

erosion, shading, detritus, buffer width, and so on, explained 46–53%, and nutrients just 15–22%.

As a bit of a comparison, surveys of zooplankton in stormwater ponds draining urban areas of Sante Fe, Argentina, counted more than 65 species, many of which exploded in number after rainfalls, but clearly represented species tolerant of pollutants.[1] Among many water quality problems, nitrate and phosphorus levels in the ponds were around 2 mg/L and 2.7 mg/L, respectively. Many species of rotifers and protozoans were found that could also survive in minor-moisture environments, such as mosses and rain gutters. Despite the seemingly high biodiversity, after 48 hours of exposure to stale stormwater pond water, 78% of test *Daphnia magna* died.

Figure 11.12 refers to a study from Maine showing water-quality problems with increasing watershed imperviousness (see Figures 4.8 and 4.9). Results here show the effect on ecological communities through a "qualitative habitat index" (QHI), which measures 10 different stream characters covering substrate condition, erosion, sedimentation, riparian buffers, and channel configuration.[2] Each measure falls on a 0–20 scale, and the summed QHI values range from 0 to 240, including two submeasures. These numbers break into four categories. Streams having optimal conditions score between 180 and 240, suboptimal streams score between 120 and 179, marginal streams score 60–119, and anything below 60 is considered in poor condition.

Riparian width of these same streams also decreases with imperviousness (see Figure 12.2). Notice that, in this particular example, no clear threshold exists at an imperviousness of 10% often seen elsewhere.

---

[1] de Paggi et al. (2008). [2] Morse (2001) describes the QHI. Barbour et al. (1999) provides the current approach.

Urbanized watersheds lose sensitive species.

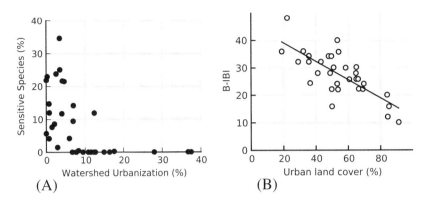

Figure 11.13: (A) Streams in southern California show that as little as 8% watershed urbanization results in the loss of sensitive aquatic insects (after Riley et al. 2005). (B) The quality of streams' benthic habitat, measured through benthic index of biological integrity (B-IBI), decreases with increasing urban land cover in the Puget Sound region of Washington State (Morley and Karr 2002).

The next few figures show that species drop out of streams with increasing watershed imperviousness. Figure 11.13A, for example, shows the results of an extensive survey of urban streams in southern California that demonstrated physical transformations.[1] Streams flowing out of watersheds with more than 8% urbanization essentially lost all of their "sensitive" aquatic insects.

Sensitive stream species included organisms from the EPT orders Ephemeroptera (mayflies), Plecoptera (stoneflies), and Trichoptera (caddisflies) that are intolerant to disturbed conditions. Granted, there's some circularity in the classification of "sensitive," but these insects have either a nymphal stage (with gradual molting into adult form) or larval stage (followed by a distinct preadult pupal stage) that lives and feeds underwater. These results pretty convincingly show that above urbanization levels of 10%, streams are quite degraded from the perspective of these animals.

A little bit to the north, a 3-year sampling effort during the late 1990s showed that stream health decreased with urban land use in the Puget Sound watershed in Washington State, shown in Figure 11.13B.[2] Researchers sampled

---

[1] Riley et al. (2005) demonstrated the rapid loss of sensitive stream species with urbanization.
[2] Morley and Karr (2002) measure IBI in Puget Sound.

Species drop out of city streams.

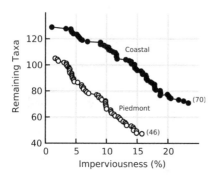

Figure 11.14: A comprehensive survey of Maryland's streams shows that sensitive aquatic invertebrates drop out of streams as watershed imperviousness increases (Utz et al. 2009). Once watershed imperviousness has reached 15–20%, all of the sensitive species are lost.

16 different subwatersheds with 45 sites in second- and third-order lowland streams in King and Snohomish Counties. Samples were taken using a so-called Surber sampler that outlines a small 0.1-m^2 quadrat of the stream bottom, which is then stirred up and as the material is carried downstream it's caught in a mesh bag attached to a vertical frame. Here, the benthic index of biological integrity (B-IBI) includes just the benthic invertebrates, with variation limited from 10 (poor condition) to 50 (excellent condition). Half of the sites were in poor or very poor condition, defined by a value less than 26.

Urban land use was the most important determinant at the scales of the subbasin as well as the complete riparian area, meaning a 200 m riparian buffer extending fully upstream, and the local scale, meaning a 200 m riparian buffer extending just 1km upstream. This dependence on multiple scales means that both subbasin and local scales are important. The local scale includes important features of riparian cover; for example, the presence of stonefly taxa (temperature-sensitive shredders) depends on leaf input to the stream. However, urban land cover was a slightly better fit than impervious surface. The study authors argue for land cover as the better determinant because it also includes compacted soils, which we've seen can be nearly as impervious as impervious surfaces (see Figure 4.2).

Figure 11.14 shows the results of yet another amazingly detailed study that looked at 180 taxa across the coastal plains, piedmont, and highlands of Maryland,[1] a state that covers 32,000 km^2, of which 95% drains into the Chesapeake Bay. The data come from the Maryland Biological Stream Survey, run by the state's Department of Natural Resources, which randomly selects 75-m stream lengths across the state for sampling—all lengths of all streams are potential sample sites. This study pulled 893 sites from the coastal

---

[1] Utz et al. (2009).

plains, 879 from the Piedmont, and 569 from the highlands, and performed a biological survey at each site.

These curves show at which level of imperviousness all of the taxa sensitive to imperviousness have been lost, and the imperviousness at which all that remains are those taxa that either have no response or a positive response to imperviousness.

Here's how to interpret the curves. First, order all of the streams according to increasing imperviousness. At zero imperviousness there's some number of species found across all of the streams of all imperviousness values, denoted by the intercepts, 129 and 105 for coastal and Piedmont regions, respectively. Now start dropping from the species tallies the streams with the lowest levels of imperviousness and ask what taxa remain in the remaining higher impervious streams. As imperviousness increases, some taxa that were present only in the low-imperviousness streams are no longer found in the remaining high-imperviousness set and the curve steps down one species at a time.

Species continue to drop out of the samples as imperviousness increases to 15% or so in the Piedmont, and 20% or so in the coastal area, after which there's no further loss of species. The remaining taxa are either tolerant of imperviousness or respond positively to the presence of imperviousness. A positive response could come about, perhaps, if a species' strongest competitor is sensitive to imperviousness, and once it's gone the insensitive species can flourish. Of course, there are some organisms that like disturbances—just think of the weeds that pop up in any disturbed patch of soil. Similar to disturbing soil, directly connected impervious surfaces cause stream disturbances that make some species thrive. These results from the U.S. East Coast clearly confirm the loss of West Coast species at very low levels of imperviousness that we saw in Figure 11.13A.

While urbanization caused the loss of many taxa, most survived high levels of agricultural use (see Figure 11.22). However, the study's authors note that it is possible to argue that agriculture did its damage decades ago and the agriculturally sensitive species simply aren't around anywhere, while urbanization is a relatively new phenomenon.

Watersheds in North Carolina also show interesting results for organismal tolerance to humans.[1] Most important land use factors were percent forest (increasing aquatic species diversity) and percent agriculture (decreasing diversity). As we've seen before, urban land was correlated with the degradation of streams. Adding a new twist, the strongest correlate with increasing species diversity was a factor called "watershed topographic complexity," essentially the variability in elevation across the watershed, which might simply reflect the notion that people shy away from developing and farming in hilly watersheds. The shape of a watershed also seemed important, with round ones being

---

[1] Potter et al. (2004) correlated diversity of aquatic organisms to stream and watershed features.

Imperviousness correlates with the loss of aquatic organisms.

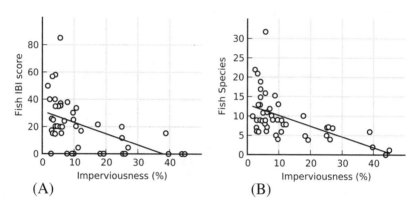

Figure 11.15: Both the fish biological index (A) and fish biodiversity (B) in southeastern Wisconsin streams decrease with increasing watershed imperviousness (Wang et al. 2001). Conditions for most fish are inhospitable above 40% imperviousness.

more at risk to losing biodiversity. Shape is important because "round" watersheds focus runoff into their streams more quickly due to shorter flow paths than elongated ones; hence, the flashy flows are just that much more flashy. What all of this means, again, is that the presence of organisms intolerant of human disturbance was best predicted by forested land use. Finally, these land use correlations varied across the state, reflecting that more agriculture is found on the coast, more forested riparian areas are found in the Piedmont, and watershed and riparian forests are correlated in the mountains.

In the study of southeastern Wisconsin streams first presented in Figure 5.2, connected imperviousness and other related urban land use variables explained the most variation in many stream quality measures, two of which are shown in Figure 11.15.[1] These variables include a combined measure of highways, streets, and parking lots, as well as commercial land use and total urban land use fractions. The next best predictors included agricultural, government, and residential land uses, environmental corridor (undisturbed land connected to the stream), woodland, vegetated land, and water/wetlands.

In figure 11.15A the IBI index comes from a detailed analysis of the community structure of the sampled fish community and represents the health of the stream on a range from 0 (bad) to 100 (good), whereas the number of

---

[1] Wang et al. (2001).

fish species in Figure 11.15B is a simple quantitative variable. These measures were variably affected by agricultural lands but mostly increased slightly or were unaffected. As we can see, however, both measures decrease strongly with increasing imperviousnessness.

On the other hand, the study authors examined a habitat score designed specifically for Wisconsin streams, which was not affected by the connected impervious fraction. As they describe the measure, its components "differentially weighted measures of channelization, instream cover, bank erosion, sinuosity, variation in thalweg depth [the lowest point in the stream], and riparian vegetation."[1] Although erosion increased with imperviousness, the other components were unaffected in these Wisconsin streams.

In the end, this study like many others concluded that relatively low levels of urbanization inevitably lead to seriously harmed fish communities. In a case like this one for IBI and fish biodiversity, we can dismiss the worries of causation versus correlation. It's more than just a correlation because causation is clearly one-directional—the loss of biodiversity can't, in any universe, be the cause of the connected impervious surface. But exactly *how* connected imperviousness reduces biodiversity has a number of possible causative mechanisms: urban stream syndrome, pollutants, reduced base flows, and so forth. Indeed, this same study looked at base-flow dependence on imperviousness (see Figure 5.2), and that effect might be explanatory on the decrease in the biological community.

Along these lines, repeated measurements from 39 cold-water streams in the same region of Wisconsin and Minnesota replicate a sharp imperviousness threshold at 7% for many types of invertebrates.[2] The study also considered temperature, showing that a 7-day average stream temperature is important for invertebrate biodiversity and about as important as impervious surface coverage. Yet, for example, watershed imperviousness explains only 32% of the temperature variation, and other things such as stream shading were also important, meaning that reducing stream damage coming from human land use must address a multifactor problem.

Figures 4.8B and 4.9 showed results of a stream study performed in Maine, demonstrating that total suspended solids and nitrate levels increased and the sediments become finer with increasing imperviousness. Data in Figure 11.16 extend those study results to show that biodiversity also decreases with increasing imperviousness.[3] A total of 93 insect taxa were collected, of which 66 were found in both seasons. Clearly, stream invertebrate diversity, whether considering all taxa or just EPT alone, decreases with increasing imperviousness. Figures 4.8B and 4.9 omit error bars for clarity, but they're only about twice the size of the plotting symbols depicted there and tend to be larger at lower imperviousness, but not terribly large in any event. A close examination

---

[1] Wang et al. (2001).   [2] Wang and Kanehl (2003).   [3] Morse et al. (2003).

Taxonomic richness decreases with increasing imperviousness.

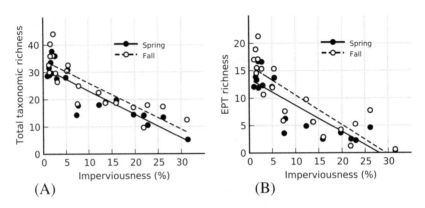

Figure 11.16: Total taxonomic richness of stream invertebrates (A) and EPT species richness (B) decrease with increasing imperviousness in Maine watersheds (Morse et al. 2003). Study authors argue for a threshold imperviousness of 6%, at which both biodiversity measures drop significantly.

---

shows a drop in biodiversity of about a third of the species between 5% and 10% and, again, indicates that 10% imperviousness is an important threshold for species richness.

We've seen above that even low levels of imperviousness cause the loss of sensitive species. An important goal of low-impact development, discussed in Chapter 13, is to reduce imperviousness through a variety of approaches, as well as disconnect impervious surfaces from streams, thereby minimizing their harm.[1]

Piping road runoff is a standard practice that brings bad consequences even to rural areas and should be minimized whenever possible. While keeping watershed imperviousness below 10% is not always feasible given the fact of cities and a large human population, these results argue for stormwater control measures (SCMs) that disconnect impervious surfaces, for example, adding rain gardens and green roofs.

Figure 11.17 demonstrates the importance of connection status on aquatic organisms. As shown in Figure 4.10, the correlation between TIA and directly connected impervious area is very strong. In that curve, within the range of 2–12% TIA the fraction connected to a stormwater drainage system and hence directly connected to a stream ranged from 0% to 80%. The study

---

[1] Walsh (2004).

Directly connected impervious surfaces flush out rare species.

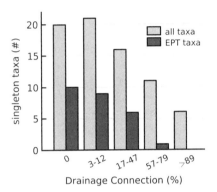

Figure 11.17: Watersheds analyzed here surround Melbourne, Australia, and have impervious surface fractions between 2% and 12%. For these watersheds, the number of rare stream taxa depends greatly on the fraction of imperviousness directly connected to streams (Walsh 2004).

results shown in Figure 11.17 use the same streams from Melbourne, Australia, with streams draining watersheds ranging from 0% to 50% imperviousness. Figure 11.17 shows how frequently rare taxa—defined as having been found only once in the survey—are found in a stream given its connection status of impervious surfaces. Whereas around 20 total rare taxa were collected from streams without any connected impervious surfaces, just 10 were found in streams having half their impervious surfaces connected. The number drops even more precipitously for EPT taxa.

Despite the correlation between imperviousness and connection status, connection was the strongest independent variable predicting species richness in the intermediate range of TIA. However, looking at the family level, biodiversity was not as strongly affected by urbanization due to a gain in disturbance-tolerant organisms.

Fish aren't immune, and their eggs are particularly vulnerable. Data in Figure 11.18 come from four different regions of the Hudson River in eastern New York State having a broad range of land use fractions for urban, agriculture, and forest.[1] Sixteen tributaries were studied from Albany to Yonkers, and watershed sizes ranged from 4 to 3,082 km². Upstream areas are more rural and agricultural, whereas downstream areas nearer New York City are more urbanized.

Although the original plot separates resident freshwater fish from ocean fish spawning in freshwater streams, called "anadromous" fish, Figure 11.18 includes all fish species together. Eggs from a total of 23 species were collected

---

[1] Limburg and Schmidt (1990) sampled fish egg densities in the Hudson River.

Urbanization reduces the density of fish eggs.

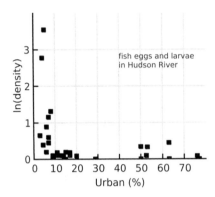

Figure 11.18: Streams flowing into New York's Hudson River from watersheds with high fractions of urban land use show reduced counts of fish eggs and larvae (Limburg and Schmidt 1990). Density involves the number of eggs and larvae per cubic meter, and the vertical axis plots the natural log, where $\ln(3) \approx 20$.

from these inland streams, but the anadromous alewife herring, *Alosa pseudoharengus*, made up 93% of the sample. Statistically, more than 70% of the egg density variation was accounted for by urban land use.

Measurements of dissolved oxygen (DO) showed that urban watersheds had lower percent saturation from mid-April on through the summer. Oxygen, of course, is important because fish need it. Water can hold a certain amount of oxygen, dependent on its temperature, with water at $0°C$ holding about twice as much oxygen as water at $30°C$. Percent saturation accounts for the temperature dependence, but a reduction like the one found in this location is typically due to high amounts of organic matter dispersed into water. That organic matter leads to the typical microbial decomposition that burns up oxygen. Urban watersheds also displayed higher variability in DO, a variability suspected to come from the variabilities in urban disturbances.

These results again conform with the conventional knowledge that stream problems crop up at around watershed impervious fractions of 10%; here at that fraction we see a tremendous drop-off in fish eggs.

Figure 11.19 shows directly how stream health decreases with urbanization in the Toronto watershed. The strongest correlate with urban land cover among the 10 components of this figure's IBI was the presence of brook trout, a species needing high-quality cold water (see Figures 9.9 and 9.10). We should think of brook trout as an indicator species, whose presence and health imply a healthy ecosystem, and whose absence from its expected range indicates some sort of environmental problem. Drilling down into the details, an increase in the proportion of fish with more than three black-spot cysts on one side, an indicator of poor fish health, correlated well with a decrease in the IBI value. Stream length with intact forest at least 20 m wide served as riparian forest

Urban land cover reduces stream health.

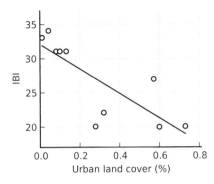

Figure 11.19: For the streams draining the large watersheds surrounding
Toronto, Ontario, a fish-oriented
measure of stream health demonstrates a rapid decline with increasing urban land use (Steedman 1988).

indicator. Indeed, the statistical model that explained 68% of stream health
was the expression IBI $= 29.47 - 19.35$ urban $+14.21$ riparian, where "urban"
and "riparian" are the watershed's urban land cover and riparian forest cover
fractions, respectively. Given that stream health corresponded to brook trout
presence, scientists concluded that good outcomes for IBI needed more than
25% riparian forest with no urban land use, or 100% riparian forest cover
when urban land use hits 50%.

We've not seen any results that show urbanization is anything other than
harmful to stream health. The picture is a bit more mixed with agricultural land
use, as we'll see some nonnegative effects in Figure 11.22. First, however,
Figure 11.20 shows that agricultural land use harms aquatic organisms in
southeastern Michigan. This study was done in the River Raisin basin, which
drains into Lake Erie, in an area that had very little urbanization but 70%
agricultural land use.[1] Over the 20-year period examined (1968–1988), the
number of farms nearly halved, while forest land use increased by 13%, a
not uncommon phenomenon seen across the U.S.[2] Land used for agriculture
also decreased while urban land use increased, but no number was mentioned.
Over this time interval riparian forest corridor area increased around 40%, in
parallel with forest cover, reflected by an increase in corridor width.

Researchers estimated stream health using habitat and IBI index scores at
23 sites, where the habitat index arises from a visual inspection of the stream
and the IBI comes from fish sampling in 100-m reaches. As shown in Figure
11.20, stream health decreases for both measures with increasing upstream
watershed agricultural land use, responsible for three-quarters of the trend in
habitat index. Other cited studies indicate that the most important scale for

---

[1] Allan et al. (1997). [2] Wilson (2011).

Agriculture harms streams in southeastern Michigan.

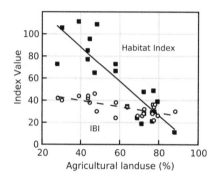

Figure 11.20: In the River Raisin basin of southern Michigan, two measures of stream health, a habitat index and a fish-based index of biological integrity, respond negatively to increasing agricultural land use (Allan et al. 1997).

stream health was the land use within a 100-m stream buffer. These contrasting scales aren't so much conflicting as they are designed, intentionally or not, to capture effects at different land use scales.[1]

In a really neat application of the study results, a statistical model incorporating a tremendous amount of land cover information estimated the consequences of converting land surrounding an existing land use by 100-m buffer increments. When the increment was applied to urban areas, runoff and sediments increased by about 5%, and nutrients increased by more than 10%. Agriculture land use expansions had a similar effect, though slightly lower except for the export of sediments, which increased by more than 15%. Only the expansion of forested land decreased these consequences by 5–10%.

In contrast to visual inspections, a hydrologic study examined 21 forested wetlands of New Jersey, all of them surrounded by urban areas, and ranging in size from 5 to 500 hectares.[2] Wetlands spanned a variety of hydrogeomorphic classes for wetlands: depression; slope; riverine; mineral flats, and mineral flat–riverine. Each class differs in how the hydrology works, where the water comes from, and how the subsurface water flows.[3] Signs of flooding were infrequent, but such indicators as the presence of moss provided evidence for the sustained presence of moisture. Classification sounds nice, but it was problematic. Researchers cautioned that qualitative indicators of wetland hydrology (root depth, water marks, moss presence) didn't reflect better and more accurate quantitative measures, such as water level in wells. Hydrological differences between wetlands classes were also obscured by differences in

---

[1] Allan et al. (1997). [2] Ehrenfeld et al. (2003). [3] A U.S. Army Corps of Engineers site describes the wetlands types more thoroughly, if not with slight differences: water.epa.gov/ lawsregs/guidance/wetlands/hgm.cfm.

Agriculture harms streams in Pennsylvania.

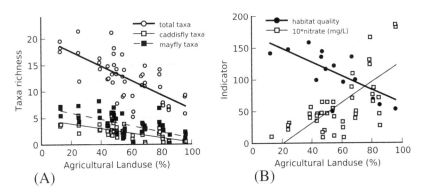

Figure 11.21: Pennsylvanian streams respond negatively to agricultural land use as indicated by several measures, including invertebrate biodiversity (A) and habitat and nutrient measures (B) (Genito et al. 2002). Nitrate concentrations represent repeated measurements over 3 years. Compare these more modest declines with the declines from urban land use in Figure 11.15.

---

disturbance regimes caused by urbanization. In short, human land use disturbs wetlands as well as streams.

Another example of decreased stream health connected to increased agricultural land use—mostly crops and pasture—comes from Pennsylvania,[1] where forested fraction decreased and agriculture increased simultaneously. As shown in Figure 11.21A, relatively steep declines take place in invertebrate taxa, averaged from 3 years (1998–2000) of data from each location for each taxa type. Part of the reason for the declines is shown in Figure 11.21B, which demonstrates increasing nitrate levels and decreasing habitat quality with increasing agricultural land use. Sediments also were thought to play an important role in the loss of taxa.

These results of declining stream condition with agriculture certainly seem to contrast with agriculture in the Southeast. As a part of the study showing organisms dropping out of Maryland's streams with greater urbanization, described in Figure 11.14, researchers also performed surveys of the species in both the coastal and piedmont regions (Figure 11.22). They then examined which species were common, or at least more common, in each of the two land use variables, impervious or agriculture.[2]

---

[1] Genito et al. (2002). [2] Utz et al. (2009) provide in their supplementary data the species positively associated with imperviousness, as well as the imperviousness at which other species drops out of the streams.

East Coast agriculture doesn't harm many aquatic species.

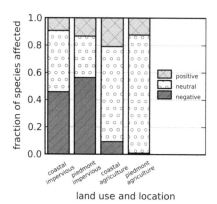

Figure 11.22: In Maryland's streams, large fractions of stream invertebrates experience negative effects from imperviousness, whereas only a very small fraction are harmed by agriculture (Utz et al. 2009).

In summary, imperviousness harms 40–50% of aquatic organisms in both coastal and Piedmont areas of the East Coast, but agriculture has a negative effect on less than 10% of species. And, as we saw before (see Figure 11.14), impervious surface sensitive taxa were generally lost below 15% cover. Agriculture's impacts on streams seems to be more mixed than those of imperviousness.

Chapter Readings:

Chadwick, M.A., D.R. Dobberfuhl, A.C. Benke, A.D. Huryn, K. Suberkropp, and J.E. Thiele. 2006. Urbanization affects stream ecosystem function by altering hydrology, chemistry, and biotic richness. *Ecol. Appl.* 16: 1796–1807.

Imberger, S.J., C.J. Walsh, and M.R. Grace. 2008. More microbial activity, not abrasive flow or shredder abundance, accelerates breakdown of labile leaf litter in urban streams. *J. N. Am. Benthol. Soc.* 27: 549–561.

Meyer, J.L., D.L. Strayer, J.B. Wallace, S.L. Eggert, G.S. Helfman, and N.E. Leonard. 2007. The contribution of headwater streams to biodiversity in river networks. *J. Am. Water Resources Assoc.* 43: 86–103.

Morse, C. C., A.D. Huryn, and C. Cronan. 2003. Impervious surface area as a predictor of the effects of urbanization on stream insect communities in Maine, U.S.A. *Environ. Monit. and Assess.* 89: 95–127.

Purcell, A.H., D.W. Bressler, M.J. Paul, M.T. Barbour, E.T. Rankin, J.L. Carter, and V.H. Resh. 2009. Assessment tools for urban catchments: Developing biological indicators based on benthic macroinvertebrates. *J. Am. Water Resources Assoc.* 45: 306–319.

Utz, R.M., R.H. Hilderbrand, and D.M. Boward. 2009. Identifying regional differences in threshold responses of aquatic invertebrates to land cover gradients. *Ecol. Indic.* 9: 556–567.

Walsh, C.J. 2004. Protection of in-stream biota from urban impacts: Minimise catchment imperviousness or improve drainage design? *Mar. Freshw. Res.* 55: 317–326.

Wang, L., J. Lyons, P. Kanehl, and R. Bannerman. 2001. Impacts of urbanization on stream habitat and fish across multiple spatial scales. *Environ. Manag.* 28: 255–266.

Part III
Solutions

Chapter 12

Streams and Trees: Riparian Ecosystems

Forested riparian buffers of at least 30-m (100 foot) width provide services that offset the impacts of urbanization. Healthy riparian ecosystems promote many factors lead to high stream biodiversity. In headwater streams, organismal diversity increases downstream, reflecting changes in flow patterns as a stream extends farther from its source. Another important factor is watershed land use, with more forests leading to higher biodiversity.

One important ecosystem service is nutrient removal. Uptake in streams reduces to one-third of its undisturbed watershed levels at just 30% high-intensity urban land use. In part, nutrient uptake requires organic matter to fuel microbial respiration, but flashy flows of urban watersheds wash this material away.

Riparian forests also help with other water quality services such as filtering sediments that enter streams with surface runoff. Trees in these streamside forests grow roots deep into the soils, helping with stormwater infiltration, and when they die they leave organic matter behind that continues the denitrification of nitrate, pulling this nutrient out of downstream surface waters. Forested riparian buffers also keep streams cool, which helps water-temperature-sensitive fish such as native trout species. These natural aspects of riparian buffers remain difficult to replicate in the restoration of degraded urban streams.

Once nutrients make their way downstream and enter reservoirs, lakes, estuaries, and beyond, their fate can take many paths. Generally, nutrients increase primary productivity in the water body, but removal processes, including sedimentation and denitrification, are quite effective. Water residence times of about a year reduce both nitrogen and phosphorus export fractions to about 20–30%.

Aquatic insect populations increase downstream from the headwaters.

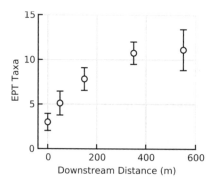

Figure 12.1: Aquatic invertebrate biodiversity increases downstream from headwater sources, here measured as the number of taxa found of the three orders Ephemeroptera, Plecoptera, and Trichoptera (EPT) (Meyer et al. 2007). Data come from West Virginia and Kentucky.

In the most natural of conditions, streams gain species as they flow downstream, which only makes sense: the first few intermittently dry meters of a stream can't possibly reflect the biodiversity of a downstream perennially wet section. Figure 12.1 shows this concept from 34 unmapped and/or intermittent headwater streams in West Virginia and Kentucky.[1] The EPT counts included 86 insect genera from 47 families.

However, these results showing a biodiversity increase downstream in no way imply that the initial portions of streams have no importance—far from it. Another study performed a long-term survey of eight headwater streams at Coweeta Hydrological Laboratory in western North Carolina, streams so small that none were shown on standard topographic maps. Results revealed organisms from 51 taxa and 145 genera of insects, some unique to first- and second-order streams. Another study from Ontario, Canada, found eight taxa unique to the first 20 m of a stream, not found anywhere downstream. Headwaters provide an important refuge from predators and competitors, provide spawning sites, and have a unique chemistry for species that occur nowhere else in the streams. And we're losing these headwater streams to development (see Figure 10.2).

These results come from natural streams, but one could potentially translate the results to daylighted urban streams. Where an outfall pipe emerges to become a daylighted "creek," that location might act like a headwater. Only after a couple hundred meters might one expect to see a significant number of aquatic insects. Of course, if that stream goes back into a pipe, the whole stream might get reset back to a depauperate headwater stream.

---

[1] Meyer et al. (2007) surveyed studies of headwater stream biodiversity.

Buffers narrow with increasing imperviousness.

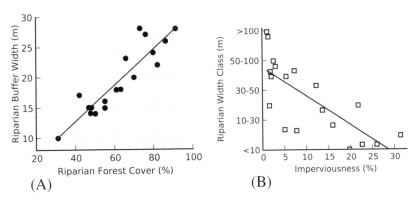

(A)   (B)

Figure 12.2: The width of riparian buffers correlate with two features: the fraction of forest cover in the buffer in Georgia streams (A) (Jones et al. 2006) and the watershed imperviousness in southern and central Maine watersheds (B) (Morse et al. 2003).

The results of Figure 12.2A demonstrate the correlation between forest cover within the buffer and buffer width for streams in Georgia where the trout study of Figure 11.4 was performed.[1] The implication is that narrow riparian buffers lack forest cover, meaning narrow buffers have decreased infiltration, reduced input of organic matter, and decreased effectiveness at removing nitrogen (see Figure 12.13).

Reflecting the land cover correlates of riparian areas of Figure 11.4, average riparian width also decreases with watershed imperviousness, as shown in Figure 12.2B. In this work, measurement of riparian width was facilitated by a categorical classification scheme indicated by the y-axis.[2] Results show that natural streams are wide, typically ranging from 30 to 100 m, but high imperviousness closes in on and reduces these buffer widths.

The entire watershed, and not just the stream, affects the biodiversity of stream invertebrates. The study shown in Figure 12.3 examined 138 streams in the central portion of the north island of New Zealand, a region called Waikato. Data shown here come from onsite ecological and biodiversity measurements, and the study used a GIS analysis of land cover. Quite a number of measures of biodiversity increased with forest cover at the catchment scale, but here I show just two plots related to EPT richness. Forest cover at the so-called

---

[1] Jones et al. (2006). [2] Detailed measurement information can be found in Morse (2001).

Aquatic insect diversity increases with forestation.

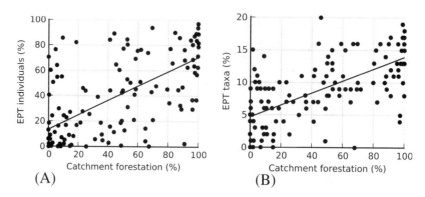

(A) (B)

Figure 12.3: Frequencies of aquatic invertebrate insects from the EPT orders increase in both number (A) and taxa (B) as catchment forest cover increases (Death and Collier 2010). Results come from New Zealand.

catchment scale was most statistically significant for EPT species richness in New Zealand watersheds (see Figure 12.3).[1] Catchment, drainage basin, and watershed are just different words for the same thing; however, there exists a clear hierarchical nature, much like stream orders. Vegetation at scales of 50–100 m reaches (with "reaches," here meaning the distance between pools) and 1.3-km segments (the distance between stream confluences) also predicted EPT richness.

The data points look quite scattered, but the statistical correlations give values of $R^2 = 0.44$ linking EPT *taxa* to catchment scale, and $R^2 = 0.37$ for the segment scale. For the percentage of EPT *individuals* in the samples, the catchment scale has $R^2 = 0.43$, and the segment scale, $R^2 = 0.35$. There's certainly not much difference between those two scales in terms of explanatory power, or between taxa richness versus individual abundances. The study authors conclude that retaining 80% of the taxa would require 40–60% of the catchment to be in natural forest cover. It's interesting to note that no really sharp transition is observed in these biodiversity plots with forestation as there is with the legendary 10% imperviousness.

In the end, higher levels of forestation means more aquatic organisms. The rub is that we might be thinking of these as aquatic organisms, not terrestrial ones, while we forget that the adult stage of many aquatic organisms spends time above water. And that's an important link.

---

[1] Death and Collier (2010) studied New Zealand stream health.

Diversities of aquatic insects and sensitive fish increase together.

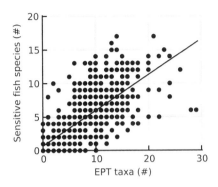

Figure 12.4: The number of sensitive fish species surveyed increase with the number of co-located EPT taxa (Miltner and Rankin 1998). Results come from a broad survey of Ohio's surface waters.

Good places for aquatic invertebrates are also good places for fish. Results of Figure 12.4 come from a large study of 1,657 sites compiled by the Ohio Environmental Protection Agency, with the data set beginning in 1981.[1] Sensitive fish mentioned here are just those species having a demonstrated harm from pollutants. The correlation explains about 50% of the variation in fish species and arises from two responses: both fish and invertebrates perish from a common exposure to pollutants, and the fish perish in response to the loss of their invertebrate prey.

It also shows the utility of sampling aquatic invertebrates. Even though many people don't care about stream insects or the regulations that preserve them, they do care about fish—if you care about fish, you must care about regulating stream health, and stream health is measured by stream invertebrate sampling.

There remains some disparity in the two aspects of biodiversity, in part because total phosphorus and total inorganic nitrogen influenced fish communities in a more consistent manner than they influenced invertebrate communities. In headwater to small streams, increasing nutrients decreased sensitive fish while increasing tolerant and omnivorous fish. In contrast, very little response was seen in the invertebrate community as measured through the invertebrate community index,[2] except for a large reduction at quite high nutrient concentrations.

Figure 12.5 reports the results of a fish species survey in 209 stream locations within 10 watersheds around Toronto in the summers of 1984 and

[1] Miltner and Rankin (1998) showed the correlation between aquatic insects and sensitive fish. [2] The Ohio Watershed website has a description of this index at tycho.knowlton.ohio state.edu/ici.html.

Bigger watersheds support more fish species.

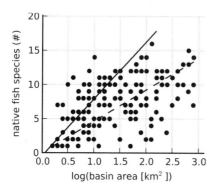

Figure 12.5: Higher fish biodiversity is found in larger stream basins around Toronto, Ontario (Steedman 1988). The solid line represents the expected maximum number of species; the dashed line is the best fit for observed species given the constraint of passing through the origin.

1985.[1] The land cover information comes from aerial photos taken in 1976—outdated by now, for sure. Repeated species sampling shows up to 8% and 24% variation in results within and between seasons, respectively, at the same location. These repetitions also, presumably, give us a handle on other studies' precision in species sampling.

The figure shows the well-known "species–area curve" for biodiversity, a phenomenon seen when encompassing all sites from the study.[2] Imagine intensively sampling 1 acre of forest or meadow or a 100-m stretch of river, and you would imagine finding some number of unique species. Certainly you'd find many individuals of some common species and just one individual of a rare species, but in the species richness measure, both count as "one." Now imagine sampling 100 acres or 10 km of river: you'd find many more individuals and more species. Maybe you'd find 100 times more individuals, but you wouldn't find 100 times more species because you've already found a lot of the common ones in the first small sample. Now imagine sampling 10,000 acres or 1,000 km of river; again, you'd find more species, but only the really rare ones that you hadn't found at the smaller scales. Thus, species richness increases taper off as the area sampled increases.

The dashed line in Figure 12.5 is a best fit to the data, constrained to go through the origin, whereas the solid line represents the maximum species richness for a watershed of a specific size. The study authors take important note that the larger basins lack species that *should* exist in their surveys, but those large basins are highly urbanized. These watersheds are the same ones that have poor biological conditions (see Figure 11.19). If those species had

---

[1] Steedman (1988). [2] Another great study showing the species–area curve is Stout and Wallace (2005).

Organic matter increases nutrient uptake.

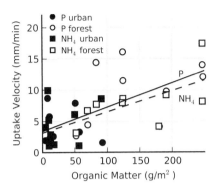

Figure 12.6: Nutrient uptake velocity increases with increasing fine benthic organic matter in streams surrounding Atlanta, GA (Meyer et al. 2005). Organic matter density is reported in units of ash-free dry mass.

been found, then the points, and the dashed best-fit line, should have even higher values at larger basin sizes.

Nutrient processing is one of the most important services provided by riparian ecosystems. However, urban streams get a double whammy in that they have higher nutrient concentrations alongside lower nutrient uptake rates, as shown in Figure 12.6. The data reveal that these uptake rates depend on the amount of fine benthic organic matter because it's thought, reasonably, that biotic demand determines uptake rate. Organic matter is like the fuel for microbial populations that need nutrients to grow. This organic matter decreased with increased urbanization, as high peak flows flushed organic matter from urban streams. A second concern might be that urban streams, with their higher nutrient levels, are so saturated with nitrogen that microbes simply can't take in anymore. Still, fine benthic organic matter was the best predictor of nutrient uptake.

Besides these dependencies, NH_4 uptake increased with the sum of gross primary productivity and "community respiration," essentially a measure of the transformation rate of organic matter to carbon dioxide. Organic matter, however, isn't the main issue since even natural streams are often depleted of organic matter, which in turn limits microbial populations, which in turn keeps dissolved oxygen (DO) levels high and provides an ideal condition for organisms.

Urban streams hold both ammonium and nitrate, and lots of the biological processes transforming these molecules happen in the stream bottom, where all of the microbes doing the work live.[1] Several important pathways exist (see Figure 6.1). First, NH_4 can be nitrified to NO_3, and we just have more nitrate.

---

[1] Peterson et al. (2001) describes the nitrogen process in stream bottoms.

NH$_4$ uptake is faster in smaller streams.

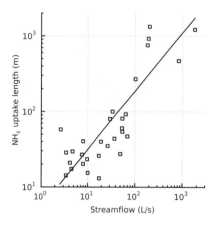

Figure 12.7: Ammonium uptake distance in U.S. rivers increases with streamflow, meaning that the uptake velocity decreases as river size increases (Peterson et al. 2001). These results demonstrate that smaller, slower-moving streams make greater contributions to nitrogen removal.

Second, both ammonium and nitrate can be taken up directly by the microbes, algae, and fungi. Once this conversion to biomass occurs, subsequent decay in the stream bottom again releases NH$_4$, which can then be nitrified back to NO$_3$. Third, some small fraction of the NO$_3$ can be denitrified to N$_2$ gas and make its way back to the atmosphere. These processes take place predominantly in headwater streams, processing more than half of imported nitrogen.

Figure 12.7 shows data on the uptake of ammonium containing labeled ^{15}N from 12 headwater streams throughout the U.S.[1] Researchers dumped a bit of this spiked ammonium in the water as a tracer and then tracked its distribution and concentration as it flowed downstream. About 70–80% of uptake takes place at the stream bottom, through the routes of photosynthesis, microbial uptake, and a simple binding to sediments. Ammonium with only common nitrogen isotopes behaves in the same way. A molecule of nitrate travels about 5–10 times longer than one of ammonium, simply reflecting the relative importance of different biological uptake pathways.[2]

The uptake distance increases when the uptake rate decreases: lower uptake rates mean longer uptake distances. In this study, ammonium uptake lengths ranged from 10–100 m up to 1 km or so. If reducing nutrient levels is important, then smaller streams—which are those with the lower discharge rates—are the most important streams, having a more rapid uptake of NH$_4$. This relation comes about since most of the nutrient uptake takes place at the interface between the water and the soil, and small streams have a greater benthic surface-to-water volume ratio. Think about a river as if it were

---

[1] Peterson et al. (2001). [2] Meyer et al. (2003).

Intact riparian buffers reduce nitrates and dissolved solids.

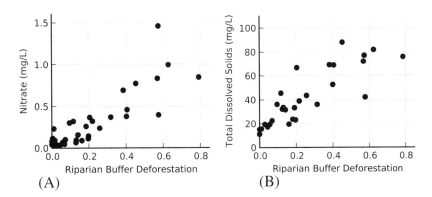

Figure 12.8: Deforestation within 100-m riparian buffers affects many stream variables, including the concentrations of nitrate (A) and dissolved solids (B) (Scott et al. 2002). These results come from western North Carolina watersheds.

a pipe: the cross-sectional area increases with the *square* of the pipe's radius, but the perimeter is directly proportional to the radius. Those two different dependencies mean that as a stream gets bigger the volume of water increases much faster with stream size than the area of the soil–water interface. In other words, big streams have relatively less stream bottom pulling nutrients from much more water.

Given the importance of headwater streams for nutrient uptake, it's unfortunate that the most endangered streams in urban areas are the small headwater streams, as we saw in Figure 10.2.[1]

Stream bottoms are important, but riparian buffers also improve stream water quality in the western North Carolina mountains.[2] Using 36 streams from the Upper Tennessee River system, several stream variables were correlated against many types of land use and watershed properties. The most important variable for water quality at a location, generally speaking, was the amount of deforestation within a 100-m buffer measured from a stream's center, extending 2 km upstream. In addition to the increase in nitrate and total dissolved solids, shown in Figure 12.8, increases in ammonium, turbidity, and temperature also occurred with increased buffer deforestation.

---

[1] A very readable introduction to the importance of headwater streams is Meyer et al. (2003).
[2] Scott et al. (2002) studied streams in the western North Carolina mountains.

There's also a historical dependence. Fine particles in the substrate increased with riparian deforestation from the 1970s rather than riparian deforestation observed in the 1990s, when the study took place. Results also showed large woody debris (LWD) quickly dropped with increasing land use intensity (meaning the density of roads and buildings) in the catchment. The correlation of LWD with riparian buffer forestation was weak, however, likely because of the time lag between reforestation and new inputs of LWD as trees die of old age.

The Chesapeake Bay watershed covers a large area and includes around 18 million people, making it an important target for environmental protection. In comparison, my entire state of North Carolina has just 10 million people. Given its environmental importance, a large group of academic and governmental researchers collaborated on a Chesapeake Bay watershed review,[1] prompted by an agreement among Maryland, Virginia, and Pennsylvania to reduce nutrient loadings to the Chesapeake Bay by 40% by 2000. The multi-state agreement requires a control on nonpoint sources, and those controls require riparian buffers to control 50–90% of the nitrogen coming from agricultural runoff.

How well do riparian buffers take up nitrogen? The conceptual model here is that microbial processes remove oxygen from nitrate-rich runoff coming from fields as the water moves from the edge of the buffer toward the stream. The lack of oxygen leads to anaerobic conditions closer toward the stream, thereby removing the nitrate via denitrification.

Figure 12.9 shows two collections of data on nitrate removal through riparian buffers in a broad collection of situations and demonstrates the importance of buffer width to assure nitrate removal.[2] Nitrogen uptake by woody vegetation in riparian buffers can be as high as 77–84 kg/ha/year, while phosphorus uptake reaches 7.5–10 kg/ha/year. Reflecting their greater importance, lower-order streams have higher potential for interactions between vegetation in riparian buffers and the nutrient-laden water. Another set of studies demonstrated that riparian buffers removed 80% of sediments, meaning riparian buffers provide control of sediments and nitrates (up to 90% removal). However, the results for phosphorus are less compelling.

An interesting thing happens in the riparian buffers adjacent to tile-drained agricultural fields northeast of London, Ontario, Canada.[3] Using a line of monitoring wells to take groundwater samples, researchers found that the riparian areas were areas of natural drainage and groundwater recharge. That recharge produced a downward plume of water sinking below the riparian soils into yet deeper groundwater. Perhaps visualize a container of evaporating liquid nitrogen sitting on a stool, with the vapor plume streaming down to the floor—a

---

[1] Lowrance et al. (1997). [2] Lowrance et al. (1997) and Van Appledorn (2009). [3] Cey et al. (1999) examined riparian zones of southern Ontario agricultural areas.

Riparian buffers remove nitrogen.

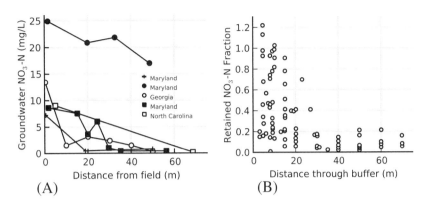

Figure 12.9: (A) Groundwater nitrogen decreases as it flows toward streams from fields (Lowrance et al. 1997). Data come from five different studies within the Chesapeake Bay watershed. (B) A compendium of seven studies (including three from A) confirms the importance of wide riparian buffers (>30 m) in nitrate removal (Van Appledorn 2009).

similar plume of recharging water falls through the riparian area into the groundwater. In the Ontario study, the recharge water flowed down through the soil and took with it the nitrogen-rich water draining from the agricultural fields. Not only did the nutrient-rich runoff not enter the stream, but the nitrate concentration and DO declined with depth, indicating that denitrification took place in the deeper groundwater. As a tracer chemical, chloride concentration in the plume was constant, showing that dilution didn't obscure denitrification. In this case the riparian zone acts as a barrier to nitrate-rich runoff, diverting it from entering streams and, rather than directly acting as a denitrification zone, pushed it into a denitrifying, hyporheic zone where surface water and groundwater mix. All in all, this study shows that one has to be careful of the two-dimensional conceptual model of lateral flows entering riparian zones, carrying nitrate to be denitrified. Groundwater, like the atmosphere, exists in three dimensions, and one must be sure what processes take place where.

Now come back to consider the 30-m scale of riparian buffers needed for denitrification, reflected in Figure 12.9. Urban buffers are too small. For example, the city of Durham, North Carolina, requires only 50-foot (15-m) stream buffers, losing out on significant nitrogen removal possibilities.

Consider another regulation of wide stream buffers designed to protect water quality: septic tank placement near streams. A study from Melbourne,

Wet, organic-rich riparian buffers remove nitrogen.

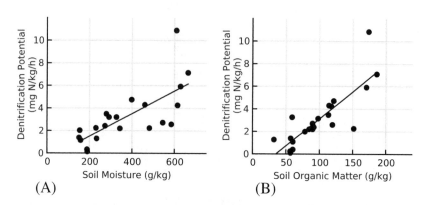

Figure 12.10: Two important factors that promote denitrification are soil moisture (A) and organic matter (B). The denitrification potential measured in soil samples taken from riparian buffers of Baltimore, MD, streams increases markedly with both factors (Groffman and Crawford 2003).

Australia, showed that *E. coli* from septic tanks were completely removed after just tens of meters of overland flow, but once they reached the streams, these pathogens took kilometers to clear.[1] The conclusion was that wide buffers help, and no septic tanks should be allowed within, say, 100 m (300 feet) of a stream.

Recall from the discussion in Chapter 6 that denitrification involves the transformation of nitrate (NO_3) into nitrogen gas (N_2). In actuality, the process involves several steps with intermediate forms of nitrogen. Accounting for these different forms, we define the ability of a soil sample to convert reactive nitrogen to NO, N_2O, or N_2 as its "denitrification potential."[2] Figure 12.10 shows how this potential varies with soil moisture and organic matter. The study involved eight Baltimore, Maryland, streams—four urban and four rural—with either herbaceous or forested riparian buffers. One of the herbaceous sites, however, was a 50-m-wide power line cut through a forested buffer instead of a long stretch of natural meadow. Nitrate concentrations ranged up to 4 mg/L, comparable to the agricultural levels shown, for example, in Figure 6.6.

---

[1] Walsh and Kunapo (2009). [2] Groffman and Crawford (2003).

Soil samples used for their measurements were taken within 5 m of the stream bank to test how much denitrification takes place in those soils, and the experiment measures denitrification enzyme activity (DEA) by measuring the amount of N_2O gas coming off these specially prepared soil samples incubated under anaerobic conditions.

The greatest activity was found in reeds and other wetland vegetation growing around a storm drain. The study authors argue that, although denitrification in groundwater and subsurface flows is often more important, surface flows have importance in infiltration-free urban streams. They also concluded that urbanization doesn't necessarily lead to a loss of denitrification function in riparian buffers and that there was no difference in denitrification potential between forested and herbaceous sites. What matters is keeping these areas wet and full of organic matter. In short, maintaining base flows in healthy riparian buffers rich in organic matter preserves the ecosystem service of denitrification.

All of the ingredients, nitrate, water, and organic matter, as well as a lack of oxygen, need to come together for denitrification. Sometimes these denitrifying interfaces are narrow strips as small as 1 m wide, occurring between nitrogen-carrying deep groundwater and oxygen-poor shallow groundwater that meet in riparian areas.[1] These small regions can also result from NO_3-rich groundwater flowing into soils with high organic content, typical of riparian zones, but also marked by wetlands and peaty areas. Finally, denitrification can also take place where surface water infiltrates oxygen-poor stream beds.

Another description is that denitrification depends on oxygen, carbon, and nitrogen, or in yet another way, it depends on flowing water that brings oxygen and nitrogen to microbes inhabiting organic-rich soils. A critical part of the picture is the need for water flow in water-saturated soils to bring in oxygen, which microbes use to metabolize organic matter and thereby create anaerobic conditions appropriate for denitrification of cotransported nitrates.

Authors of the Baltimore, Maryland, study shown in Figure 12.11 compared denitrification in degraded, restored, and natural, reference streams. The study showed that denitrification really takes place in the top 10 cm of their riparian soils, in contrast to the discussion of Figure 12.9. In this situation the restored streams raise the water table, and it's at that new interface where denitrification takes place. In Figure 12.11, everything decreased with depth, as shown in Figure 12.12, and root biomass bringing organic matter into the soil profile is the link between organic matter and denitrification.

All these results confirm the developing picture that stormwater erodes streams (see Figure 5.14A); stream channeling lowers the water table, dries out the organic riparian soils, and reduces the anaerobic process of denitrification

---

[1] McClain et al. (2003).

Good riparian buffers have roots.

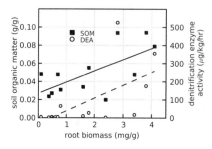

Figure 12.11: Within the riparian buffers of Baltimore, MD, tree roots increase soil organic matter (SOM) and the denitrification potential as measured by microbial denitrification enzyme activity (DEA) (Gift et al. 2010).

Good riparian buffers have deep roots.

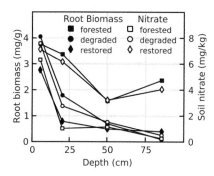

Figure 12.12: A comparison of forested, degraded, and restored riparian buffers in Baltimore, MD, shows that only forested buffers have deep roots, and only restored buffers have high nitrate levels throughout the soil profile (Gift et al. 2010). Restored buffers take time to increase the needed denitrification-promoting root biomass levels.

that turns nitrate into atmospheric nitrogen gas.[1] Stream restoration projects attempt to correct these conditions.

Riparian zones of natural, forested streams that serve as reference sites for comparing with restored streams were wet with good soils but had little nitrogen because these streams drained natural watersheds, not degraded ones. The obvious differences among degraded, restored, and forested streams in Figure 12.12 are that root biomass drops sharply between 20 and 50 cm in degraded and urban streams but stays high in forested streams from years of root growth in intact riparian buffers. Indeed, even though the restored sites were 10 years old, they still had low root biomass compared with the reference sites. Clearly, true restoration takes time.

---

[1] Gift et al. (2010) studied denitrification in urban riparian areas.

However, another measure could be misleading. These natural streams have long been nitrogen limited, which means is that they have low denitrification potential because they haven't been as primed with denitrifying microbes. Thus, despite higher moisture levels and higher organic matter, the reference streams had low nitrate inputs and showed very little difference in denitrification compared with degraded and restored streams.

If stream restoration is going to work in nutrient-rich urban streams, it needs to raise the water table in the riparian zone and bring organic matter into the restored riparian areas. In degraded streams the groundwater contains the nitrogen while the riparian soils hold some organic matter, but the water table is lowered. Restoration then requires a modification of stream hydrology to keep the riparian areas wet. To keep the stream restored, the problems that led to the initial degradation also need to be addressed. In restored sites biomass availability can come from filling deep trenches with sawdust, a so-called high-carbon hot spot,[1] through which nitrate-rich water must flow. Over the long term, stable riparian buffers must promote the vegetation that leads to deep-rooted vegetation that increases soil organic biomass, which we've already seen increases denitrification, as well as increasing the depth—closer to the water table—where denitrification takes place.

Figure 12.9 showed that wide riparian buffers decrease nitrate levels through denitrification in both deep and shallow soils (see also Figure 12.11). A USEPA review of 40 studies of riparian buffers examined the role of width for a variety of buffer conditions, including forested and grassland, as well as those surrounding wetlands.[2] This review separated nitrate removal studies into those examining surface flows, for example, the runoff flowing across the surface of a grassy filter strip, and those examining subsurface flows of groundwater where the nitrate reaches denitrification zones. These results are displayed in Figure 12.13.

Dividing the studies in this way showed that buffer width was less important than making sure the buffers pushed nitrate-rich water into subsurface flows. Considering surface flows, buffer width explained 30% of nitrate removal. This importance makes sense because these are flows through grassy strips designed to remove sediments, organic matter, and such. In this situation it's completely reasonable that wider is better. In the case of subsurface flows, buffer width didn't affect nitrate removal at all, but removal effectiveness was nearly 90% compared with just 33% for surface flows. Infiltration of nitrate-bearing stormwater near streams is very important for removal of nitrate.

The general point of Figures 12.9 and 12.13 is that narrow buffers display great variability in nitrogen removal, so much so that in some cases narrow buffers actually contribute to the nitrogen-loading problem. On the other hand, buffers wider than 50 m more consistently removed high amounts of nitrogen

---

[1] Gift et al. (2010). [2] Mayer et al. (2005).

Wide riparian buffers best remove nitrogen.

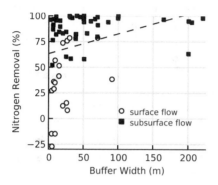

Figure 12.13: Wider buffers more consistently remove nitrogen (Mayer et al. 2005). However, this compendium of 40 worldwide studies also shows that subsurface flows through riparian buffers are critically important to nitrogen removal. The dashed line presents the linear regression to both sets.

from the riparian zone, with the greatest percentage of nitrogen removed by wide buffers with low flows. Situations exist where 85% of nitrates are removed within an initial 5 m of a buffer, but in other situations a 31-m buffer removed only 40%. Grassy strips also help, with examples of 7-m-wide strips removing 62% of nitrates but other examples of similar widths being ineffective.

Nitrogen removal effectiveness also decreased as nitrogen load increased, meaning the buffer can only do so much, and adding more pollutants reduces its percentage removal. In this case of high nitrogen loading, wider buffers were slightly more effective than narrow buffers. The study also notes high variability in nitrogen reductions due to seasonality and conditions.

In part because of different infiltration rates, different types of vegetative cover also led to different levels of nitrogen removal effectiveness. In order of most effective buffer cover are forest (90% removal), forested wetland (85%), grass/forest (81%), grass (22%), and wetland (7%).

As shown in Figure 12.7, low-flow headwater streams take up nitrogen quicker than higher flowing streams. A striking example of the importance of a low-flow stream is that for a 10th-order stream like the Mississippi River, 90% of its cumulative length consists of ephemeral, first-, and second-order streams. Stream buffers on these small streams, most often destroyed by urbanization (see Figure 10.2), demand protection because that's where most of the nutrient processing takes place.

The USEPA study recommended protecting buffers against soil compaction from all sources, such as vehicles, livestock, and impervious surfaces, because the subsurface flows require good infiltration. Some of Durham's riparian buffers, for example, serve multiple purposes, including paved trails and sewer/ power line corridors, both of which require periodic truck traffic for maintenance. Since vehicles compact soils (see Figure 4.2), these uses are in direct conflict with this recommendation. The study also recommends not removing vegetation and leaf litter, or even thinning trees, from riparian buffers so as to enhance the input of vital organic matter. Finally, buffers must be protected from the urban development that promotes the urban stream syndrome of channelization and erosion that drains and dries out riparian areas.

In stark contrast to the results demonstrating the importance of wide riparian buffers, in a non-peer-reviewed article written by industry consultants in the November–December 2009 issue of *Stormwater* magazine,[1] the authors conclude that "for streambank stability, temperature control, minimizing degradation from direct impacts, and pollutant removal capacities, substantial benefits are achieved within the first 50 feet of vegetated buffer width. Marginal increases in benefits may accrue when buffer widths are increased beyond 50 feet." Fifty feet is just 15 m, well below the more than 50 m (150 feet) recommended by the USEPA authors. One quote from this *Stormwater* article[2] was especially appreciated by a homebuilders website, which stated that "well-planned developments incorporating 'averaged' vegetated buffers of 50 feet or less, combined with shared BMP [best management practice] treatment trains, may be more protective of riparian and wetland ecosystem values than the much larger buffers required by some regional and local regulations"[3]—a very different conclusion than that of the scientific literature.

These discussions bring up the idea that riparian buffers have many ecosystem functions that involve some of the more structural stream variables that change with riparian buffer deforestation.[4] One such function is filtering sediments from runoff before it reaches streams, the consequences of which are shown in Figure 12.14. This particular measured variable involves the coarse particle fraction of the stream bottom substrate, which becomes finer with increased deforestation. As shown in Figure 10.6, fine sediments inhibit stream–groundwater interchange, an exchange that's important for denitrification.

Despite the fact that suspended solids don't require very wide buffers for removal—for example, a mere 10-m (30-foot) buffer yields 80% or more reduction (though steep slopes require wider buffers)—stormwater control is more than just suspended sediments.

Stream buffers reduce stream temperatures, too, as Figure 12.15 demonstrates. The stream temperature results of Figure 12.15B come from the same study as Figure 11.4 that used thousands of stream segments for a GIS analysis but did careful examination of 18 segments to check against aerial photos.[5]

---

[1] Rupprecht et al. (2009). [2] Rupprecht et al. (2009). [3] hbact.org screenshot, January 22, 2013. [4] Scott et al. (2002). [5] Jones et al. (2006).

Intact buffers remove fine sediments.

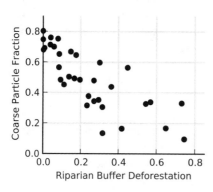

Figure 12.14: Deforestation within 100-m riparian buffers affects the substrate of streams in western North Carolina (Scott et al. 2002). A high fraction of coarse particles (>6.4 cm, or 3 inches) means that streams have bottoms covered by gravel, providing an ideal aquatic insect habitat.

Forested riparian buffers are cool.

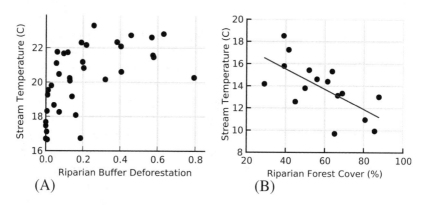

Figure 12.15: (A) Streamside vegetation in forested riparian buffers critically affects stream temperatures in western North Carolina (Scott et al. 2002). (B) Deforestation in riparian buffers increases stream temperatures (maximum 7-day average) by as much as 4–6°C (Jones et al. 2006). These temperature differences can determine the presence or absence of trout.

This plot uses their secondary sites and shows that stream temperature depends on riparian forest cover, with deforestation being responsible for as much as a 5–6°C temperature increase. This change is huge for temperature-sensitive trout (see Figure 9.9), meaning the difference between presence or absence of trout. The study authors also demonstrated that stream temperature does not depend on watershed-scale forest cover, just this riparian-scale cover. Overall, narrow riparian buffers of 50 feet, from a water quality perspective, are "risky."[1]

Forested riparian buffers yield the cooler stream temperatures that support trout, and the question becomes, how much do we like trout and healthy streams? One measure is economic, and a study more than a decade old found that the six federal hatcheries in the Southeast region, located in Arkansas, Tennessee, Kentucky, and Georgia, cost $2.1 million annually but generated $107 million in direct returns and $212 million in indirect returns and supports 2,800 jobs. Rather astonishing, additional estimates show that, for every federal budget dollar spent, trout fishing yields $109 to $141 of economic return, but this industry needs healthy streams to support the trout.[2]

The importance of riparian buffers discussed with the last few figures brings up the idea of stacked ecosystem services, in which a riparian buffer affects many issues, including stream temperatures, denitrification, suspended solids removal, recreational fishing, wildlife habitat, stream stabilization, flood control, and so on. Each of these issues have independent benefits, but they are simultaneously lost, at least to some extent, when riparian buffers stop functioning. Comparing one benefit alone against total costs in a cost–benefit calculation is shortsighted when many components are important.

Once natural riparian buffers are lost, does stream restoration work to restore the buffers' multifunction role? One study from Maryland used agricultural streams with preexisting buffers ranging from 10% to 48% intact forest. On top of this amount was added 1–30% more forest in stream restoration projects, giving a total of 1,123 hectares of riparian forest alongside 1,102 km of stream. That restoration gave a final riparian forestation range of 12–61%.[3] The restoration projects were performed between 1998 and 2005, but no effect on nutrient levels had been observed through 2006. Researchers suspect that the reasons simply reflect a lack of time for a response, a dominance of nutrients by agriculture, or insufficient restoration. Given some of the decades-long time delays in nutrient transport through groundwater (see Figure 10.10), that lack of an observed effect may not be terribly surprising.

One 2005 study that considered more than 37,000 stream restoration projects estimated that more than $1 billion per year has been spent on restoration projects. Most common goals for restoration, in order of decreasing frequency, are water quality enhancement, riparian zone improvement, stream habitat improvement, fish movement, and stream bank stabilization. Only 10% of the projects examined whether or not improvements actually happened after the restorations.[4] Another study summarized the results of 78 restoration

---

[1] Wenger (1999) is an earlier paper than the USEPA paper defending wide buffers and provides excellent descriptions of riparian buffer functions. [2] U.S. Fish and Wildlife Service (2001). [3] Sutton et al. (2010). [4] Bernhardt et al. (2005).

Lakes and estuaries remove nutrients.

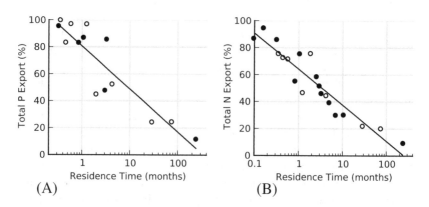

Figure 12.16: Export of phosphorus (A) and nitrogen (B) from estuaries (filled circles) and lakes (open circles) decreases with an increase in the time those nutrients spend within the water body (after Nixon et al. 1996). Regression lines include both lakes and estuaries.

projects presented in 18 other studies that concentrated on restoring the physical degradation of urban and agricultural streams by adding habitat structure.[1] These changes included grading the banks, adding meanders to the stream, adding LWD and rocks, carving out pools, and so on, with heavy machinery. Sadly, only *two* of the 78 projects found an increase in the biodiversity of stream invertebrates. The researchers noted that perhaps not enough time had passed to see an effect, even though it had been a decade for many studies. Given the cost and apparent lack of results, they recommended that watershed-level approaches to stormwater be emphasized instead of expensive stream restoration projects.

As we move from streams, rivers, and lakes, we know that surface waters drain to oceans. Some rivers discharge directly to continental shelves. Estimates are that nearly 1 billion kg of nitrogen is buried annually in deltas, and an equal amount on the continental slopes.[2] These amounts represent 29% of the nitrogen discharge from the five major rivers to the North Atlantic, the Amazon, Orinoco, Mississippi, Magdelena, and Grijalva-Usumacinta, which together account for about 45% of nitrogen exported.

The rest of the nitrogen exported to oceans moves through estuaries. Off the coast of North Carolina, for example, the Tar and Neuse Rivers empty out into the Pamlico Sound, an estuary where a large volume of brackish water

---

[1] Palmer et al. (2010). [2] Nixon et al. (1996).

Denitrification takes time in estuaries.

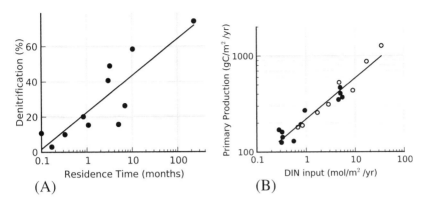

Figure 12.17: (A) Denitrification taking place in estuaries increases with water residence time. Larger water bodies fed by smaller streams have longer water residence times. (B) An increasing amount of dissolved inorganic nitrogen (DIN) leads to greater primary productivity. A mole of nitrogen has a mass of 14 g. Data are from Nixon et al. (1996).

has a limited connection to the ocean. The Pamlico Sound joins the Albemarle Sound in a large region separated from the Atlantic Ocean by the Outer Banks. The amount of nitrogen that leaves an estuary, after having been transported there by rivers or air, depends in large part on how long water stays in the estuary before exiting into the ocean. That time depends on the estuary's water volume and flow characteristics, estimated much like the residence time of water in the atmosphere (see Chapter 2).

Rough estimates exist for the nutrient cycling that takes place in large water bodies, both freshwater and saltwater, around the North Atlantic from Florida to Denmark. As we see in Figure 12.16A, phosphorus export from estuaries and lakes decreases with increasing residence time in the water body, while the loss of phosphorus occurs as burial in sediments.

Nitrogen export also shows a decrease with increasing residence time, and as Figure 12.16B shows, after roughly a year only about one-third of the nitrogen remains for export. Denitrification causes the loss. The longer nitrogen spends in a water body, the more that's denitrified and released as N_2 into the atmosphere, shown in Figure 12.17A.[1] Overall, estuaries remove 30–65% of nitrogen, slightly less of the phosphorus, and 90–95% of the sediments transported by rivers that would otherwise be transported out onto the shelf.

---

[1] Nixon et al. (1996).

River-transported nitrogen exported directly to oceans also experiences denitrification. For these areas off the coasts, nitrogen export by rivers exceeds atmospheric deposition by 3.5–4.7 times, though on the U.S. Atlantic shelf atmospheric deposition is higher. In fact, denitrification taking place on the continental shelves of North America involves an amount of nitrogen that equals about 1.4–2.3 times the amount of nitrogen exported there by rivers and the atmosphere—the rest comes from nutrients carried onto the shelves by water coming from the deep ocean.

Increased nitrogen inputs also increase primary productivity in estuaries, as shown in Figure 12.17B. These estuaries take in nutrients, and, as shown in Figure 6.19, the longer the nutrients remain in the estuary, the longer photosynthesizing organisms have to take it up and produce more biomass. Averaged across the oceanic shelves of the North Atlantic, primary productivity yields 165 gC/m^2/year. Then, like farmers pulling crops off of fields, commercial ocean fisheries take an amount of nitrogen with their catch that amounts to about 5% of the flux from terrestrial and atmospheric sources.

As mentioned above, most phosphorus gets buried in sediments. Sedimentation rates on the continental shelves run about 25 cm per 1,000 years, or roughly an inch per century. However, as reported earlier in Figure 5.20, some areas off the southeastern U.S. experience deposition rates of nearly 1 cm/year, or 39 inches per century. The disparity arises because even with large amounts of sediment coming down some rivers (see Figure 5.18), such rivers are widely separated on the coast and there's a lot of ocean bottom.

Chapter Readings

Groffman, P.M., and M.K. Crawford. 2003. Denitrification potential in urban riparian zones. *J. Environ. Quali.* 32: 1144–1149.

Jones, K.L., G.C. Poole, J.L. Meyer, W. Bumback, and E.A. Kramer. 2006. Quantifying expected ecological response to natural resource legislation: A case study of riparian buffers, aquatic habitat, and trout populations. *Ecol. Soc.* 11: 15.

Lowrance, R., L.S. Altier, J.D. Newbold, R.R. Schnabel, P.M. Groffman, J.M. Denver, D.L. Correll, J.W. Gilliam, J.L. Robinson, R.B. Brinsfield, K.W. Staver, W. Lucas, and A.H. Todd. 1997. Water quality functions of riparian forest buffers in Chesapeake Bay watersheds. *Environ. Manag.* 21: 687–712.

Mayer, P.M., S.K. Reynolds, T.J. Canfield, and M.D. McCutchen. 2005. Riparian buffer width, vegetative cover, and nitrogen removal effectiveness: A review of current science and regulations. EPA/600/R-05/118. USEPA (Cincinnati, OH).

Meyer, J.L., M.J. Paul, and W.K. Taulbee. 2005. Stream ecosystem function in urbanizing landscapes. *J. N. Am. Benthol. Soc.* 24: 602–612.

Moore, A.A., and M.A. Palmer. 2005. Invertebrate biodiversity in agricultural and urban headwater streams: Implications for conservation and management. *Ecol. Appl.* 15: 1169–1177.

Peterson, B.J., W.M. Wollheim, P.J. Mulholland, J.R. Webster, J.L. Meyer, J.L. Tank, E. Marti, W.B. Bowden, H.M. Valett, A.E. Hershey, W.H. McDowell, W.K. Dodds, S.K. Hamilton, S. Gregory, and D.D. Morrall. 2001. Control of nitrogen export from watersheds by headwater streams. *Science* 292: 86–90.

Sutton, A.J., T.R. Fisher, and A.B. Gustafson. 2010. Effects of restored stream buffers on water quality in non-tidal streams in the Choptank River basin. *Water Air Soil Pollut.* 208: 101–118.

Sweeney, B.W., T.L. Bott, J.K. Jackson, L.A. Kaplan, J.D. Newbold, L.J. Standley, W.C. Hession, and R.J. Horwitz. 2004. Riparian deforestation, stream narrowing, and loss of stream ecosystem services. *Proc. Natl. Acad. Sci. USA* 101: 14132–14137.

Chapter 13

Control Measures

Stormwater control measures (SCMs) must have different objectives dependent on the amount of rainfall. SCMs often go by the term BMP, which stands for "best management practice." A hurricane stalled out over the eastern U.S. can dump a dozen inches of rain in a day or two—there can be no hope for SCMs that rely on groundwater infiltration to take care of the flooding.

SCMs must involve both privately held property, holding the most impervious surfaces, and publicly held property, holding 80% of the connected imperviousness. Streets and driveways account for more than half of all impervious surfaces but more than 80% of connected imperviousness. Buildings make up another 40% of the imperviousness but only a very small fraction of connected imperviousness.

Prevention is better than remediation. Little dolphins stenciled onto stormwater drains typify nonstructural SCMs. Other nonstructural approaches include riparian buffer preservation, SCM maintenance, road and ditch maintenance, sediment removal, vegetation maintenance, and litter removal. Street sweeping and road salt reductions both intercept pollutants before they enter runoff.

Another broad class of SCMs includes low-impact development, which prevents precipitation from become runoff in the first place. This approach requires clever site design features ranging from rain barrels to porous pavement, street construction with an eye to stormwater concerns, and careful construction practices.

Once runoff picks up pollutants, pollution removal and flow control involve infiltration, filtration, detention, and retention of stormwater runoff, making use of chemical, biological, and physical processes, using structures that include constructed wetlands, biofilters and bioretention areas, and ponds.

Publicly owned impervious surfaces are connected.

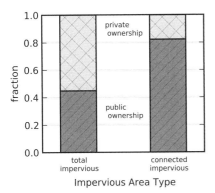

Figure 13.1: Data from a Boulder, CO, study shows that just over half of all impervious surfaces sit on private land, but the connected fraction is overwhelmingly on public land (Lee and Heaney 2003). Surfaces like roads and parking lots contribute the highest proportion (see Figure 13.2).

We all own the stormwater problems of connected imperviousness. Figure 13.1 comes from a study of only a dozen acres or so of one city, Boulder, Colorado, but what it lacks in size it makes up for with an exhaustive study of those parcels.[1]

Various studies discussed throughout this book make the argument that directly connected impervious area causes great harm, for example, to stream biodiversity (see Figure 11.17). This study demonstrated that the measurement of connected imperviousness is complicated, and getting to Figure 13.1 involved the following steps: First, get aerial photo data from city hall, which may or may not have additional levels of analysis. Next, identify all impervious surfaces—streets, roofs, sidewalks, and roads. Next, pull out all isolated impervious surfaces on the reasonable assumption that there's no piping of small impervious areas. Next, have people get out of doors and perform field investigations of the drainage conditions of roads. Next, have more people perform field investigations of the entire area's remaining impervious surfaces, including driveways and homes, focusing on such details as whether or not downspouts drain water to stormwater systems.

As the level of analysis progressed, the various classifications changed dramatically. These researchers tallied the effort required at each stage, and the complete, careful analysis of these 5.8 hectares took 195 hours, much of it being the field analysis. A primary conclusion is that a curb and gutter analysis is critically important for determining connection status: are roads bordered by curbs? If so, it's connected, but if not, then water runs off into the ditch. Aerial photos don't readily distinguish the difference.

---

[1] Lee and Heaney (2003).

Streets dominate connected impervious surfaces.

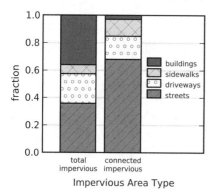

Figure 13.2: Additional data from the Boulder, CO, study of Figure 13.1 shows the relative amounts of different types of impervious surfaces and their partitioning into total and connected areas (Lee and Heaney 2003).

This site provided a remarkably sobering example. Before any field investigations of the area, with a total imperviousness of 35.9%, roads were estimated to contribute 7,478 m^2 of directly connected impervious area. However, after going out to look for curbs, researchers concluded that roads contributed just 5,152 m^2. Similarly, directly connected buildings dropped from 5,022 m^2 to 215 m^2, contributing only 2.8% of directly connected imperviousness after the final analysis! Clearly, connection in a photo doesn't mean connection on the ground.

Thus, there's an implication of tremendous uncertainty in the reported levels of directly connected imperviousness cited in studies that don't carry out such a careful classification. This caution doesn't impugn other studies; rather, it warrants an appreciation of the difficulties. Of course, the general message remains that imperviousness, and directly connected imperviousness in particular, is clearly a problem.

Continuing on with this careful analysis of imperviousness, Figure 13.2 details both total and directly connected impervious area by origin.[1] We've already seen that for small storms a very high proportion of the runoff comes from connected imperviousness (see Figure 2.7), meaning nearly all of that runoff comes from directly connected streets, whereas buildings and privately held property likely infiltrate their share of runoff.

These estimates are helpful in guiding how to ameliorate the flashy runoff from impervious surfaces. For example, estimating peak discharge through the "rational method," originating in the mid-1800s,[2] involves the product $C \times i \times A$, where C is the runoff coefficient, i is the rainfall intensity (estimated

[1] Lee and Heaney (2003). [2] Lee and Heaney (2003).

Low-impact development reduces imperviousness.

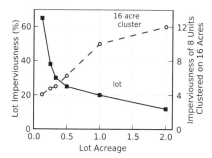

Figure 13.3: Clustering houses within a larger parcel reduces total imperviousness. Smaller lots increase each lot's imperviousness, but fewer roads lead to less overall imperviousness (after Arnold and Gibbons 1996; Richards 2006). This example considers eight houses within a 16-acre subdivision.

from the time taken for runoff to flow from the farthest reaches of the stormwater system and the frequency of rainfalls in a region[1]), and A is the drainage area. As this study shows, estimating the parameters requires calculating directly connected impervious area. Another expression determined from data, $D = 0.15T^{1.41}$, gives the percentage of directly connected impervious area, D, depending on the amount of total impervious area, T.[2]

Low-impact development (LID) seems like an inherent contradiction, but the term summarizes a broad set of strategies that mitigate stormwater problems at the source: using open space, maximizing infiltration, reducing impervious surfaces, and using rain gardens to build stormwater-sensitive developments.[3]

Clustering houses within a development is one such LID strategy.[4] Using some oft-cited numbers produced by the Soil Conservation branch of the USDA[5] for imperviousness per lot for house lots of various lot sizes, Figure 13.3 shows that clustering houses within a larger parcel greatly reduces overall imperviousness.[6] The emphasis is on a choice between low-density development throughout the watershed or a concentrated high-density development in one spot with the rest of the watershed protected from development. Importantly, it is *not* high density development *everywhere*!

Here's the scenario: Suppose a developer can place eight houses on a 16-acre parcel, with roads going to each house. One approach divides the parcel into eight 2-acre lots with a house on each. At this lot acreage, the imperviousness per lot is just 12%, but there's a lot of roads, which means the whole

[1] A nice description can be found at water.me.vccs.edu/courses/CIV246/lesson11_3.htm.   [2] From Lee and Heaney (2003), but the expression itself is attributed to Alley and Veenhuis (1983), cited within.   [3] A USEPA publication, Richards (2006), discusses LID in lay terms.   [4] Jacob and Lopez (2009).   [5] USDA (1986: Table 2-2a).   [6] Arnold and Gibbons (1996).

16-acre cluster has a maximum imperviousness of 12%. Another development approach builds eight townhouses on 1 acre, leaving the other 15 acres as undeveloped open space. In this case impervious surfaces account for 65% of the developed acre, but the entire 16 acre parcel is just slightly more than 4% impervious.

In addition, that open space can be used for surface infiltration through control measures such as grassy swales, which I think of as a roadside ditch. Other approaches include rain gardens that infiltrate water below the soil surface in a trench or open-bottomed tank, called subsurface infiltration, that goes nearly direct to groundwater. These SCMs redirect surface flow to subsurface flow, where the main problem can be groundwater contamination, especially when there is deep infiltration. Overall, that's low impact.

Here's a second way to envision the benefits of clustering, according to the example of the USDA report.[1] Suppose a 10,000-acre watershed will be developed. Consider a sprawling approach that builds at either one, four, or eight houses per acre for a total of 10,000, 40,000, and 80,000 houses. Runoff volumes *per house* at these different densities are 18,700, 6,200, and 4,950 cubic feet (cfu) per year, while the entire watershed's runoff more than doubles and the impervious surface more than triples across these densities.

Now consider the effect of clustering, with 10,000 houses built on 10,000, 2,500, and 1,250 acres at respective cluster densities of one, four, and eight houses per acre, with the remaining land "preserving critical ecological and buffer areas." Runoff volumes per house again equal 18,700, 6,200, and 4,950 cfu/year, but from the low to high cluster densities the entire watershed's runoff decreases from 187 million to 49.5 million cfu/year and from 20% to 8.1% imperviousness for the same number of houses.

However, years after constructing an LID community, the open space set aside by this clustering might start to look awfully enticing to development, from the perspective of both elected officials and housing developers. Tough legal protections must be put in place to protect that land from development; otherwise, the watershed suffers from the overall high imperviousness of high-density development everywhere, and that high density has bad stormwater consequences.

Why does clustering reduce runoff so well? Figure 13.4A shows why, from the subdivision scale, through a study done in Waterford, Connecticut.[2] The lines compare two traditional developments built in 1988 and 2003 in separate small watersheds (2 and 5.5 hectares) with an LID of 1.7 hectares built in 2002. The numbers of lots were 12, 43, and 17, respectively. Downspouts from the houses in the traditional developments eject their runoff onto single-lot driveways that connect to roads with curbs and drains. In contrast, each house in the LID has a rain garden for roof and lot runoff, shared driveways,

---

[1] USDA (1986). [2] Hood et al. (2007) compared an LID with two standard developments.

Low-impact development reduces runoff.

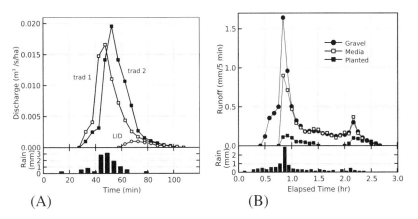

Figure 13.4: (A) Two traditional developments in Waterford, CT, have much greater runoff discharge than a similarly sized LID under identical rainfalls (Hood et al. 2007). (B) At a much smaller scale, a roofs having media plus plants have much less runoff than flat roofs with gravel or planting media, which show a similar runoff reduction (VanWoert et al. 2005).

and a bioretention area in the cul-de-sac. The roads were built with concrete pavers with runoff entering infiltration trenches. LID imperviousness was 22%, with 26% open space, while the two traditional developments had 29% and 32% imperviousness, and neither had any open space. As the curves in Figure 13.4A show, the traditional development's peak discharge was around 11 times greater than the peak discharge from the LID. Also notice the roughly 20-min lag time of the clustered development's peak flow compared with the peak flow from the traditional developments. LID slows water down and lets it infiltrate—an impressive stormwater volume reduction.

LID is a part of reversing the mindset of getting rid of urban water as quickly as possible. Several infiltration devices were used in a retrofit of Tokyo, where trenches along roads and in parks reduced peak flows by 60% and volume by 50%, at one-third the cost of other detention approaches.[1] These successes also helped Tokyo's combined stormwater and sewer systems by reducing overflows from 36 to 7 per year.

The same phenomenon of slowing down water and reducing runoff can be greatly scaled down, as shown in Figure 13.4B. These curves come from tiny

---

[1] Pitt (2005).

Low-impact development retains stormwater.

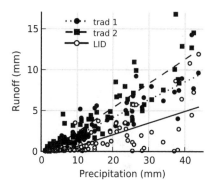

Figure 13.5: Runoff fractions for traditionally built neighborhoods in Waterford, CT, greatly exceed those of LIDs (Hood et al. 2007). These data represent many rainfall events, each of which reflects the curves of Figure 13.4A.

experimental roofs just 0.67×2.44 m (2×8 feet), giving results remarkably similar to the subdivision runoff results.[1] The three roofs were mimicking options for commercial roofs. Above a waterproof layer, one roof had 2 cm of gravel, representing a traditional commercial roof. The other two placed a drainage layer, filter fabric, and 1 inch (2.5 cm) of growing medium over the waterproof membrane. The green roof added a thick layer of *Sedum* species, placed as seed and allowed to germinate and grow for nearly 4 months before the 430-day-long experiments started. Stormwater retention for medium rains (2–6 mm) were 34%, 82%, and 83% for the gravel, media-only, and planted roofs, respectively. Higher rainfalls (>6 mm) produced stormwater retentions of 22%, 39%, and 52%, respectively. A green roof is much better at holding back water than a normal gravel roof, but a "media-only" roof does quite well, too. Plants, however, also provide an important media retention function.

Back at the several-hectare scale of Figure 13.4A comparing traditional and clustered neighborhoods, Figure 13.5 shows the rainfall–runoff curves for the same developments.[2] The clustered development produced no runoff for rains <6 mm, whereas the traditional neighborhoods had a threshold rainfall value of around 3 mm. Likewise, the LID lag time for small rains of short duration was triple that of traditional developments, which, when averaged over all of the neighborhoods in an entire watershed, greatly reduces downstream peak flows. However, for large rains of more than 1 inch or of long duration the lag time was not significantly different from that of traditional neighborhoods. Still, averaged over all rains, LID runoff fraction was 0.067, versus 0.19 and 0.24 for the two traditional developments. That reduction is a tremendous benefit to stream health!

---

[1] VanWoert et al. (2005) examined runoff from green roofs. [2] Hood et al. (2007).

In the U.K., LID is often called "sustainable urban drainage systems" (SUDS). In Australia it's called "water-sensitive urban design" (WSUD) and has a much broader conceptual foundation. Still other folks call LID by the term "sustainable urban water management" (SUWM).[1] The thinking extends to using water as a resource, even providing using stored stormwater for urban cooling. As an example, one study centered on a small town near Canberra, Australia, and examined several WSUD retrofitting scenarios using various modeling approaches.[2] The study's baseline was the existing traditional urban design, complete with garden and lawn irrigation using potable water. In the southern Australian climate, with 63 cm annual rainfall, that baseline irrigation led to summertime urban cooling estimates of 4.6°C compared with the surrounding desert. This irrigation-induced cooling contrasts with the urban heat islands (UHIs) reported for many cities.[3] Fully implementing WSUD principles further enhanced evapotranspiration and led to a cooling of 5.1°C and a 20% stormwater flow reduction, most of which came from the use of green roofs irrigated with stored stormwater, and the rest from swales and a constructed wetland. Terminating irrigation via potable water while retaining the WSUD components led to a cooling of 4.2°C.

Along these lines of cooling summer urban temperatures, a comparison of green roofs with white reflective and black nonreflective roofs in Austin, Texas, showed that green roofs reduced interior temperatures by as much as 18°C.[4] The additional temperature reduction from WSUD implementation implies an energy use reduction of about 1.5% in this modeled residential neighborhood.

Green roofs provide a wonderful opportunity to reduce stormwater volume, provide urban cooling, and reduce the associated summertime energy costs. A green roof is a straightforward concept, although it requires a careful construction process. Essential features include an impermeable drainage layer covered by a root and soil barrier, and above that soil and vegetation. Start by thinking of a normal roof as the impermeable layer—though it has to be a continuous barrier, not like a shingled roof through which roots could easily find a path and wreak costly havoc. The structure under the roof also needs the sturdiness to support the added weight of a water-saturated green roof. Given those conditions, green roofs can reduce runoff fractions down to 12% for rainfall events below 1 inch, but it certainly depends on features like media type, medium depth, and season.[5] Some evidence demonstrates that green roofs can also get better with age as the plants establish and increase organic matter in their soils. Though not universal, after 5 years one green

---

[1] Van de Meene et al. (2011). [2] Mitchell et al. (2008) modeled various levels of WSUD for Canberra, Australia. [3] Wilson (2011) discusses UHIs where city temperatures exceed nearby rural temperatures. [4] Simmons et al. (2008). [5] Berndtsson (2010) provides an excellent, detailed review of green roofs with citations of analyses of specific roofs.

Water retention of green roofs depends on thickness.

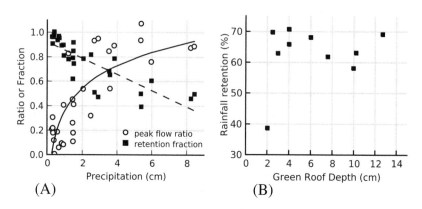

(A)   (B)

Figure 13.6: (A) The ratio of peak flow and water retention for a green roof compared with a normal roof indicates superior green roof performance, especially for small rainfalls (Carter and Rasmussen 2006). (B) A variety of green roofs show that a relatively thin, 1-inch (2.5-cm) green roof can retain about 60% of a rainfall's water (Dietz 2007).

roof increased its water retention capability from 17% to 67%, which means more water is available for evapotranspiration.[1]

Figure 13.6A presents results from a study comparing small sections (42 m²) of a normal and green flat roof on a building at the University of Georgia, in Athens.[2] The normal roof consisted of about 2 inches of gravel on top of a waterproof membrane, whereas the green roof had a complicated layering of root barrier, water-retaining layers, and growing medium, much like the scenario of Figure 13.4B. Water retention layers could store roughly 9 L/m² (a 9-mm water storage depth); the medium was 3 inches (7.5 cm) deep, and it was planted with various *Sedum* and *Delosperma* species. Compared with the gravel roof, the green roof demonstrated excellent retention and peak flow reductions across 31 rainfalls—including a couple of hurricane remnants—spread over 13 months.

A study from just south of Oxford, U.K., examined roof runoff from flat, moderate (22°), and steeply (50°) sloped roofs.[3] Conclusions include that low-sloped roofs get higher precipitation, runoff from 50° roofs is 40% greater than 22° roofs, and runoff from roofs facing north or east is 30% greater than

---

[1] Getter et al. (2007). [2] Carter and Rasmussen (2006). [3] Ragab et al. (2003).

that from south- and west-facing roofs. Finally, roofs facing the prevailing wind receive more precipitation but have higher evaporation, conditions that lead to less runoff.

Flat roofs that could be converted green are all around us. One report concluded that, of the 53.8% imperviousness of the 237-hectare Tanyard Branch watershed of Athens, Georgia, roofs alone cover 15.9% of the area and constitute 29.5% of the impervious surface.[1] The share of flat roofs is highest in commercial, downtown, and university areas, ranging from 25 to 85% of the roof area. The very large effort of turning all of the roofs green, and removing that area from the impervious surface coverage, would reduce total impervious area from 37% to 20%. Converting just the flat roofs would reduce total runoff from a half-inch rainfall by 40% and peak flows by 13%.

Figure 13.6B summarizes green roofs' rainfall retention from myriad locations and shows that a green roof needs to be only an inch or so deep to control most precipitation.[2] In part, the result may reflect that retention involves both storage and subsequent evapotranspiration. As a detention pond must empty before it can store more water, a green roof must either drain away its stored water or evapotranspire it away. Plants can transpire only a certain amount of water per day, and deeper media may act like a pond that can't drain before the next rainfall. Whatever the mechanism, these results strongly suggest that thin, 4-cm roofs are good, and that's good news in that thin green roofs certainly need less structural support than thicker ones. Of course, there are situations where thicker is better; for example, thin roofs (5 cm thick) in cold climates can result in more freezing damage to plants than thicker roofs (10–15 cm). That latter point leads to the caveat that one should use at least a 4-inch deep green roof in cold climates.[3]

Typical plants for green roofs include the *Sedum* species, a drought- and freeze-tolerant group of succulents. In a project testing their hardiness to establishment, a variety of medium depths were used with plantings in East Lansing, Michigan.[4] Figure 13.7 shows that moisture and absolute cover are enhanced by having media 7 cm deep or deeper, the depth needed for an optimally functioning green roof. For the shallower 4-cm depth, the two species that grew well were *S. sarmentosum* and *S. stefco*. Those two grew well at other depths, too, along with *S. floriferum*, *S. sexangulare*, and *S. spurium* John Creech.

A North Carolina study compared a flat green roof, 70 m² area with medium 7.5 cm deep, and a smaller 27 m² green roof with 10-cm depth set at a 3% pitch. Both roofs retained 64% of total rainfall, meaning that evapotranspiration was doing its job of reducing stormwater runoff. As a result, the roofs reduced

---

[1] Carter and Jackson (2007). [2] Dietz (2007) summarizes a several studies involving green roofs. [3] Boivin et al. (2001) examined freezing on green roof plants. [4] Getter and Rowe (2008, 2009) examined *Sedum* establishment in green roofs.

Happiness is a 7-cm green roof.

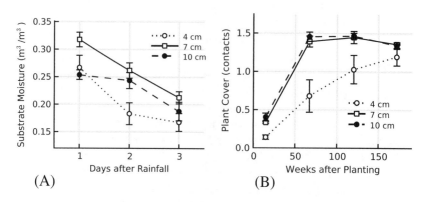

Figure 13.7: (A) A comparison of various green roof thicknesses shows that a 4-cm-thick roof dries out rather quickly compared with thicker roofs. (B) A 7-cm or thicker green roof provides better moisture levels in a Michigan installation, important when considering the establishment of *Sedum* seedlings (Getter and Rowe 2009).

the peak flow by 75% compared with non-green-roof controls. Unfortunately, total nitrogen and phosphorus export increased compared with the nutrients from the inflow, coming from compost in media.[1]

Another study compared four green roof drainage systems, two of which use matlike layers for drainage and retention, produced by American Hydrotech and Xeroflor, and two other systems that use voids to retain water, Sarnafil and Siplast.[2] Using 2.4 × 2.4 m sections of artificial roofs at a 2% slope with substrate depths ranging up to 10 cm, researchers examined three planting types: sedums from seeds, sedums from plugs, and plugs of 18 species native to the area of Michigan where the study took place. Consistent with the studies we've seen here, the roofs retained about half the rainfall, ranging from 38.6% for 2-cm substrate Xeroflor and 58.1% for 10-cm substrate Siplast. Nitrate levels for all roof types were higher on day 314 than on day 140, except in the case of native plants, but phosphorus levels decreased.

Storing and using stormwater at the parcel scale, perhaps for irrigation of green roofs, rain gardens, or lawns, can help reduce stormwater flows and reduce municipal water supply demands.[3] Harvesting stormwater involves capturing precipitation from an impervious surface, storing it in tanks, and

---

[1] Hathaway et al. 2008 studied two North Carolina green roofs. [2] Monterusso et al. (2004) compared commercially available (in 2004) green roofs. [3] Sample and Heaney (2006).

pumping it out for irrigation or nonpotable household needs. If the main goal is to reduce peak flows, then a "dewatering" of the tank to some predetermined level takes place within a few days in preparation for the next rainfall. If water use is the primary objective, then all of the water could be slowly drawn down as the various water use needs arise. This draw-down, however, must reflect the varying degrees of necessary reliability (will the tank go empty?) dependent on water supply availability and costs during droughts. One modeling project based on a small area of Boulder, Colorado, concluded that using cisterns sized at 10,000 gallons was optimal for that particular situation. It's not a cheap benefit for a homeowner: a quick Internet search shows that 10,000-gallon cisterns run about $5,000. Other harvesting approaches for very large stormwater volumes (millions of gallons) include plastic or concrete retention chambers placed beneath parking lots.[1] Sadly, a study based on areas of Virginia concluded that there was no positive financial benefit over a 50-year period, with a need for a doubling of water rates and stormwater fees to reach the breakeven point.[2]

Can stormwater harvesting possibly reduce the huge stormwater flows seen in urban streams?[3] Alternatively, might harvesting reduce streamflows so much that streams dry up? A study using a couple of Australian cities as modeling examples for stormwater harvesting assumed ponds that equaled 1% of the catchment area in size, with a 1-m-deep permanent pool, a 1-m detention depth, an infiltration rate of 0.18 mm/hour (typical of heavy clay), an evaporative loss equal to the local potential evapotranspiration, and, finally, a detention time of 70.5 hours. The two cities, Melbourne and Brisbane, have annual rainfalls of 65.3 cm and 177 cm, respectively. The study also examined three different impervious situations of 14%, 42%, and 70%. Harvesting under these conditions helped restore streamflows toward natural conditions and reduce pollution in all cases except suburban and urban Brisbane scenarios. In those cases, although peak flow was reduced, it came at the cost of a greater frequency of more high flows over a few days—in other words, several days of high flow rather than just one day of really high flow. Still, that situation helps stream health by reducing flashy flows. Under the suburban and urban Melbourne scenarios, these conditions resulted in stormwater "overharvesting," a situation in which less water gets released downstream compared with predevelopment. The other situations resulted in underharvesting. So the answers are yes and yes, meaning careful stormwater harvesting design is warranted if the consideration of downstream neighbors is important.

Setting a green roof on its edge yields a green facade.[4] Think of the ivy-covered ivory towers: rather than thermally massive stone absorbing the sun's

---

[1] Peruse any recent issue of *Stormwater* at www.stormh2o.com for examples. [2] Sample and Liu (2014). [3] Fletcher et al. (2007) modeled stormwater "harvesting" effects on streamflow. [4] Köhler (2008) reviews green facades.

rays, green leaves take them in, photosynthesize, and quickly transfer heat to the surrounding air. Irrigating this vegetation could reduce stormwater flows. These vertical, vegetated structures also collect dust and purify air, as well as provide some animal habitat. Though the style goes back at least 2,000 years, modern updates since the 1970s adapt the approach for today's cities. Ivy cover also provides winter insulation, increasing wall temperatures by as much as 3°C, thereby reducing heating costs. Green facades are more apparent to building occupants, and an interview of residents of green facade buildings indicated positive responses that mentioned general desires for natural spaces. On the other hand, negative responses focused on specific concerns like winter leaf fall. Many of the negative attitudes resulted from the early use of inappropriate plant species, and recent manuals on green facade practices take these problems into account.

The counterpart to a green roof on the ground is grassy paving (Figure 13.8). It's an alternative to an impervious surface that heats up and discharges polluted runoff to nearby stream. Grassy paving satisfies many needs simultaneously. Its permeability allows water to pass through, reducing stormwater runoff into retention ponds, infiltration trenches, and streams. Filtration through the soil at the parking lot source reduces downstream needs for water treatment, particularly when the stormwater becomes drinking water.[1] As shown earlier, many problems arise directly from so much stormwater being efficiently and rapidly transported from impermeable surfaces and into stormwater systems, which usually empty directly into urban streams.[2] Grassy paving would, at worst, slow down this transport and, at best, eliminate most runoff and pollutants during light rains.

Permeable pavement, whether concrete blocks, plastic grids, or permeable asphalt or concrete, involving undersurface storage within a gravel base, generally reduced runoff from precipitation as desired, but also controlled heavy metals. The control of nutrients was more variable, and permeable paving has problems with clogging and freezing. The damage from snowplows could be intense. Despite assertions to the contrary, permeable pavement installed over clay soils works well if installed correctly with a deep gravel base and underdrain systems.[3] Here in North Carolina, on sandier soils developers get credit for permeable paving at a 60% permeability fraction; everywhere else the regulations assume that permeable paving acts just like impermeable paving, especially when fitted with underdrain systems.[4]

Impermeable and paved surfaces also reduce onsite water retention, reducing water availability to trees and other vegetation. As we know, grass

---

[1] Brattebo and Booth (2003) studied permeable paving's water retention and pollutant-filtering abilities over several years, finding extremely favorable results. [2] For example, Walsh et al. (2005a). [3] Dietz (2007) also summarizes a variety of LID approaches. [4] Hunt and Bean (2006) present North Carolina permeable paving results.

Paving and grass can be combined.

Figure 13.8: Parking can coexist with permeable, grassy surfaces. At top are two examples of low-tech grassy paving surfaces that provide compaction-free grassy spaces (photos by Paula Bailey). Photos at bottom display modern grassy paving using a plastic substructure that provides the vehicular support (photos by Invisible Structures, Inc.).

---

cools the air through transpiration,[1] but the water has to stick around to be transpired. Grassy paving alleviates this problem by increasing water infiltration, recharging soil moisture levels, and helping trees grow shadier over parking lots.

Let's not forget that grassy paving, being partly alive, needs consistent upkeep, perhaps killing tree seedlings as they recruit into the grass of underused spots, and some watering during droughts. This maintenance could be costly for a small business, whereas the public gains most of the citywide air and water quality benefits.

Rain gardens have much the same functional goals of their green roof and pavement partners. Figure 13.9 complements the green roof plot of Figure 13.4B, showing the inflow and outflow for a rain garden built in Haddam,

---

[1] Ebdon et al. (1999).

Rain gardens hold back water.

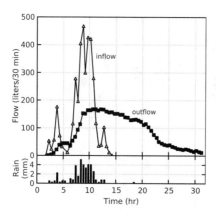

Figure 13.9: A rain garden in Haddam, CT, delays runoff, reducing peak flows at the outlet (Dietz and Clausen 2005). Although the control measure delays runoff, its performance in reducing pollutants isn't stellar, with only NH_3-N removed to any extent (not shown).

Connecticut, one example of a broader class of bioretention control measures.[1] This rain garden is really just a big puddle with mulch, having a surface dimension of 2.7 × 3.4 m and a depth of about 0.6 m (about 2 feet), holding about 2.5 m^3 of soil, sized to hold the first inch of rain water collected from a nearby roof. Though not usually done with bioretention ponds, this one was lined with an impermeable membrane and a pipe at the bottom that drains water from the rain garden. In this case the underflow drain was only for outflow measurement purposes, but some bioretention ponds use underdrains as a design feature. That membrane obviously reduced groundwater infiltration, so the performance of a "real" rain garden would be better than that measured in this study.

This specific plot shows a single rainfall from October 1, 2003. Compared with the inflow, peak outflow decreased, outflow duration increased, and 98.8% of the inflow water left through the underdrain pipe. No really serious changes in water quality or temperature of the outflow were observed compared with the inflow, though there was some reduction of NH_3^-. There was even a net export of phosphorus from the ponds, primarily due to the disturbed soils from construction; however, that phosphorus export settled down after the rain garden aged for a year. The main benefit from this noninfiltrating garden was the peak flow reduction that helps mitigate flashy streamflows.

Figure 13.10 shows results from an extensive examination of bioretention cells (which include the less formal rain gardens) carried out at four sites in North Carolina and two in Maryland.[2] Call them ponds, cells, or rain gardens,

---

[1] Dietz and Clausen (2005) studied runoff from Connecticut rain gardens. [2] Li et al. (2009) studied bioretention cells in North Carolina and Maryland.

Bioretention ponds retain water.

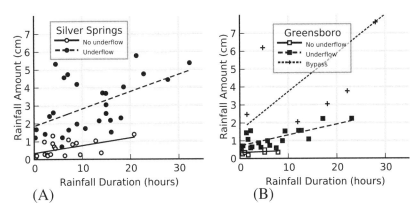

Figure 13.10: These two bioretention ponds from Silver Springs, MD (A), and Greensboro, NC (B), have drains beneath their soils, but only heavy or extended rainfalls lead to "underflow" through these drains (Li et al. 2009). Open symbols represent rainfalls that had no underflow, filled symbols represent those with underflow, and the plus marks are extremely heavy rainfalls (in Greensboro) that led to flows through bypass channels.

but really what they have in common is that they're just a hole partly filled with mulch. Imagine having a shallow stormwater pond, but the bottom is flat and covered with grass and shrubs. That depth is the ponding depth. Somewhere in that bottom sits another large but shallow hole filled with mulch or soil media. From the bottom of that hole potentially sits drain pipes of various types. Certainly much better engineering goes into them than described here, but the captured water then either percolates into the ground or seeps out of the pipe with a surface flow somewhere downhill.

In this study, pond surface areas covered about 5% of drainage area, meaning something like a 20 × 100 foot area per acre, with a ponding depth ranging from 15 to 30 cm, and between 0.5 and 1 m of soil media. All ponds had trees, shrubs, and mulch. Ideally, such a cell would be sized to deal with all of the water from its basin. Also ideally, all water would infiltrate into the ground, and no "underflow" would come out from the drain pipe. However, at least that water has been filtered of sediments, but the water still runs off site and, presumably, downstream as a part of a flashy flow. An even more undesired third situation, bypass flow, takes place when so much rain falls that water starts bypassing the bioretention pond. Bypass flow can't be avoided during those

6-inch in 24-hour deluges, and in fact, these structures need to be carefully engineered to prevent damage in those situations.

These plots highlight one feature from a very detailed study about the water retained in and water flowing out of bioretention ponds. Rainfall characteristics of events that either did or did not produce flow out of the underflow pipes are plotted in Figure 13.10 for ponds in Silver Springs, Maryland, and Greensboro, North Carolina. As one might expect, slow, light rains stay mainly in the bioretention pond. Although a bigger and deeper cell is better for evapotranspiration and infiltration, this wasn't borne out in these two plots: the Greensboro site was roughly 2.5 times larger and 30% deeper than the Silver Springs pond yet showed more underflow drainage. Still, both ponds delayed and reduced peak flows to a very great extent (not shown here). Evapotranspiration removed 19% of the water, and seepage into the ground at the bottom of the cell (called "exfiltration") accounted for just 8% of the water. That latter number was so small due to clayey soils.

Deep cells work great for small rainfalls, and simple cells seemingly work almost as well as complicated ones with internal water storage.[1] Beyond stormwater flow and volume reduction, bioretention facilities work well at controlling heavy metals, and perhaps temperature pulses given the right conditions, though nutrient control seems problematic.[2]

As shown above, rain gardens and bioretention ponds reduce peak stormwater flows, but the pollution-reduction benefits are harder to measure. Figure 13.11 shows the results of one such study on biofilters, which are yet another name for bioretention pond, or rain garden, or vegetated strips. As the name suggests, the essential concept of a biofilter involves filtration, retention, and biological uptake. Whatever its name, if it does those things, then it can be called a biofilter.

These experiments were done in a greenhouse using 125 two-foot-diameter containers.[3] Technically going by the term "mesocosms," these studies provide important information when the real thing is just too hard to work with or too expensive to replicate. These laboratory units were small versions of a biofilter—think of cores drilled out of the real thing—with many variables tested. Each container had a depth of 30, 50, or 70 cm, filled with a sandy loam soil. Figure 13.11A emphasizes the role of plant species. As we see here, the plants *Carex appressa* (tall sedge) and *Melaleuca ericifolia* (an Australian shrub called "swamp paperbark") removed total nitrogen better than the other species, and in fact, other species, as well as soil alone, contributed nitrogen to the runoff. Results for nitrates reflect these data. Total suspended solids were removed almost entirely by all configurations, and around 80% of phosphorus was removed, though *Carex appressa* was particularly effective at reducing total phosphorus. However, in experiments that included organic matter in the

---

[1] Li et al. (2009).   [2] Dietz (2007).   [3] Bratieres et al. (2008); see their supplementary data.

Nitrogen removal in biofilters.

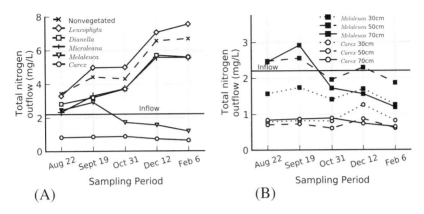

Figure 13.11: Nitrogen removal in greenhouse-scale experimental biofilters depends on plant species, with *Carex appressa* and *Melaleuca ericifolia* removing nitrogen (A), and filter media depth, with deeper performing better (B) (Bratieres et al. 2008).

soil media, and planted with *Carex appressa*, biological breakdown in the soil medium led to the export of phosphorus.

Overall conclusions of this extensive study were that biofilters should have a sandy loam filter and that the size of the biofilter should be at least 2% of its watershed; for example, a 500-acre watershed should have 10 acres of biofilter. At a smaller scale, a 1-acre lot should have about 800 square feet, a 20×40 ft area. Besides pollution, of course, one must consider aesthetics and drought survival if the biofilter is to serve as an urban open space amenity, and in those situations, plant species in biofilters need careful consideration.

Another study of two bioretention areas in Greensboro and Chapel Hill, North Carolina, showed highly variable reductions in nitrate export, one with 13% and the other with 75%.[1] Total phosphorus varied greatly, including the situation of the effluent holding more phosphorus than the influent. In fact, concentrations of both nitrogen and phosphorus were higher in the outflow, but because the bioretention and infiltration worked so well, there was a greatly reduced outflow—by half, but often to about a tenth of the inflow—which reduced total nutrient outflow. The SCMs were also highly effective at reducing metals, by at least 80%, but up to 99%. However, iron export increased, as did total suspended solids.

---

[1] Hunt et al. (2006) studied nutrient removal in bioretention areas.

Good things can happen within simple stormwater ponds, too. For example, denitrification increased with increasing infiltration rate beneath an infiltration pond off of Monterey Bay in California.[1] An interesting point with this pond is that it sits well above the water table, but, somewhat like the light rays emanating from a light bulb, with high intensity near the bulb and low intensity farther away as the rays spread out, as the water infiltrates downward from the pond there's a water-saturated zone of 1–2 m thickness just below the pond, which gives way to an 10- to 20-m-thick unsaturated vadose zone above the aquifer.

The infiltrating water holds nitrate and oxygen as it passes through the pond's bottom. Microbes living in the soil consume organic material aerobically as the oxygen in the passing water is consumed. At some depth, however, the oxygen has been used up by the respiring microbes in the shallower layers above, but denitrifying bacteria continue consuming organic carbon in the absence of oxygen. The so-called redoxocline is the boundary between the oxygen-poor and oxygen-rich, or anoxic and oxic, or anaerobic and aerobic regions of the soil (see discussion in Chapter 6).

If infiltration rates are too high, so much oxygen makes it deep into the soil such that no denitrification can be observed. With some variability, denitrification stops when infiltration rates are higher than 0.7 m/day. Hence, too much infiltration hinders nutrient processing.

Recall that microbes prefer particular oxygen and nitrogen isotopes in the nitrates they consume while denitrifying, leaving behind nitrate with higher fractions of $\delta^{18}O$ and $\delta^{15}N$, providing a fingerprint of denitrification, as shown in Figure 6.4. Here, in Figure 13.12, that preference was used to reveal how denitrification depends on nitrate concentration. At low concentrations, complete denitrification of the infiltrated nitrates took place, meaning that net denitrification was limited by nitrate concentration. Hence, more denitrification could have taken place, so the pond wasn't living up to its potential. Thus, there's a balance between not too much oxygen making its way into the ground, having enough nitrate, and stimulating microbial activity.

In a study from northern Virginia, simple flood-control detention ponds did not sequester phosphorus compared with natural riparian wetlands, whereas ponds designed to both control floods and reduce pollution—let's call them "enhanced ponds"—did sequester phosphorus.[2] The latter ponds were, essentially, detention ponds with some topography and wetlands vegetation. Part of the explanation for the retention of phosphorus in the enhanced ponds was the iron brought in with urban sediments, a role taken on by aluminum in the wetlands.

---

[1] Schmidt et al. (2011) examine infiltration and denitrification rates. [2] Hogan and Walbridge (2007).

Denitrification under stormwater ponds.

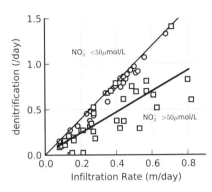

Figure 13.12: Relative denitrification rate (scaled by initial concentration) increases beneath an infiltration pond as more water flows into the Monterey Bay, CA, ground (Schmidt et al. 2011). Thin solid line represents complete denitrification. At infiltration rates greater than those shown here, denitrification shuts down because sufficient oxygen is provided to sustain aerobic respiration.

Stormwater ponds need to be empty when a large rain comes, but a quick-draining pond loses its peak flow reduction function, and as just discussed, infiltration rates that are too high lose out on subsurface denitrification benefits. Holding onto water in a pond provides another benefit: reducing the toxicity of pollutants. Figure 13.13 shows data from 87 stormwater runoff samples from residential, industrial, and commercial/institutional areas in Birmingham, Alabama.[1] Surfaces producing runoff included roofs, parking areas, storage locations, and landscaping. Researchers also sampled creeks and stormwater ponds. Of all these various samples, standard toxicology methods rated 9% of them "extremely toxic" and 32% "moderately toxic" for compounds that included pesticides, metals, and PAHs. Parking lots, service stations, and storage areas were the worst places for toxicity and associated with particulates. Pollutants coming from gas stations situated around Tijuana, Mexico, had concentrations ranging from 18 to 57 times higher than similar pollutants in the U.S.[2] Even one of six landscaped areas was highly toxic, too, with some heavy metals found, but most landscaped areas were fine.

Taking the samples with high toxicity levels, various treatment approaches were tested using small bench-scale devices—about a liter of fluid in a large beaker—to recreate specific aspects of the conditions within a stormwater pond.

Filtering the water through extremely fine screens greatly reduced toxicity, shown in Figure 13.13A, and a 0.45-μm filter essentially fully detoxified the runoff. That kind of treatment for huge stormwater volumes is not prac-

---

[1] Pitt et al. (1995) measured stormwater toxicity and various degradation approaches. [2] Mijangos-Montiel et al. (2010).

Pollution removal takes time.

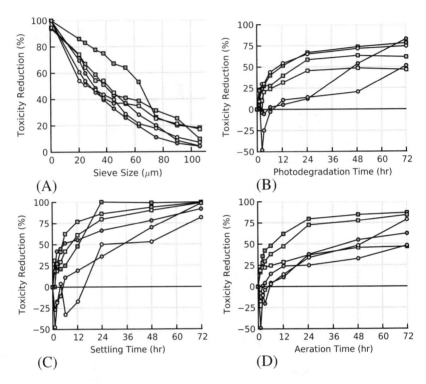

Figure 13.13: Effects on toxicity in Birmingham, AB, runoff using different control measures, (A) filtration, (B) photodegradation, (C) settling, and (D) aeration (Pitt et al. 1995). Different lines represent individual stormwater samples coming from automobile service centers, industrial loading and parking areas, and automobile salvage yards.

tical, but it demonstrates that most of the problems with toxicity are bound to particulates. A 40-μm filter represents sort of a minimum effective filtration size. As a benchmark comparison, a study that examined particle sizes in stormwater passing through vegetated filter strips measured influent particles sizes ranging from 30 to 140 μm, whereas the effluent particles ranged from 13 to 56 μm.[1] This result brings us close to the 40-μm filtration, but an order of magnitude or more finer filtration will be difficult. Perhaps comparable or not, air quality issues occur when small, 2.5-μm particles are inhaled, causing cardiovascular problems, and the bacterium *E. coli* is roughly that size, too.

---

[1] Knight et al. (2013).

Since we can't superfilter all of our stormwater, all of the other practical approaches take a couple of days, during which the water must be held for treatment. Photodegradation—meaning drenching by sunlight—reduced toxicity levels by half after 24 hours or so (Figure 13.13B). Simply letting things settle for a few days works well, too (Figure 13.13c), as does aeration for a few days (Figure 13.13D). It's important to note that control samples set aside in dark, sealed containers at room temperature did not display these same consistent toxicity degradations, as might be anticipated if toxin-consuming bacteria were sweeping up everything in sight. Final variation in toxicity of these set-aside samples ranged across a 20% increase or decrease.

It seems, then, that letting stormwater sit in a settling pond with good sunlight and aeration for a few days before it drains away to a stream might help improve stormwater quality while also reducing peak flows.

Settling is an important part of this solution, but settled things can be re-suspended. A complicated three-pond detention system in Rhode Island comprised an initial settling pond, followed by a constructed wetland, followed by a deep pond to trap fine sediments.[1] In this system, measurements showed that influent and effluent PAH concentrations were about the same. There was a first flush in PAHs, but the concentrations were about the same as or lower than the PAHs in the stagnant ponds, meaning that PAHs sitting in the pond sediments get resuspended as new water flushed in.

In another such example, a Swedish study examined using stormwater ponds to treat road runoff.[2] Many ponds didn't work because the water's turbulence mixed up the sediments and carried them downstream, short-circuiting the whole point of a settlement pond. Bigger ponds were better, and, sadly, using more complicated elliptical pond shapes didn't help increase sedimentation.

Figures 13.14 and 13.15 describe many types of stormwater treatment approaches,[3] showing that constructed wetlands are the best solution for cost and the reduction of solids and phosphorus. These types include "dry extended detention basins" (a barren hole) that hold water for less than 2 days; "wet retention basins" (a pond) that might hold water almost continuously; "constructed wetlands" (pond and meadows) that hold water for an intermediate time; and "infiltration trenches" (parking lot strips), which are pretty much anything that filters stormwater and prevents it from entering streams.

Cost is an important factor, too. Many competing aspects favor one type of stormwater system or the other. Construction cost data in Figure 13.15A show that bigger is cheaper for all stormwater systems, and annual maintenance costs, shown in Figure 13.15B, run about 3–5% of construction costs for large systems. One notable exception is that dry extended basins require very low

---

[1] Neary and Boving (2011) studied PAHs in a Rhode Island detention three-pond system. [2] Starzec et al. (2005). [3] Weiss et al. (2007).

Bioretention and infiltration work best.

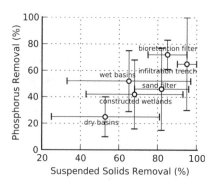

Figure 13.14: A surveyed across the U.S. shows that infiltration trenches (grassy strips surrounding parking lots) and bioretention filters (water-holding medium filled pits covered with grass and trees) work well in the reduction of solids and phosphorus (after Weiss et al. 2007). Dry basins are least effective.

Dry and wet ponds are cheapest.

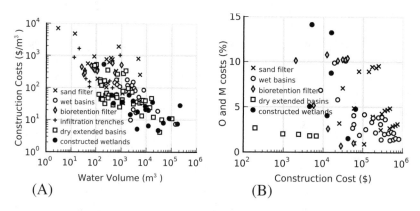

(A)   (B)

Figure 13.15: (A) Construction costs for U.S. stormwater control measures decrease greatly with increasing stormwater volume handled. Dry extended basins (ponds) and constructed wetlands are cheapest to build. (B) Dry and wet ponds have the lowest operating and maintenance (O&M) costs (Weiss et al. 2007).

maintenance costs whatever the size. Ponds are cheap to build and cheap to maintain, and that might be why stormwater ponds are ubiquitous.

Constructed wetlands are an overall good solution, nicely solving the stormwater filtration problem while promoting wildlife. Constructed wetlands that include trees can serve many purposes: purify drinking water by filtering stormwater, cool the city, support wildlife, and provide recreation and environmental education.

Each of the streams shown in Figure 9.7 also had a stormwater control structure, and these various SCMs were ranked from best to worst for increased outflow temperature: dry infiltration pond (2.3°F increase), constructed wetlands (2.4°F), extended detention dry pond (5.2°F), and wet pond (8.5°F).[1] Note that all four structure types *increased* temperatures between inflow and outflow. In cold stream areas, the latter two SCMs could eliminate temperature-sensitive species like trout, but SCMs occur in impervious areas that already are hot, and the unshaded outflows with rip-rap alone produce temperature increases of 8.5°F. This increase takes place as the high thermal mass of rip-rap collects solar energy and then transfers the heat to stormwater. Most important factors for temperature problems were undersized structures, the presence of a large wet pond, poorly shaded channels, poorly shaded wet ponds, and extended water detention periods.

We can ask similar questions of all types of stormwater systems, including constructed wetlands and infiltration trenches. Whatever option cities choose, trade-offs exist between upfront and maintenance costs, costs borne by developers and costs borne by citizens, and local stormwater costs versus nonlocal downstream pollution and sediment costs (see Figure 13.14).

Let's put these water volumes in scale. Together, my fellow Durhamites and I use about 30 million gallons, or 114,000 m³, of potable water per day. The main portion of downtown Durham covers about 1,000 acres, and a 1-inch rainfall over that area is just shy of 30 million gallons. A single constructed wetland built to filter that amount of water comes to about $5 per m³ (see Figure 13.14), or just half a million dollars (excluding land acquisition). Of course, not all of Durham's water supply would go through a single wetland, and a single rainfall yields far greater water volumes rushing through stormwater systems, but daily water use provides a convenient human scale. So, construction costs for ten of these wetlands leads to a $5 million price tag, with $250,000 annual maintenance costs. That doesn't seem outrageously expensive for cleaner water, especially when reported willingness-to-pay values for clean water range from $38 to $273 per household in 1988 dollars,[2] which, theoretically, should easily provide $10 from Durham for such infrastructure development.

Figure 13.16 shows another trade-off: a low-cost, fenced-off, single-purpose retention pond at a public library chosen over a more expensive constructed wetland where trees, birds, frogs, and children may have played. Certainly an avoidance of accident liability is the driving factor.

Two concepts drive "modern" ideas behind stormwater control: the newer concept of ecosystem services, a compendium of all the things the ecosystem provides people; and the even more wieldly but older term, "integrated

---

[1] Galli (1990: 113). [2] Bergstrom (1990) reports willingness to-pay values for clean water.

Single-purpose stormwater solutions.

Figure 13.16: A retention basin draining the North Durham Library's parking lot only serves a stormwater function: If designed more carefully, it could provide a fine location for frog-chasing, pond-stomping educational opportunities (Wilson 2011).

water resource management," a coordinated development of land and water resources to maximize economic and social welfare.[1]

Whether or not the two concepts represent fundamentally distinct ideas can be debated, but in any event, there is a so-called implementation gap—people who *do* stormwater work aren't following its principles. Implementation probably has as its primary challenge that it requires a broad synthesis of concepts and governmental departments, a task complicated by competing goals of different city and county departments. How do we get stormwater departments and the social services departments to work together to alleviate stormwater flows by diverting them to urban farms that grow food for undernourished populations?

Many researchers study that issue, however. Australia, where "low-impact development" is known "water-sensitive urban design," is one place that has questioned what's inhibiting adoption. The inherent privileges within stormwater management represents a strong inertia that prevents change.[2] Yet, alongside this administrative situation, also hindering adoption were funding limitations to try new approaches, responsibilities being split up among many departments (water supply, stormwater, and surface water quality), and a lack of legal accountability. A survey of more than 800 Australian stormwater professionals revealed strong support (>80%) for protecting stream health and for innovative approaches, but they had the attitude that only about half of the public supported change and only about one-third of politicians supported the needed changes.[3] These are called "socioinstitutional" barriers related to

---

[1] Cook and Spray (2012) summarize the difficulties behind matching implementation and the concepts of ecosystem services and integrated water resource management. [2] Brown (2005) finds inertia in stormwater management. [3] Brown and Farrelly (2009a) surveyed Australian stormwater professionals.

vision, coordination, responsibility, and resources.[1] Implementation may need a new look at government structures.[2] Lack of community engagement also factored into the lack of change.

They concluded that adoption of these changes represents a four-step process for an individual or an organization.[3] First, awareness of a problem and the recognition of a need for a solution— if people don't perceive a problem, why fix what doesn't need fixing? Second, there must be an understanding of the potential benefits of a solution. In this step, an accurate accounting of the costs and benefits may help address a financial gain in changing an approach. Third, an organization must possess the needed skills to carry out the solution. Once a problem or the existence of a benefit to be gained is recognized, are there available people that can do what needs to be done? Fourth, do sufficient incentives exist to carry out the solution? If the actors have nothing to gain, then there's a high likelihood that nothing will get done. For example, money is always a great incentive. Having high stormwater fees that could be abated by good behavior provides a stick and a carrot.[4] However, in the U.S., monthly stormwater fees are around a couple of dollars (and often not tied to stormwater runoff), but fees would need to be around \$1,000 to provide useful incentives. Furthermore, organizational culture has to value learning and integration.

Consider the difficulties of addressing a well-understood problem like human-caused climate change: many U.S. decision makers refuse to perceive a problem. Stormwater problems have much less media visibility than climate change, but making information available is the first step to a solution.

Chapter Readings

Berndtsson, J.C. 2010. Green roof performance towards management of runoff water quantity and quality: A review. *Ecol. Engin.* 36: 351–360.

Brown, R.R., and M.A. Farrelly. 2009. Challenges ahead: Social and institutional factors influencing sustainable urban stormwater management in Australia. *Water Sci. Technol.* 59: 653–660.

Claytor, R A., and T.R. Schueler. 1996. Design of stormwater filtering systems. Center for Watershed Protection (Silver Spring, MD).

Dietz, M.E., and J.C. Clausen. 2005. A field evaluation of rain garden flow and pollutant treatment. *Water, Air, Soil Pollut.* 167: 123–138.

---

[1] Brown and Farrelly (2009b) surveyed literature on institutional barriers to innovative stormwater approaches. [2] Van de Meene et al. (2011). [3] Brown and Farrelly (2009a). [4] Roy et al. (2008) compared notes on the impediments between the U.S. and Australia.

Hood, M.J., J.C. Clausen, and G.S. Warner. 2007. Comparison of stormwater lag times for low impact and traditional residential development. *J. Am. Water Resources Assoc.* 43: 1036–1046.

Li, H., L.J. Sharkey, W.F. Hunt, and A.P. Davis. 2009. Mitigation of impervious surface hydrology using bioretention in North Carolina and Maryland. *J. Hydrol. Engine.* 14: 407–415.

Pitt, R., R. Field, M. Lalor, and M. Brown. 1995. Urban stormwater toxic pollutants: Assessment, sources, and treatability. *Water Environ. Res.* 67: 260–275.

Roy, A.H., S.J. Wenger, T.D. Fletcher, C.J. Walsh, A.R. Ladson, W.D. Shuster, H.W. Thurston, and R.R. Brown. 2008. Impediments and solutions to sustainable, watershed-scale urban stormwater management: Lessons from Australia and the United States. *Environ. Manag.* 42: 344–359.

VanWoert, N.D., D.B. Rowe, J.A. Andresen, C.L. Rugh, R.T. Fernandez, and L. Xiao. 2005. Green roof stormwater retention: Effects of roof surface, slope, and media depth. *J. Environ. Qual.* 34: 1036–1044.

Weiss, P.T., J.S. Gulliver, and A.J. Erickson. 2007. Cost and pollutant removal of storm-water treatment practices. *J. Water Resources Plann. Manag.* 133: 218–229.

Appendix

U.S. Stormwater Laws

Complete discussion of the Clean Water Act (CWA) is provided by the US EPA,[1] and this brief history closely follows the summary in NRC (2009: Table 2-1).

Protection of U.S. water goes back to the end of the 1800s, when the U.S. Congress empowered the U.S. Army Corps of Engineers to deal with "obstructions to navigation," including the ability to deal with industries dumping pollutants into rivers and lakes.[2]

After WWII, Congress passed the Federal Water Pollution Control Act of 1948,[3] which held cities and industries responsible for the problems caused to the owners of riparian lands. Much of the motivation was to protect private property rights from the harms of pollution by others. The act also emphasized state and federal responsibilities in water-quality protection as a part of promoting public health and welfare. Comprehensive goals included giving voice to all of the important uses for water, such as public water supplies, protecting aquatic life, recreation, agriculture, and industrial uses. All waters were included, both surface and groundwater.

In 1965 Congress passed the Water Quality Act to deal with interstate waters. This legislation was followed closely by the 1972 Federal Water Pollution Control Act that stopped dumping pollutants into streams without permits. Amid a budget argument the legislation was vetoed by President Nixon,[4] stating "that we [should] attack pollution in a way that does not ignore other very real threats to the quality of life, such as spiraling prices and increasingly onerous taxes." Congress quickly overrode the president's veto. It's now

---

[1] See cfpub.epa.gov/watertrain/index.cfm. [2] See www.usace.army.mil/About/-History/-BriefHist ory-oftheCorps/-Environmental-Activities.aspx. [3] Schwob (1953) describes the goals, findings, and accomplishments of the Federal Water Pollution Control Act. Schwob was the first head of the Water Pollution Control division within the Department of Health, Education, and Welfare. [4] *New York Times*, Oct. 19, 1972.

best known as the Clean Water Act (CWA) after major revisions in Title 33, Chapter 26, Sections 1251–1387, in the U.S. Code,[1] which is the compendium of federal laws.[2]

In 1972, the USEPA exempted stormwater from the permitting require-ments of the CWA regulations due to its complexity.[3] The Natural Resource Defense Council (NRDC) sued. In 1977, it won in a decision known as *NRDC v. Costle*, and the courts required the USEPA to include stormwater pollution.[4]

In 1987, following *NRDC v. Costle*, CWA sections 301 and 402 were modified to deal with stormwater and adopted as law over President Reagan's veto.[5] This legislation set limits and starting stormwater permitting processes for urban areas. With this legislation, the USEPA put forth phase I rules in 1990 for construction projects exceeding 5 acres and big municipalities (more than 100,000 people). Phase II rules were put forth in 1999 for projects exceeding 1 acre and smaller cities, the latter being determined in a more complicated manner. For both sets of rules several years elapsed before their enactment by the states.

In 2007, Congress passed the Energy Independence and Security Act that required federal projects above 5,000 square feet match predevelopment hydrology as much as practically possible.

---

[1] See www.law.cornell.edu/uscode/text/33/chapter-26 [2] A one-to-one translation of the two documents, the CWA and the U.S. Code, is provided by Copeland (2010). [3] See www.nrdc.org /water/pollution/storm/chap2.asp. [4] See openjurist.org/568/f2d/1369 [5] *New York Times*, Feb. 5, 1987.

References

Aber, J.D., C.L. Goodale, S.V. Ollinger, M.-L. Smith, A.H. Magill, M.E. Martin, R.A. Hallett, and J.L. Stoddard. 2003. Is nitrogen deposition altering the nitrogen status of northeastern forests? *Bioscience* 53: 375–389.

Ahmed, F.E. 2001. Toxicology and human health effects following exposure to oxygenated or reformulated gasoline. *Toxicology Letters* 123: 89–113.

Alexander, R.B., E.W. Boyer, R.A. Smith, G.E. Schwarz, and R.B. Moore. 2007. The role of headwater streams in downstream water quality. *Journal of the American Water Resources Association* 43: 41–59.

Allan, J.D., D.L. Erickson, and J. Fay. 1997. The influence of catchment land use on stream integrity across multiple spatial scales. *Freshwater Biology* 37: 149–161.

Allen, A., D. Milenic, and P. Sikora. 2003. Shallow gravel aquifers and the urban "heat island" effect: A source of low enthalpy geothermal energy. *Geothermics* 32: 569–578.

Alvarez-Cobelas, M., D.G. Angeler, and S. Sánchez-Carrillo. 2008. Export of nitrogen from catchments: A worldwide analysis. *Environmental Pollution* 156: 261–269.

Ankley, G.T., B.W. Brooks, D.B. Huggett, and J.P. Sumpter. 2007. Repeating history: Pharmaceuticals in the environment. *Environmental Science and Technology* 41: 8211–8217.

Arnold, C.L., Jr., and C.J. Gibbons. 1996. Impervious surface coverage: The emergence of a key environmental indicator. *Journal of the American Planning Association* 64: 243–258.

Athanasiadis, K., H. Horn, and B. Helmreich. 2010. A field study on the first flush effect of copper roof runoff. *Corrosion Science* 52: 21–29.

Ball, J.E., R. Jenks, and D. Aubourg. 1998. An assessment of the availability of pollutant constituents on road surfaces. *Science of the Total Environment* 209: 243–254.

Barbour, M.T., J. Gerritsen, B.D. Snyder, and J.B. Stribling. 1999. Rapid bioassessment protocols for use in streams and wadeable rivers: Periphyton, benthic macroinvertebrates and fish. 2nd ed. EPA 841-B-99-002. U.S. Environmental Protection Agency, Office of Water (Washington, DC).

Barnes, K.K., D.W. Kolpin, E.T. Furlong, S.D. Zaugg, M.T. Meyer, and L.B. Barber. 2008. A national reconnaissance of pharmaceuticals and other organic wastewater contaminants in the United States: I) Groundwater. *Science of the Total Environment* 402: 192–200.

Baronti, C., R. Curini, G. D'Ascenzo, A. Di Corcia, A. Gentili, and R. Samperi. 2000. Monitoring natural and synthetic estrogens at activated sludge sewage treatment plants and in a receiving river water. *Environmental Science and Technology* 34: 5059–5066.

Bartens, J., S.D. Day, J.R. Harris, J.E. Dove, and T.M. Wynn. 2008. Can urban tree roots improve infiltration through compacted subsoils for stormwater management? *Journal of Environmental Quality* 37: 2048–2057.

Bartens, J., S.D. Day, J.R. Harris, T.M. Wynn, and J.E. Dove. 2009. Transpiration and root development of urban trees in structural soil stormwater reservoirs. *Environmental Management* 44: 646–657.

Bartlett, R.J., and B.R. James. 1993. Redox chemistry of soils. *Advances in Agronomy* 50: 151–208.

Beasley, G., and P. Kneale. 2002. Reviewing the impact of metals and PAHs on macroinvertebrates in urban watercourses. *Progress in Physical Geography* 26: 236–270.

Beaulac, M.N., and K.H. Reckhow. 1982. An examination of land use–nutrient export relationships. *Water Resources Bulletin* 18: 1013–1024.

Beckett, K.P., P.H. Freer-Smith, and G. Taylor. 2000. Particulate pollution capture by urban trees: Effect of species and windspeed. *Global Change Biology* 6: 995–1003.

Bender, G.M., and M.L. Terstriep. 1984. Effectiveness of street sweeping in urban runoff pollution control. *Science of the Total Environment* 33: 185–192.

Bennett, E.R., K.D. Linstedt, V. Nilsgard, G.M. Battaglia, and F.W. Pontius. 1981. Urban snowmelt–characteristics and treatment. *Journal of the Water Pollution Control Federation* 53: 119–125.

Berger, T., and C.M. Horner. 2003. In vivo exposure of female rats to toxicants may affect oocyte quality. *Reproductive Toxicology* 17: 273–281.

Bergstrom, J.C. 1990. Concepts and measures of the economic value of environmental quality: A review. *Journal of Environmental Management* 31: 215–228.

Berman, D., E.W. Rice, and J.C. Hoff. 1988. Inactivation of particle-associated coliforms by chlorine and monochloramine. *Applied and Environmental Microbiology* 54: 507–512.

Berndtsson, J.C. 2010. Green roof performance towards management of runoff water quantity and quality: A review. *Ecological Engineering* 36: 351–360.

Bernhardt, E., M.A. Palmer, J.D. Allan, G. Alexander, K. Barnas, S. Brooks, J. Carr, S. Clayton, C. Dahm, J. Follstad-Shah, D. Galat, S. Gloss, P. Goodwin, D. Hart, B. Hassett, R. Jenkinson, S. Katz, G.M. Kondolf, P.S. Lake, R. Lave, J.L. Meyer, T.K. O'Donnell, L. Pagano, B. Powell, and E. Sudduth. 2005. Synthesizing U.S. river restoration efforts. *Science* 308: 636–637.

Bernhardt, E.S., and M.A. Palmer. 2007. Restoring streams in an urbanizing world. *Freshwater Biology* 52: 738–751.

Bierman, P.M., B.P. Horgan, C.J. Rosen, A.B. Hollman, and P.H. Pagliari. 2010. Phosphorus runoff from turfgrass as affected by phosphorus fertilization and clipping management. *Journal of Environmental Quality* 39: 282–292.

Biggs, B.J.F. 2000. Eutrophication of streams and rivers: Dissolved nutrient–chlorophyll relationships for benthic algae. *Journal of the North American Benthological Society* 19: 17–31.

Blanck, H.M., M. Marcus, P.E. Tolbert, C. Rubin, A.K. Henderson, V.S. Hertzberg, R.H. Zhang, and L. Cameron. 2000. Age at menarche and tanner stage in girls exposed *in utero* and postnatally to polybrominated biphenyl. *Epidemiology* 11: 641–647.

Blann, K.L., J.L. Anderson, G.R. Sands, and B. Vondracek. 2009. Effects of agricultural drainage on aquatic ecosystems: A review. *Critical Reviews in Environmental Science and Technology* 30: 909–1001.

Bochis-Micu, C., and R.E. Pitt. 2005. Impervious surfaces in urban watersheds. Presented at the 78th annual Water Environment Federation Technical Exposition and Conference, Washington, DC, Oct. 29–Nov. 2, 2005.

Boeder, M., and H. Chang. 2008. Multi-scale analysis of oxygen demand trends in an urbanizing Oregon watershed, USA. *Journal of Environmental Management* 87: 567–581.

Böhlke, J.-K. 2002. Groundwater recharge and agricultural contamination. *Hydrogeology Journal* 10: 153–179.

Bohn, H.L., B.L. McNeal, and G.A. O'Connor. 2001. *Soil chemistry.* 3rd ed. New York: Wiley.

Boivin, M., M. Lamy, A. Gosselin, and B. Dansereau. 2001. Effect of artificial substrate depth on freezing injury of six herbaceous perennials grown in a green roof system. *Horticulture Technology* 11: 409–412.

Bonte, M., B.M. van Breukelen, and P.J. Stuyfzand. 2013. Temperature-induced impacts on groundwater quality and arsenic mobility in anoxic aquifer sediments used for both drinking water and shallow geothermal energy production. *Water Research* 47: 5088–5100.

Bookman, R., C.T. Driscoll, S.W. Effler, and D.R. Engstrom. 2010. Anthropogenic impacts recorded in recent sediments from Otisco Lake, New York, USA. *Journal of Paleolimnology* 43: 449–462.

Bookman, R., C.T. Driscoll, D.R. Engstrom, and S.W. Effler. 2008. Local to regional emission sources affecting mercury fluxes to New York lakes. *Atmospheric Environment* 42: 6088–6097.

Booth, D.B. 1990. Stream-channel incision following drainage-basin urbanization. *Water Resources Bulletin* 26: 407–417.

Booth, D.B. 1991. Urbanization and the natural drainage system: Impacts, solutions, and prognoses. *Northwest Environmental Journal* 7: 93–118.

Booth, D.B., and L.E. Reinelt. 1993. Consequences of urbanization on aquatic systems—measured effects, degradation thresholds, and corrective strategies. In: *Watershed 93 Conference Proceedings, March 21–24.* Tetra Tech (Alexandria, VA).

Bornstein, R., and Q. Lin. 2000. Urban heat islands and summertime convective thunderstorms in Atlanta: Three case studies. *Atmospheric Environment* 34: 507–516.

Boyer, E.W., R.W. Howarth, J.N. Galloway, F.J. Dentener, P.A. Green, and C.J. Vörösmarty. 2006. Riverine nitrogen export from the continents to the coasts. *Global Biogeochemical Cycles* 20: GB1S91.

Bragazza, L., C. Freeman, T. Jones, H. Rydin, J. Limpens, N. Fenner, T. Ellis, R. Gerdol, M. Hájek, T. Hájek, P. Iacumin, L. Kutnar, T. Tahvanainen, and H. Toberman. 2006. Atmospheric nitrogen deposition promotes carbon loss from peat bogs. *Proceedings of the National Academy of Sciences of the USA* 103: 19386–19389.

Bratieres, K., T.D. Fletcher, A. Deletic, and Y. Zinger. 2008. Nutrient and sediment removal by stormwater biofilters: A large-scale design optimisation study. *Water Research* 42: 3930–3940.

Brattebo, B.O., and D.B. Booth. 2003. Long-term stormwater quantity and quality performance of permeable pavement systems. *Water Research* 37: 4369–4376.

Braun, J.M., K. Yolton, K.N. Dietrich, R. Hornung, X. Ye, A.M. Calafat, and B.P. Lanphear. 2009. Prenatal bisphenol a exposure and early childhood behavior. *Environmental Health Perspectives* 117: 1945–1952.

Breivik, K., A. Sweetman, J.M. Pacyna, and K.C. Jones. 2007. Towards a global historical emission inventory for selected PCB congeners—a mass balance approach: 3. An update. *Science of the Total Environment* 377: 296–307.

Brent, R.N., and Herricks, E.E. 1998. Postexposure effects of brief cadmium, zinc, and phenol exposures on freshwater organisms. *Environmental Toxicology and Chemistry* 17: 2091–2099.

Bressler, D.W., M.J. Paul, A.H. Purcell, M.T. Barbour, E.T. Rankin, and V.H. Resh. 2009. Assessment tools for urban catchments: Developing stressor gradients. *Journal of the American Water Resources Association* 45: 291–305.

Broussard, W., and R.E. Turner. 2009. A century of changing land-use and water quality relationships in the continental US. *Frontiers in Ecology and the Environment* 7: 302–307.

Brown, P.J., R.S. Bradley, and F.T. Keimig. 2010. Changes in extreme climate indices for the northeastern United States, 1870–2005. *Journal of Climate* 23: 6555–6572.

Brown, R.A., and W.F. Hunt. 2010. Impacts of construction activity on bioretention performance. *Journal of Hydrologic Engineering* 15: 386–394.

Brown, R.R. 2005. Impediments to integrated urban stormwater management: The need for institutional reform. *Environmental Management* 36: 455–468.

Brown, R.R., and M.A. Farrelly. 2009a. Challenges ahead: Social and institutional factors influencing sustainable urban stormwater management in Australia. *Water Science and Technology* 59: 653–660.

Brown, R.R., and M.A. Farrelly. 2009b. Delivering sustainable urban water management: A review of the hurdles we face. *Water Science and Technology* 59: 839–846.

Burns, D.A., and C. Kendall. 2002. Analysis of $\delta^{15}N$ $\delta^{18}O$ to differentiate NO_3^- sources in runoff at two watersheds in the Catskill Mountains of New York. *Water Resources Research* 38: 10.1029/2001WR000292.

Burton, G.A., Jr., R. Pitt, and S. Clark. 2000. The role of traditional and novel toxicity test methods in assessing stormwater and sediment contamination. *Critical Reviews in Environmental Science and Technology* 30: 413–447.

Burton, T.M., and J.E. Hook. 1979. Non-point source pollution from abandoned agricultural land in the Great Lakes basin. *Journal of Great Lakes Research* 5: 99–104.

Bushey, J.T., A.G. Nallana, M.R. Montesdeoca, and C.T. Driscoll. 2008. Mercury dynamics of a northern hardwood canopy. *Atmospheric Environment* 42: 6905–6914.

Butler, T.J., M.D. Cohen, F.M. Vermeylen, G.E. Likens, D. Schmeltz, and R.S. Artz. 2008. Regional precipitation mercury trends in the eastern USA, 1998–2005: Declines in the Northeast and Midwest, no trend in the Southeast. *Atmospheric Environment* 42: 1582–1592.

Caissie, D. 2006. The thermal regime of rivers: A review. *Freshwater Biology* 51: 1389–1406.

Caltrans (California Department of Transportation). 2009. Compost and low impact development technical memorandum. Final report. CTSW-TM-07-172.51.4 (Sacramento, CA).

Camponelli, K.M., R.E. Casey, J.W. Snodgrass, S.M. Lev, and E.R. Landa. 2009. Impacts of weathered tire debris on the development of *Rana sylvatica* larvae. *Chemosphere* 74: 717–722.

Carter, T., and C.R. Jackson. 2007. Vegetated roofs for stormwater management at multiple spatial scales. *Landscape and Urban Planning* 80: 84–94.

Carter, T.L., and T.C. Rasmussen. 2006. Hydrologic behavior of vegetated roofs. *Journal of the American Water Resources Association* 42: 1261–1274.

Cassman, K.G., A. Dobermann, D.T. Walters, and H. Yang. 2003. Meeting cereal demand while protecting natural resources and improving environmental quality. *Annual Review of Environment and Resources* 28: 315–358.

Cey, E.E., D.L. Rudolph, R. Aravena, and G. Parkin. 1999. Role of the riparian zone in controlling the distribution and fate of agricultural nitrogen near a small stream in southern Ontario. *Journal of Contaminant Hydrology* 37: 45–67.

Chadwick, M.A., D.R. Dobberfuhl, A.C. Benke, A.D. Huryn, K. Suberkropp, and J.E. Thiele. 2006. Urbanization affects stream ecosystem function by altering hydrology, chemistry, and biotic richness. *Ecological Applications* 16: 1796–1807.

Chang, H., and T.N. Carlson. 2005. Water quality during winter storm events in Spring Creek, Pennsylvania USA. *Hydrobiologia* 544: 321–332.

Chee-Sanford, J.C., R.I. Aminov, I.J. Krapac, N. Garrigues-Jeanjean, and R.I. Mackie. 2001. Occurrence and diversity of tetracycline resistance genes in lagoons and groundwater underlying two swine production facilities. *Applied and Environmental Microbiology* 67: 1494–1502.

Chen, C.Y., and C.L. Folt. 2005. High plankton densities reduce mercury biomagnification. *Environmental Science and Technology* 39: 115–121.

Chen, C.Y., P.C. Pickhardt, M.Q. Xu, and C.L. Folt. 2008a. Mercury and arsenic bioaccumulation and eutrophication in Baiyangdian Lake, China. *Water, Air and Soil Pollution* 190: 115–127.

Chen, C.Y., N. Serrell, D.C. Evers, B.J. Fleishman, K.F. Lambert, J. Weiss, R.P. Mason, and M.S. Bank. 2008b. Methylmercury in marine ecosystems: From sources to seafood consumers. *Environmental Health Perspectives* 116: 1706–1712.

Chen, C.Y., R.S. Stemberger, N.C. Kamman, B.M. Mayes, and C.L. Folt. 2005. Patterns of Hg bioaccumulation and transfer in aquatic food webs across multi-lake studies in the Northeast US. *Ecotoxicology* 14: 135–147.

Chen, Y., R.C. Viadero Jr., X. Wei, R. Fortney, L.B. Hedrick, S.A. Welsh, J.T. Anderson, and L.-S. Lin. 2009. Effects of highway construction on stream water quality and macroinvertebrate condition in a mid-Atlantic highlands watershed, USA. *Journal of Environmental Quality* 38: 1672–1682.

Church, J.A., N.J. White, L.F. Konikow, C.M. Domingues, J.G. Cogley, E. Rignot, J.M. Gregory, M.R. van den Broeke, A.J. Monaghan, and I. Velicogna. 2011. Revisiting the earth's sea-level and energy budgets from 1961 to 2008. *Geophysical Research Letters* 38: L18601.

Clark, S.E., K.A. Steele, J. Spicher, C.Y.S. Siu, M.M. Lalor, R. Pitt, and J.T. Kirby. 2008. Roofing materials' contributions to storm-water runoff pollution. *Journal of Irrigation and Drainage Engineering* 134: 638–645.

Claytor, R.A., and T.R. Schueler. 1996. *Design of stormwater filtering systems.* Center for Watershed Protection (Silver Spring, MD).

Cleveland, C.C., A.R. Townsend, D.S. Schimel, H. Fisher, R.W. Howarth, L.O. Hedin, S.S. Perakis, E.F. Latty, J.C. von Fischer, A. Elseroad, and M.F. Wasson. 1999. Global patterns of terrestrial biological nitrogen (N_2) fixation in natural systems. *Global Biogeochemical Cycles* 13: 623–645.

Coes, A.L., T.B. Spruill, and M.J. Thomasson. 2007. Multiple-method estimation of recharge rates at diverse locations in the North Carolina coastal plain, USA. *Hydrogeology Journal* 15: 773–788.

Cook, B.R., and C.J. Spray. 2012. Ecosystem services and integrated water resource management: Different paths to the same end? *Journal of Environmental Management* 109: 93–100.

Cooper, S.R., S.K. McGlothlin, M. Madritch, and D.L. Jones. 2004. Paleoecological evidence of human impacts on the Neuse and Pamlico estuaries of North Carolina, USA. *Estuaries* 27: 617–633.

Copeland, C. 2010. Clean water act: A summary of the law. Congressional Research Service Report for Congress RL30030 (Washington, DC).

Copolovici, L., A. K annaste, T. Remmel, V. Vislap, and U. Niinemets. 2011. Volatile emissions from *Alnus glutionosa* induced by herbivory are quantitatively related to the extent of damage. *Journal of Chemical Ecology* 37: 18–28.

Corsi, S.R., L.A. De Cicco, M.A. Lutz, and R.M. Hirsch. 2015. River chloride trends in snow-affected urban watersheds: Increasing concentrations outpace urban growth rate and are common among all seasons. *Science of the Total Environment* 508: 488–497.

Councell, T.B., K.U. Duckenfield, E.R. Landa, and E. Callender. 2004. Tire-wear particles as a source of zinc to the environment. *Environmental Science and Technology* 38: 4206–4214.

Cox, C. 2001. Atrazine: Toxicology. *Journal of Pesticide Reform* 21: 12–20.

Cristol, D.A., R.L. Brasso, A.M. Condon, R.E. Fovargue, S.L. Friedman, K.K. Hallinger, A.P. Monroe, and A.E. White. 2008. The movement of aquatic mercury through terrestrial food webs. *Science* 320: 335.

Curren, J., S. Bush, S. Ha, M.K. Stenstrom, S.-L. Lau, and I.H.M. Suffet. 2011. Identification of subwatershed sources for chlorinated pesticides and polychlorinated biphenyls in the Ballona Creek watershed. *Science of the Total Environment* 409: 2525–2533.

Curtis, M.C. 2002. Street sweeping for pollutant removal. Report by the Department of Environmental Protection (Montgomery County, MD).

Damásio, J., D. Barceló, R. Brix, C. Postigo, M. Gros, M. Petrovic, S. Sabater, H. Guasch, M.L. de Alda, and C. Barata. 2011. Are pharmaceuticals more harmful than other pollutants to aquatic invertebrate species: A hypothesis tested using multi-biomarker and multi-species responses in field collected and transplanted organisms. *Chemosphere* 85: 1548–1554.

Davies, S.P., and S.K. Jackson. 2006. The biological condition gradient: A descriptive model for interpreting change in aquatic ecosystems. *Ecological Applications* 16: 1251–1266.

Death, R.G., and K.J. Collier. 2010. Measuring stream macroinvertebrate responses to gradients of vegetation cover: When is enough enough? *Freshwater Biology* 55: 1447–1464.

Delucchi, M.A. 2000. Environmental externalities of motor-vehicle use in the US. *Journal of Transport Economics and Policy* 34: 135–168.

DeMott, R.P., and T.D. Gauthier. 2014. Comment on "PAH concentrations in lake sediment decline following ban on coal-tar-based pavement sealants in Austin, Texas." *Environmental Science and Technology* 48: 14061–14062.

de Paggi, S.J., P. Juan, C. Pablo, C. Jorge, and G. Bernal. 2008. Water quality and zooplankton composition in a receiving pond of the stormwater runoff from an urban catchment. *Journal of Environmental Biology* 29: 693–700.

Descloux, S., T. Datry, M. Philippe, and P. Marmonier. 2010. Comparison of different techniques to assess surface and subsurface streambed colmation with fine sediments. *International Review of Hydrobiology* 95: 520–540.

de Vries, J.J., and I. Simmers. 2002. Groundwater recharge: An overview of processes and challenges. *Hydrogeology Journal* 10: 5–17.

Dietz, M.E. 2007. Low impact development practices: A review of current research and recommendations for future directions. *Water, Air and Soil Pollution* 186: 351–363.

Dietz, M.E., and J.C. Clausen. 2005. A field evaluation of rain garden flow and pollutant treatment. *Water, Air, and Soil Pollution* 167: 123–138.

Dixon, P.G., and T.L. Mote. 2003. Patterns and causes of atlanta's urban heat island-initiated precipitation. *Journal of Applied Meteorology* 42: 1273–1284.

Dodds, W.K. 2003. Misuse of inorganic N and soluble reactive P concentrations to indicate nutrient status of surface waters. *Journal of the North American Benthological Society* 22: 171–181.

Dodds, W.K., W.W. Bouska, J.L. Eitzmann, T.J. Pilger, K.L. Pitts, A.J. Riley, J.T. Schloesser, and D.J. Thornbrugh. 2009. Eutrophication of U.S. freshwaters: Analysis of potential economic damages. *Environmental Science and Technology* 43: 12–19.

Dodds, W.K., J.R. Jones, and E.B. Welch. 1998. Suggested classification of stream trophic state: Distributions of temperate stream types by chlorophyll, total nitrogen, and phosphorus. *Water Research* 32: 1455–1462.

Donohue, I., and J.G. Molinos. 2009. Impacts of increased sediment loads on the ecology of lakes. *Biological Reviews* 84: 517–531.

Downing, J.A., and E. McCauley. 1992. The nitrogen:phosphorus relationship in lakes. *Limnology and Oceanography* 37: 936–945.

Drapper, D., R. Tomlinson, and P. Williams. 2000. Pollutant concentrations in road runoff: Southeast Queensland case study. *Journal of Environmental Engineering* 126: 313–320.

Driscoll, C.T., Y-J. Han, C.Y. Chen, D.C. Evers, K.F. Lambert, T.M. Holsen, N.C. Kamman, and R.K. Munson. 2007. Mercury contamination in forest and freshwater ecosystems in the northeastern United States. *Bioscience* 57: 17–28.

Ebdon, J.S., A.M. Petrovic, and R.A. White. 1999. Interaction of nitrogen, phosphorus, and potassium on evapotranspiration rate and growth of Kentucky bluegrass. *Crop Science* 39: 209–218.

Eertman, R.H.M., C.L.F.M.G. Groenink, B. Sandee, H. Hummel, and A.C. Smaal. 1995. Response of the blue mussel *Mytilus edulis* L. following exposure to PAHs or contaminated sediment. *Marine Environmental Research* 39: 169–173.

Eganhouse, R.P., and P.M. Sherblom. 2001. Anthropogenic organic contaminants in the effluent of a combined sewer overflow: Impact on Boston Harbor. *Marine Environmental Research* 51: 51–74.

Egodawatta, P., E. Thomas, and A. Goonetilleke. 2009. Understanding the physical processes of pollutant build-up and wash-off on roof surfaces. *Science of the Total Environment* 407: 1834–1841.

Ehrenfeld, J.G., H.B. Cutway, R. Hamilton IV, and E. Stander. 2003. Hydrologic description of forested wetlands in northeastern New Jersey, USA: An urban/suburban region. *Wetlands* 23: 685–700.

Ely, J.C., C.R. Neal, C.F. Kulpa, M.A. Schneegurt, J.A. Seidler, and J.C. Jain. 2001. Implications of platinum-group element accumulation along U.S. roads from catalytic-converter attrition. *Environmental Science and Technology* 35: 3816–3822.

Eskenazi, B., J. Chevrier, S.A. Rauch, K. Kogut, K.G. Harley, C. Johnson, C. Trujillo, A. Sjodin, and A. Bradman. 2012. In utero and childhood polybrominated diphenyl ether (PBDE) exposures and neurodevelopment in the CHAMACOS study. *Environmental Health Perspectives* 121: 257–262.

Evers, D.C., N.M. Burgess, L. Champoux, B. Hoskins, A. Major, W.M. Goodale, R.J. Taylor, R. Poppenga, and T. Daigle. 2005. Patterns and interpretation of mercury exposure in freshwater avian communities in northeastern North America. *Ecotoxicology* 14: 193–221.

Evers, D.C., Y.-J. Han, C.T. Driscoll, N.C. Kamman, M.W. Goodale, K.F. Lambert, T. Holsen, C.Y. Chen, T.A. Clair, and T. Butler. 2007. Biological mercury hotspots in the northeastern United States and southeastern Canada. *Bioscience* 57: 29–43.

Evett, J.B., M.A. Love, and J.M. Gordon. 1994. Effects of urbanization and land-use changes on low streamflow. UNC-WRRI-94-284. Water Resources Research Institute of the University of North Carolina (Raleigh, NC).

Fent, K., A.A. Weston, and D. Caminada. 2006. Ecotoxicology of human pharmaceuticals. *Aquatic Toxicology* 76: 122–159.

Fletcher, T.D., V.G. Mitchell, A. Deletic, T.R. Ladson, and A. Séven. 2007. Is stormwater harvesting beneficial to urban waterway environmental flows? *Water Science and Technology* 55: 265–272.

Focazio, M.J., D.W. Kolpin, K.K. Barnes, E.T. Furlong, M.T. Meyerd, S.D. Zaugg, L.B. Barber, and M.E. Thurman. 2008. A national reconnaissance for pharmaceuticals and other organic wastewater contaminants in the United States: II) Untreated drinking water sources. *Science of the Total Environment* 402: 201–216.

Förster, J. 1996. Patterns of roof runoff contamination and their potential implications on practice and regulation of treatment and local infiltration. *Water Science and Technology* 33: 39–48.

Foulquier, A., F. Malard, S. Barraud, and J. Gibert. 2009. Thermal influence of urban groundwater recharge from stormwater infiltration basins. *Hydrological Processes* 23: 1701–1713.

Foulquier, A., F. Malard, F. Mermillod-Blondin, T. Datry, L. Simon, B. Montuelle, and J. Gibert. 2010. Vertical change in dissolved organic carbon and oxygen at the water table region of an aquifer recharged with stormwater: Biological uptake or mixing? *Biogeochemistry* 99: 31–47.

Frank, R., H. Lumsden, J.E. Barrt, and H.E. Braun. 1983. Residues of organochlorine insecticides, industrial chemicals, and mercury in eggs and in tissues taken from healthy and emaciated common loons, Ontario, Canada, 1968–1980. *Archives of Environmental Contamination and Toxicology* 12: 641–654.

Fries, G.F. 1985. The PBB episode in Michigan: An overall appraisal. *Critical Reviews in Toxicology* 16: 105–156.

Furumai, H., H. Balmer, and M. Boller. 2002. Dynamic behavior of suspended pollutants and particle size distribution in highway runoff. *Water Science and Technology* 46: 413–418.

Gallagher, M.W., E. Nemitz, J.R. Dorsey, D. Fowler, M.A. Sutton, M. Flynn, and J. Duyzer. 2002. Measurements and parameterizations of small aerosol deposition velocities to grassland, arable crops, and forest: Influence of surface roughness length on deposition. *Journal of Geophysical Research* 113: D12201.

Galli, J. 1990. *Thermal impacts associated with urbanization and stormwater management best management practices.* Metropolitan Washington Council of Governments (Washington, DC).

Galloway, J.N., J.D. Aber, J.W. Erisman, S.P. Seitzinger, R.W. Howarth, E.B. Cowling, and J. Cosby. 2003. The nitrogen cascade. *Bioscience* 53: 341–356.

Galloway, J.N., and E.B. Cowling. 2002. Reactive nitrogen and the world: 200 years of change. *Ambio* 31: 64–71.

Gebbers, R., and V.I. Adamchuk. 2010. Precision agriculture and food security. *Science* 327: 828–831.

Genito, D., W.J. Gburek, and A.N. Sharpley. 2002. Response of stream macroinvertebrates to agricultural land cover in a small watershed. *Journal of Freshwater Ecology* 17: 109–119.

German, J., and G. Svensson. 2002. Metal content and particle size distribution of street sediments and street sweeping waste. *Water Science and Technology* 46: 191–198.

Getter, K.L., and D.B. Rowe. 2008. Media depth influences *Sedum* green roof establishment. *Urban Ecosystems* 11: 361–372.

Getter, K.L., and D.B. Rowe. 2009. Substrate depth influences Sedum plant community on a green roof. *Hortscience* 44: 401–407.

Getter, K.L., D.B. Rowe, and J.A. Andresen. 2007. Quantifying the effect of slope on extensive green roof stormwater retention. *Ecological Engineering* 31: 225–231.

Gewurtz, S.B., S.P. Bhavsar, E. Awad, J.G. Winter, E.J. Reiner, R. Moody, and R. Fletcher. 2011. Trends of legacy and emerging-issue contaminants in Lake Simcoe fishes. *Journal of Great Lakes Research* 37: 148–159.

Gewurtz, S.B., R. Lazar, and G.D. Haffner. 2000. Comparison of polycyclic aromatic hydrocarbon and polychlorinated biphenyl dynamics in benthic invertebrates of Lake Erie, USA. *Environmental Toxicology and Chemistry* 19: 2943–2950.

Gianguzza, A., and S. Orecchio. 2006. The PAH composition in limpets (*Patella vulgata L.*) from the coasts of Sicily (Italy). *Polycyclic Aromatic Compounds* 26: 35–57.

Gift, D.M., P.M. Groffman, S.S. Kaushal, and P.M. Mayer. 2010. Denitrification potential, root biomass, and organic matter in degraded and restored urban riparian zones. *Restoration Ecology* 18: 113–120.

Glenn, D.W., and J.J. Sansalone. 2002. Accretion and partitioning of heavy metals associated with snow exposed to urban traffic and winter storm maintenance activities. II. *Journal Of Environmental Engineering* 128: 167–185.

Göbel, P., C. Dierkes, and W.G. Coldewey. 2007. Stormwater runoff concentration matrix for urban areas. *Journal of Contaminant Hydrology* 91: 26–42.

Goetz, S., and G. Fiske. 2008. Linking the diversity and abundance of stream biota to landscapes in the mid-Atlantic USA. *Remote Sensing of Environment* 112: 4075–4085.

Golden, H.E., E.W. Boyer, M.G. Brown, S.T. Purucker, and R.H. Germain. 2009. Spatial variability of nitrate concentrations under diverse conditions in tributaries to a lake watershed. *Journal of the American Water Resources Association* 45: 945–962.

Gray, J.R., G.D. Glysson, L.M. Turcios, and G.E. Schwarz. 2000. Comparability of suspended-sediment concentration and total suspended solids data. USGS Water Resources Investigations Report 00-4191 (Washington, DC).

Greaves, A.K., R.J. Letcher, C. Sonne, and R. Dietz. 2013. Brain region distribution and patterns of bioaccumulative perfluoroalkyl carboxylates and sulfonates in east Greenland polar bears (*Ursus maritimus*). *Environmental Toxicology and Chemistry* 32: 713–722.

Green, T.R., M. Taniguchi, H. Kooi, J.J. Gurdak, D.M. Allen, K.M. Hiscock, H. Treidel, and A. Aureli. 2011. Beneath the surface of global change: Impacts of climate change on groundwater. *Journal of Hydrology* 405: 532–560.

Gregory, J.H., M.D. Dukes, P.H. Jones, and G.L. Miller. 2006. Effect of urban soil compaction on infiltration rate. *Journal of Soil and Water Conservation* 61: 117–124.

Griffiths, M.L., and R.S. Bradley. 2007. Variations of twentieth century temperature and precipitation extreme indicators in the Northeast United States. *Journal of Climate* 20: 5401–5417.

Grimmond, C.S.B., and T.R. Oke. 1986. Urban water balance: 2. Results from a suburb of Vancouver, British Columbia. *Water Resources Research* 22: 1404–1412.

Grimmond, C.S.B., and T.R. Oke. 1991. An evapotranspiration–interception model for urban areas. *Water Resources Research* 27: 1739–1755.

Groffman, P.M., and M.K. Crawford. 2003. Denitrification potential in urban riparian zones. *Journal of Environmental Quality* 32: 1144–1149.

Hageman, K.J., S.L. Simonich, D.H. Campbell, G.R. Wilson, and D.H. Landers. 2006. Atmospheric deposition of current-use and historic-use pesticides in snow at national parks in the western United States. *Environmental Science and Technology* 40: 3174–3180.

Hale, R.C., M.J. La Guardia, E. Harvey, and T.M. Mainor. 2002. Potential role of fire retardant-treated polyurethane foam as a source of brominated diphenyl ethers to the US environment. *Chemosphere* 46: 729–735.

Han, H., N. Bosch, and J.D. Allan. 2011. Spatial and temporal variation in phosphorus budgets for 24 watersheds in the Lake Erie and Lake Michigan basins. *Biogeochemistry* 102: 45–58.

Hand, J.L., B.A. Schichtel, W.C. Malm, and M. L. Pitchford. 2012. Particulate sulfate ion concentration and SO_2 emission trends in the United States from the early 1990s through 2010. *Atmosheric Chemistry and Physics* 12: 10353–10365.

Haque, S.J., S. Onodera, and Y. Shimizu. 2013. An overview of the effects of urbanization on the quantity and quality of groundwater in South Asian megacities. *Limnology* 14: 135–145.

Harmel, D., S. Potter, P. Casebolt, K. Reckhow, C. Green, and R. Haney. 2006. Compilation of measured nutrient load data for agricultural land uses in the United States. *Journal of the American Water Resources Association* 42: 1163–1178.

Harrad, S., and S. Hunter. 2006. Concentrations of polybrominated diphenyl ethers in air and soil on a rural–urban transect across a major UK conurbation. *Environmental Science and Technology* 40: 4548–4553.

Hathaway, A.M., W.F. Hunt, and G.D. Jennings. 2008. A field study of green roof hydrologic and water quality performance. *Transactions of the American Society of Agricultural and Biological Engineers* 51: 37–44.

Hatt, B.E., T.D. Fletcher, C.J. Walsh, and S.L. Taylor. 2004. The influence of urban density and drainage infrastructure on the concentrations and loads of pollutants in small streams. *Environmental Management* 34: 112–124.

Heberer, T. 2002a. Occurrence, fate, and removal of pharmaceutical residues in the aquatic environment: A review of recent research data. *Toxicology Letters* 131: 5–17.

Heberer, T. 2002b. Tracking persistent pharmaceutical residues from municipal sewage to drinking water. *Journal of Hydrology* 266: 175–189.

Heiskary, S., and H. Markus. 2001. Establishing relationships among nutrient concentrations, phytoplankton abundance, and biochemical oxygen demand in Minnesota, USA, rivers. *Journal of Lake and Reservoir Management* 17: 251–262.

Herb, W.R., B. Janke, O. Mohseni, and H.G. Stefan. 2008. Thermal pollution of streams by runoff from paved surfaces. *Hydrological Processes* 22: 987–999.

Herb, W.R., B. Janke, O. Mohseni, and H.G. Stefan. 2009. Runoff temperature model for paved surfaces. *Journal of Hydrologic Engineering* 14: 1146–1155.

Hilyard, E.J., J.M. Jones-Meehan, B.J. Spargo, and R.T. Hill. 2008. Enrichment, isolation, and phylogenetic identification of polycyclic aromatic hydrocarbon-degrading bacteria from Elizabeth River sediments. *Applied and Environmental Microbiology* 74: 1176–1182.

Hites, R.A. 2004. Polybrominated diphenyl ethers in the environment and in people: A meta-analysis of concentrations. *Environmental Science and Technology* 38: 945–956.

Hites, R.A. 2005. Brominated flame retardants in the environment. *Journal of Environmental Monitoring* 7: 1033–1036.

Hogan, D.M., and M.R. Walbridge. 2007. Best management practices for nutrient and sediment retention in urban stormwater runoff. *Journal of Environmental Quality* 36: 386–395.

Hollis, G.E. 1988. Rain, roads, roofs and runoff: Hydrology in cities. *Geography* 73: 9–18.

Hood, M.J., J.C. Clausen, and G.S. Warner. 2007. Comparison of stormwater lag times for low impact and traditional residential development. *Journal of the American Water Resources Association* 43: 1036–1046.

Hoover, M.D. 1944. Effect of removal of forest vegetation upon water-yields. *Transactions of the American Geophysical Union* 25: 969–977.

Hopkins, B.C., J.D. Willson, and W.A. Hopkins. 2013. Mercury exposure is associated with negative effects on turtle reproduction. *Environmental Science and Technology* 47: 2416–2422.

Howarth, R.W., G. Billen, D. Swaney, A. Twonsend, N. Jaworski, K. Lajtha, J.A. Downing, R. Elmgren, N. Caraco, T. Jordan, F. Ferendse, J. Freney, V. Kudeyarov, P. Murdoch, and Z.-L. Zhu. 1996. Regional nitrogen budgets and riverine N&P fluxes for the drainages to the North Atlantic Ocean: Natural and human influences. *Biogeochemistry* 35: 75–139.

Howarth, R.W., D.P. Swaney, E.W. Boyer, R. Marino, N. Jaworski, and C. Goodale. 2006. The influence of climate on average nitrogen export from large watersheds in the northeastern United States. *Biogeochemistry* 79: 163–186.

Hrodey, P.J., T.M. Sutton, and E.A. Frimpong. 2009. Land-use impacts on watershed health and integrity in Indiana warmwater streams. *American Midland Naturalist* 161: 76–95.

Hughes, R.M., P.R. Kaufmann, and M.H. Weber. 2011. National and regional comparisons between Strahler order and stream size. *Journal of the North American Benthological Society* 30: 103–121.

Hunt, W.F., III, and E.Z. Bean. 2006. NC State University permeable pavement research, and changes to the State of NC runoff credit system. Presented at the 8th International Conference on Concrete Block Paving, San Francisco, CA, Nov. 6–8, 2006.

Hunt, W.F., A.R. Jarrett, J.T. Smith, and L.J. Sharkey. 2006. Evaluating bioretention hydrology and nutrient removal at three field sites in North Carolina. *Journal of Irrigation and Drainage Engineering* 132: 600–608.

Hwang, H.-M., and T.L. Wade. 2008. Aerial distribution, temperature-dependent seasonal variation, and sources of polycyclic aromatic hydrocarbons in pine needles from the Houston metropolitan area, Texas, USA. *Journal of Environmental Science and Health A* 43: 1243–1251.

Hwang, H.-M., T.L. Wade, and J.L. Sericano. 2003. Concentrations and source characterization of polycyclic aromatic hydrocarbons in pine needles from Korea, Mexico, and United States. *Atmospheric Environment* 37: 2259–2267.

Hwang, H.-M., T.L. Wade, and J.L. Sericano. 2008. Residue-response relationship between PAH body burdens and lysosomal membrane destabilization in eastern oysters (*Crassostrea virginica*) and toxicokinetics of PAHs. *Journal of Environmental Science and Health A* 43: 1373–1380.

Hyatt, T.L., and R.J. Naiman. 2001. The residence time of large woody debris in the Queets River, Washington, USA. *Ecological Applications* 11: 191–202.

Hylander, L.D., and M. Meili. 2003. 500 years of mercury production: Global annual inventory by region until 2000 and associated emissions. *Science of the Total Environment* 304: 13–27.

Imberger, S.J., C.J. Walsh, and M.R. Grace. 2008. More microbial activity, not abrasive flow or shredder abundance, accelerates breakdown of labile leaf litter in urban streams. *Journal of the North American Benthological Society* 27: 549–561.

Isaak, D.J., S. Wollrab, D. Horan, and G. Chandler. 2012. Climate change effects on stream and river temperatures across the Northwest U.S. from 1980–2009 and implications for salmonid fishes. *Climatic Change* 113: 499–524.

Jackson, A.K., D.C. Evers, S.B. Folsom, A.M. Condon, J. Diener, L.F. Goodrick, A.J. McGann, J. Schmerfeld, and D.A. Cristol. 2011. Mercury exposure in terrestrial birds far downstream of an historical point source. *Environmental Pollution* 159: 3302–3308.

Jackson, J., and R. Sutton. 2008. Sources of endocrine-disrupting chemicals in urban wastewater, Oakland, CA. *Science of the Total Environment* 405: 153–160.

Jacob, J.S., and R. Lopez. 2009. Is denser greener? An evaluation of higher density development as an urban stormwater-quality best management practice. *Journal of the American Water Resources Association* 45: 687–701.

Janke, B.D., O. Mohseni, W.R. Herb, and H.G. Stefan. 2011. Heat release from rooftops during rainstorms in the Minneapolis/St. Paul metropolitan area, USA. *Hydrological Processes* 25: 2018–2031.

Jennings, D.B., and S.T. Jarnagin. 2002. Changes in anthropogenic impervious surfaces, precipitation and daily streamflow discharge: A historical perspective in a mid-Atlantic subwatershed. *Landscape Ecology* 17: 471–489.

Jeremiason, J.D., D.R. Engstrom, E.B. Swain, E.A. Nater, B.M. Johnson, J.E. Almendinger, B.A. Monson, and R.K. Kolka. 2006. Sulfate addition increases methylmercury production in an experimental wetland. *Environmental Science and Technology* 40: 3800–3806.

Johansson, A.-K., U. Sellström, P. Lindberg, A. Bignert, and C.A. de Wit. 2009. Polybrominated diphenyl ether congener patterns, hexabromocyclododecane, and brominated biphenyl 153 in eggs of peregrine falcons (*Falco peregrinus*) breeding in Sweden. *Environmental Toxicology and Chemistry* 28: 9–17.

Jones, K.L., G.C. Poole, J.L. Meyer, W. Bumback, and E.A. Kramer. 2006. Quantifying expected ecological response to natural resource legislation: A case study of riparian buffers, aquatic habitat, and trout populations. *Ecology and Society* 11: 15.

Justić, D., N.N. Rabalais, R.E. Turner, and Q. Dortch. 1995. Changes in nutrient structure of river-dominated coastal waters: Stoichiometric nutrient balance and its consequences. *Estuarine, Coastal and Shelf Science* 40: 339–356.

Kalantzi, O.I., A.J. Hall, G.O. Thomas, and K.C. Jones. 2005. Polybrominated diphenyl ethers and selected organochlorine chemicals in grey seals (*Halichoerus grypus*) in the North Sea. *Chemosphere* 58: 345–354.

Kamman, N.C., N.M. Burgess, C.T. Driscoll, H.A. Simonin, W. Goodale, J. Linehan, R. Estabrook, M. Hutcheson, A. Major, A.M. Scheuhammer, and D.A. Scruton. 2005. Mercury in freshwater fish of northeast North America—A geographic perspective based on fish tissue monitoring databases. *Ecotoxicology* 14: 163–180.

Kanaly, R.A., and S. Harayama. 2000. Biodegradation of high-molecular-weight polycyclic aromatic hydrocarbons by bacteria. *Journal of Bacteriology* 182: 2059–2067.

Kaushal, S.S., P.M. Groffman, L.E. Band, C.A. Shields, R.P. Morgan, M.A. Palmer, K.T. Belt, C.M. Swan, S.E.G. Findlay, and G.T. Fisher. 2008. Interaction between urbanization and climate variability amplifies watershed nitrate export in Maryland. *Environmental Science and Technology* 42: 5872–5878.

Kaushal, S.S., P.M. Groffman, G.E. Likens, K.T. Belt, W.P. Stack, V.R. Kelly, L.E. Band, and G.T. Fisher. 2005. Increased salinization of fresh water in the northeastern United States. *Proceedings of the National Academy of Sciences of the USA* 102: 13517–13520.

Kaushal, S.S., G.E. Likens, N.A. Jaworski, M.L. Pace, A.M. Sides, D. Seekell, K.T. Belt, D.H. Secor, and R.L. Wingate. 2010. Rising stream and river temperatures in the United States. *Frontiers in Ecology and the Environment* 8: 461–466.

Kays, B.L. 1980. Relationship of forest destruction and soil disturbance to increased flooding in the suburban North Carolina Piedmont. *Metropolitan Tree Improvement Alliance Proceedings* 3: 118–1125.

Keim, R.F., A.E. Skaugset, and D.S. Bateman. 2000. Dynamics of coarse woody debris placed in three Oregon streams. *Forest Science* 46: 13–22.

Kennedy, C.D., D.P. Genereux, D.R. Corbett, and H. Mitasova. 2009. Relationships among groundwater age, denitrification, and the coupled groundwater and nitrogen fluxes through a streambed. *Water Resources Research* 45: W09402.

Khetan, S.K., and T.J. Collins. 2007. Human pharmaceuticals in the aquatic environment: A challenge to green chemistry. *Chemical Reviews* 107: 2319–2364.

Kim, L.-H., K.-D. Zoh, S. Jeong, M. Kayhanian, and M.K. Stenstrom. 2006. Estimating pollutant mass accumulation on highways during dry periods. *Journal of Environmental Engineering* 132: 985–993.

King, K.W., J.C. Balogh, K.L. Hughes, and R.D. Harmel. 2007. Nutrient load generated by storm event runoff from a golf course watershed. *Journal of Environmental Quality* 36: 1021–1030.

Kinouchi, T., H. Yagi, and M. Miyamoto. 2007. Increase in stream temperature related to anthropogenic heat input from urban wastewater. *Journal of Hydrology* 335: 78–88.

Knight, E.M.P., W.F. Hunt III, and R.J. Winston. 2013. Side-by-side evaluation of four level spreader-vegetated filter strips and a swale in eastern North Carolina. *Journal of Soil and Water Conservation* 68: 60–72.

Köhler, M. 2008. Green facades: A view back and some visions. *Urban Ecosystems* 11: 423–436.

Kok, J.F., and N.O. Renno. 2009. A comprehensive numerical model of steady state saltation (COMSALT). *Journal of Geophysical Research* 114: D17204.

Kolpin, D.W., E.T. Furlong, M.T. Meyer, E.M. Thurman, S.D. Zaugg, L.B. Barber, and H.T. Buxton. 2002. Pharmaceuticals, hormones, and other organic wastewater contaminants in U.S. streams, 1999–2000: A national reconnaissance. *Environmental Science and Technology* 36: 1202–1211.

Konrad, C.P., and D.B. Booth. 2002. Hydrologic trends associated with urban development in western Washington streams. USGS Water-Resources Investigations Report 02-4040 (Tacoma, WA).

Konrad, C.P., and D.B. Booth. 2005. Hydrologic changes in urban streams and their ecological significance. *American Fisheries Society Symposium* 47: 157–177.

Koster, M.D., D.P. Ryckman, D.V.C. Weseloh, and J. Struger. 1996. Mercury levels in Great Lakes herring gull (*Larus argentatus*) eggs, 1972–1992. *Environmental Pollution* 93: 261–270.

Kumar, S.V., Merwade, J. Kam, and K. Thurner. 2009. Streamflow trends in Indiana: Effects of long term persistence, precipitation and subsurface drains. *Journal of Hydrology* 374: 171–183.

Kunkel, K.E., K. Andsager, and D.R. Easterling. 1999. Long-term trends in extreme precipitation events over the conterminous United States and Canada. *Journal of Climate* 12: 2515–2527.

Kunkel, K.E., M.A. Palecki, L. Ensor, D. Easterling, K.G. Hubbard, D. Robinson, and K. Redmond. 2009. Trends in twentieth-century U.S. extreme snowfall seasons. *Journal of Climate* 22: 6204–6216.

Lal, R. 2003. Global potential of soil carbon sequestration to mitigate the greenhouse effect. *Critical Reviews in Plant Sciences* 22: 151–184.

LaPara, T.M., T.R. Burch, P.J. McNamara, D.T. Tan, M. Yan, J.J. Eichmiller. 2011. Tertiary-treated municipal wastewater is a significant point source of antibiotic resistance genes into Duluth-Superior Harbor. *Environmental Science and Technology* 45: 9543–9549.

Larson, M.G., D.B. Booth, and S.A. Morley. 2001. Effectiveness of large woody debris in stream rehabilitation projects in urban basins. *Ecological Engineering* 18: 211–226.

Law, R.J., C.R. Allchin, J. de Boer, A. Covaci, D. Herzke, P. Lepom, S. Morris, J. Tronczynski, and C.A. de Wit. 2006. Levels and trends of brominated flame retardants in the European environment. *Chemosphere* 64: 187–208.

Lawrence, G.B., G.M. Lovett, and Y.H. Baevsky. 2000. Atmospheric deposition and watershed nitrogen export along an elevational gradient in the Catskill Mountains, New York. *Biogeochemistry* 50: 21–43.

Lee, J.G., and J.P. Heaney. 2003. Estimation of urban imperviousness and its impacts on storm water systems. *Journal of Water Resources Planning and Management* 129: 419–426.

Leopold, L.B. 1968. *Hydrology for urban land planning: A guidebook on the hydrologic effects of land use.* USGS Circular 554 (Reston, VA).

Leopold, L.B., M.G. Wolman, and J.P. Miller. 1964. *Fluvial processes in geomorphology.* Freeman and Company (San Francisco, CA).

Lerner, D.N. 2002. Identifying and quantifying urban recharge: A review. *Hydrogeology Journal* 10: 143–152.

Lester, R.E., and A.J. Boulton. 2008. Rehabilitating agricultural streams in Australia with wood: A review. *Environmental Management* 42: 310–326.

Lester, R.E., W. Wright, and M. Jones-Lennon. 2007. Does adding wood to agricultural streams enhance biodiversity? An experimental approach. *Marine and Freshwater Research* 58: 687–698.

Li, H., L.J. Sharkey, W.F. Hunt, and A.P. Davis. 2009. Mitigation of impervious surface hydrology using bioretention in North Carolina and Maryland. *Journal of Hydrologic Engineering* 14: 407–415.

Likens, G.E., D.C. Buso, and T.J. Butler. 2005. Long-term relationships between SO_2 and NO_x emissions and $SO_4{}^{2-}$ and $NO_3{}^-$ concentration in bulk deposition at the Hubbard Brook Experimental Forest, NH. *Journal of Environmental Monitoring* 7: 964–968.

Limburg, K.E., and R.E. Schmidt. 1990. Patterns of fish spawning in Hudson River tributaries: Response to an urban gradient? *Ecology* 71: 1238–1245.

Lindberg, S., R. Bullock, R. Ebinghaus, D. Engstrom, X. Feng, W. Fitzgerald, N. Pirrone, E. Prestbo, and C. Seigneur. 2007. A synthesis of progress and uncertainties in attributing the sources of mercury in deposition. *Ambio* 36: 19–32.

Line, D.E., and N.M. White. 2007. Effects of development on runoff and pollutant export. *Water Environment Research* 79: 185–189.

Liu, Z.-J., D.E. Weller, D.L. Correll, and T.E. Jordan. 2000. Effects of land cover and geology on stream chemistry in watersheds of Chesapeake Bay. *Journal of the American Water Resources Association* 36: 1349–1365.

Lovett, G.M., T.H. Tear, D.C. Evers, S.E.G. Findlay, B.J. Cosby, J.K. Dunscomb, C.T. Driscoll, and K.C. Weathers. 2009. Effects of air pollution on ecosystems and biological diversity in the eastern United States. *The Year in Ecology and Conservation Biology* 1162: 99–135.

Lovett, G.M., M.M. Traynor, R.V. Pouyat, M.M. Carreiro, W.-X. Zhu, and J.W. Baxter. 2000. Atmospheric deposition to oak forests along an urban–rural gradient. *Environmental Science and Technology* 34: 4294–4300.

Lowrance, R., L.S. Altier, J.D. Newbold, R.R. Schnabel, P.M. Groffman, J.M. Denver, D.L. Correll, J.W. Gilliam, J.L. Robinson, R.B. Brinsfield, K.W. Staver, W. Lucas, and A.H. Todd. 1997. Water quality functions of riparian forest buffers in Chesapeake Bay watersheds. *Environmental Management* 21: 687–712.

Lowrance, R., R. Leonard, and J. Sheridan. 1985. Managing riparian ecosystems to control nonpoint pollution. *Journal of Soil and Water Conservation* 40: 87–91.

Lu, R., R.P. Turco, K. Stolzenbach, S.K. Friedlander, C. Xiong, K. Schiff, L. Tiefenthaler, and G. Wang. 2003. Dry deposition of airborne trace metals on the Los Angeles Basin and adjacent coastal waters. *Journal of Geophysical Research* 108: 4074.

MacDonald, D.D., C.G. Ingersoll, and T.A. Berger. 2000. Development and evaluation of consensus-based sediment quality guidelines for freshwater ecosystems. *Archives of Environmental Contamination and Toxicology* 39: 20–31.

Mahler, B.J., P.C. Van Metre, T.J. Bashara, J.T. Wilson, and D.A. Johns. 2005. Parking lot sealcoat: An unrecognized source of urban polycyclic aromatic hydrocarbons *Environmental Science and Technology* 39: 5560–5566.

Mahler, B.J., P.C. Van Metre, and E. Callender. 2006. Trends in metals in urban and reference lake sediments across the United States, 1970 to 2001. *Environmental Toxicology and Chemistry* 25: 1698–1709.

Mahler, B.J., P.C. Van Metre, J.T. Wilson, M. Musgrove, T.L. Burbank, T.E. Ennis, and T.J. Bashara. 2010. Coal-tar-based parking lot sealcoat: An unrecognized source of PAH to settled house dust. *Environmental Science and Technology* 44: 894–900.

Manikkam, M., R. Tracey, C. Guerrero-Bosagna, and M.K. Skinner. 2013. Plastics derived endocrine disruptors (BPA, DEHP and DBP) induce epigenetic transgenerational inheritance of obesity, reproductive disease and sperm epimutations. *PLoS ONE* 8: e55387.

Maniquiz, M.C., J. Choi, S. Lee, H.J. Cho, and L.-H. Kim. 2010. Appropriate methods in determining the event mean concentration and pollutant removal efficiency of a best management practice. *Environmental Engineering Research* 15: 215–223.

Marcus, W.A. 2002. Mapping of stream microhabitats with high spatial resolution hyperspectral imagery. *Journal of Geographical Systems* 4: 113–126.

Marmonier, P., Y. Delettre, S. Lefebvre, J. Guyon, and A.J. Boulton. 2004. A simple technique using wooden stakes to estimate vertical patterns of interstitial oxygenation in the beds of rivers. *Archive für Hydrobiologie* 160: 133–143.

Martin, D.J., and L.E. Benda. 2001. Patterns of instream wood recruitment and transport at the watershed scale. *Transactions of the American Fisheries Society* 130: 940–958.

Martínez-Cortizas, A., X. Pontevedra-Pombal, E. García-Rodeja, J.C. Nóvoa-Muñoz, and W. Shotyk. 1999. Mercury in a Spanish peat bog: Archive of climate change and atmospheric metal deposition. *Science* 284: 939–942.

Masterson, J.P., and R.T. Bannerman. 1994. Impacts of storm water runoff on urban streams in Milwaukee Co., Wisconsin. In: *Proceedings of the American Water Resources Association, National Symposium on Water Quality*, pp. 123–133. AWRA (Herndon, VA).

Mayer, P.M., S.K. Reynolds, T.J. Canfield, and M.D. McCutchen. 2005. Riparian buffer width, vegetative cover, and nitrogen removal effectiveness: A review of current science and regulations. EPA/600/R-05/118 (Cincinnati, OH).

McArthur, J.V., J.M. Aho, R.B. Rader, and G.L. Mills. 1994. Interspecific leaf interactions during decomposition in aquatic and floodplain ecosystems. *Journal of the North American Benthological Society* 13: 57–67.

McClain, M.E., E.W. Boyer, C.L. Dent, S.E. Gergel, N.B. Grimm, P.M. Groffman, S.C. Hart, J.W. Harvey, C.A. Johnston, E. Mayorga, W.H. McDowell, and G. Pinay. 2003. Biogeochemical hot spots and hot moments at the interface of terrestrial and aquatic ecosystems. *Ecosystems* 6: 301–312.

McDaniel, T.V, P.A. Martin, N. Ross, S. Brown, S. Lesage, and B.D. Pauli. 2004. Effects of chlorinated solvents on four species of North American amphibians. *Archives of Environmental Contamination and Toxicology* 47: 101–109.

McDonald, A.G., W.J. Bealey, D. Fowler, U. Dragosits, U. Skiba, R.I. Smith, R.G. Donovan, H.E. Brett, C.N. Hewitt, and E. Nemitz. 2007. Quantifying the effect of urban tree planting on concentrations and depositions of PM_{10} in two UK conurbations. *Atmospheric Environment* 41: 8455–8467.

McHenry, M.L., E. Shott, R.H. Conrad, and G.B. Grette. 1998. Changes in the quantity and characteristics of large woody debris in streams of the Olympic Peninsula, Washington, U.S.A. (1982–1993). *Canadian Journal of Fisheries and Aquatic Sciences* 55: 1395–1407.

Meyer, J.L., L.A. Kaplan, D. Newbold, D.L. Strayer, C.J. Woltemade, J.B. Zedler, R. Beilfuss, Q.Carpenter, R. Semlitsch, M.C. Watzin, and P.H. Zedler. 2003. Where rivers are born: The scientific imperative for defending small streams and wetlands. American Rivers and Sierra Club (www.americanrivers.org/WhereRiversAreBorn).

Meyer, J.L., M.J. Paul, and W.K. Taulbee. 2005. Stream ecosystem function in urbanizing landscapes. *Journal of the North American Benthological Society* 24: 602–612.

Meyer, J.L., D.L. Strayer, J.B. Wallace, S.L. Eggert, G.S. Helfman, and N.E. Leonard. 2007. The contribution of headwater streams to biodiversity in river networks. *Journal of the American Water Resources Association* 43: 86–103.

Meyer, S.C. 2005. Analysis of base flow trends in urban streams, northeastern Illinois, USA. *Hydrogeology Journal* 13: 871–885.

Mijangos-Montiel, J.L., F.T. Wakida, and J. Temores-Peña. 2010. Stormwater quality from gas stations in Tijuana, Mexico. *International Journal of Environmental Research* 4: 777–784.

Miller, C.V., J.M. Denis, S.W. Ator, and J.W. Brakebill. 1997. Nutrients in streams during baseflow in selected environmental settings of the Potomac River basin. *Journal of the American Water Resources Association* 33: 1155–1171.

Miltner, R.J., and E.T. Rankin. 1998. Primary nutrients and the biotic integrity of rivers and streams. *Freshwater Biology* 40: 145–158.

Mitchell, V.G., H.A. Cleugh, C.S.B. Grimmond, and J. Xu. 2008. Linking urban water balance and energy balance models to analyse urban design options. *Hydrological Processes* 22: 2891–2900.

Monteith, D.T., J.L. Stoddard, C.D. Evans, H.A. de Wit, M. Forsius, T. Høgåsen, A. Wilander, B.L. Skjelkvåle, D.S. Jeffries, J. Vuorenmaa, B. Keller, J. Kopácek, and J. Vesely. 2007. Dissolved organic carbon trends resulting from changes in atmospheric deposition chemistry. *Nature* 450: 537–540.

Monterusso, M.A., D.B. Rowe, C.L. Rugh, and D.K. Russell. 2004. Runoff water quantity and quality from green roof systems. *Acta Horticulturae* 639: 369–376.

Montreuil, O., P. Merot, and P. Marmonier. 2010. Estimation of nitrate removal by riparian wetlands and streams in agricultural catchments: Effect of discharge and stream order. *Freshwater Biology* 55: 2305–2318.

Moore, A.A., and M.A. Palmer. 2005. Invertebrate biodiversity in agricultural and urban headwater streams: Implications for conservation and management. *Ecological Applications* 15: 1169–1177.

Moore, M.N. 1990. Lysosomal monitoring cytochemistry in marine environmental monitoring. *Histochemical Journal* 22: 187–191.

Moran, M.J., J.S. Zogorski, and P.J. Squillace. 2005. MTBE and gasoline hydrocarbons in ground water of the United States. *Groundwater* 43: 615–627.

Morley, S.A., and J.R. Karr. 2002. Assessing and restoring the health of urban streams in the Puget Sound basin. *Conservation Biology* 16: 1498–1509.

Morse, C.C. 2001. *The response of first and second order streams to urban land-use in Maine, U.S.A.* M.Sc. thesis, Ecology and Environmental Science Program, University of Maine, Orono.

Morse, C.C., A.D. Huryn, and C. Cronan. 2003. Impervious surface area as a predictor of the effects of urbanization on stream insect communities in Maine, U.S.A. *Environmental Monitoring and Assessment* 89: 95–127.

Nadeau, T.-L., and M.C. Rains. 2007. Hydrological connectivity between headwater streams and downstream waters: How science can inform policy. *Journal of the American Water Resources Association* 43: 118–133.

Napier, F., B. D'Arcy, and C. Jefferies. 2008. A review of vehicle related metals and polycyclic aromatic hydrocarbons in the UK environment. *Desalination* 226: 143–150.

Neary, K., and T.B. Boving. 2011. The fate of the aqueous phase polycyclic aromatic hydrocarbon fraction in a detention pond system. *Environmental Pollution* 159: 2882–2890.

Nelson, E.J., and D.B. Booth. 2002. Sediment sources in an urbanizing, mixed land-use watershed. *Journal of Hydrology* 264: 51–68.

Nelson, K.C., and M.A. Palmer. 2007. Stream temperature surges under urbanization and climate change: Data, models, and responses. *Journal of the American Water Resources Association* 43: 440–452.

Nilsson, E., G. Larsen, M. Manikkam, C. Guerrero-Bosagna, M.I. Savenkova, and M.K. Skinner. 2012. Environmentally induced epigenetic transgenerational inheritance of ovarian disease. *PLoS ONE* 7: e36129.

Nixon, H., and J.-D. Saphores. 2007. Impacts of motor vehicle operation on water quality in the US: Cleanup costs and policies. *Transportation Research D* 12: 564–576.

Nixon, S.W., J.W. Ammerman, L.P. Atkinson, V.M. Berounsky, G. Billen, W.C. Boicourt, W.R. Boynton, T.M. Church, D.M. Ditoro, R. Elmgren, J.H. Garber, A.E. Giblin, R.A. Jahnke, N.J.P. Owens, M.E.Q. Pilson, and S.P. Seitzinger. 1996. The fate of nitrogen and phosphorus at the land-sea margin of the North Atlantic Ocean. *Biogeochemistry* 35: 141–180.

Nnadi, F.N., F. Kline, L. Wray, and M.P. Wanielista. 1999. Comparison of critical design storm concepts using continuous simulation with short duration storms. *Journal of American Water Resources Association* 35: 61–72.

Novotny, E.V., and H.G. Stefan. 2007. Stream flow in Minnesota: Indicator of climate change. *Journal of Hydrology* 334: 319–333.

NRC (National Research Council). 2000. *Toxicological effects of methylmercury.* National Academies Press (Washington, DC).

NRC (National Research Council). 2009. *Urban stormwater management in the United States.* National Academies Press (Washington, DC).

NRC (National Research Council). 2010. *Review of the Environmental Protection Agency's draft IRIS assessment of tetrachloroethylene.* National Academies Press (Washington, DC).

Oke, T.R. 1973. City size and the urban heat island. *Atmospheric Environment* 7: 769–779.

Ollinger, S.V., J.D. Aber, G.M. Lovett, S.E. Millham, R.G. Lathrop, and J.M. Ellis. 1993. A spatial model of atmospheric deposition for the northeastern U.S. *Ecological Applications* 3: 459–472.

Olszyna, K.J., M. Luria, and J.F. Meagher. 1997. The correlation of temperature and rural ozone levels in southeastern U.S.A. *Atmospheric Environment* 31: 3011–3022.

Orecchio, S. 2007. PAHs associated with the leaves of *Quercus ilex* L.: Extraction, GC-MS analysis, distribution and sources. Assessment of air quality in the Palermo (Italy) area. *Atmospheric Environment* 41: 8669–8680.

O'Reilly, K., J. Pietari, and P. Boehm. 2010. A review of PAHs: Polycyclic aromatic hydrocarbons in stormwater and urban sediments. *Stormwater* 11: 10–21.

Osmond, D.L., and D.H. Hardy. 2004. Characterization of turf practices in five North Carolina communities. *Journal of Environmental Quality* 33: 565–575.

Paerl, H.W., and J. Huisman. 2009. Climate change: A catalyst for global expansion of harmful cyanobacterial blooms. *Environmental Microbiology Reports* 1: 27–37.

Page, D., P. Dillon, S. Toze, D. Bixio, B. Genthe, B.E. Jiménez Cisneros, and T. Wintgens. 2010. Valuing the subsurface pathogen treatment barrier in water recycling via aquifers for drinking supplies. *Water Research* 44: 1841–1852.

Palmer, M.A., H.L. Menninger, and E. Bernhardt. 2010. River restoration, habitat heterogeneity and biodiversity: A failure of theory or practice? *Freshwater Biology* 55: 205–222.

Pardo, L.H., C. Kendall, J. Pett-Ridge and C.C.Y. Chang. 2004. Evaluating the source of streamwater nitrate using $\delta^{15}N$ and $\delta^{18}O$, in nitrate in two watersheds in New Hampshire, USA. *Hydrological Processes* 18: 2699–2712.

Paulsen, S.G., A. Mayio, D.V. Peck, J.L. Stoddard, E. Tarquinio, S.M. Holdsworth, J. Van Sickle, L.L. Yuan, C.P. Hawkins, A.T. Herlihy, P.R. Kaufmann, M.T. Barbour, D.P. Larsen, and A.R. Olsen. 2008. Condition of stream ecosystems in the US: An overview of the first national assessment. *Journal of the North American Benthological Society* 27: 812–821.

Peierls, B.L., N.F. Caraco, M.L. Pace, and J.J. Cole. 1991. Human influence on river nitrogen. *Nature* 350: 386–387.

Perry, E., S.A. Norton, N.C. Kamman, P.M. Lorey, and C.T. Driscoll. 2005. Deconstruction of historic mercury accumulation in lake sediments, northeastern United States. *Ecotoxicology* 14: 85–99.

Peters, N.E. 2009. Effects of urbanization on stream water quality in the city of Atlanta, Georgia, USA. *Hydrological Processes* 23: 2860–2878.

Peterson, B.J., W.M. Wollheim, P.J. Mulholland, J.R. Webster, J.L. Meyer, J.L. Tank, E. Marti, W.B. Bowden, H.M. Valett, A.E. Hershey, W.H. McDowell, W.K. Dodds, S.K. Hamilton, S. Gregory, and D.D. Morrall. 2001. Control of nitrogen export from watersheds by headwater streams. *Science* 292: 86–90.

Peterson, D.M., and J.L. Wilson. 1988. *Variably saturated flow between streams and aquifers.* Technical Completion Report 233. New Mexico Water Resources Research Institute (Socorro, NM).

Peterson, T.C., X. Zhang, M. Brunet-India, and J. L. Vázquez-Aguirre. 2008. Changes in North American extremes derived from daily weather data. *Journal of Geophysical Research* 113: D07113.

Pickett, S.T.A., M.L. Cadenasso, J.M. Grove, C.H. Nilon, R.V. Pouyat, W.C. Zipperer, and R. Costanza. 2001. Urban ecological systems: Linking terrestrial ecological, physical, and socioeconomic components of metropolitan areas. *Annual Reviews in Ecology and Systematics* 32: 127–57.

Pickhardt, P.C., C.L. Folt, C.Y. Chen, B. Klaue, and J.D. Blum. 2002. Algal blooms reduce the uptake of toxic methylmercury in freshwater food webs. *Proceedings of the National Academy of Sciences of the USA* 99: 4419–4423.

Pirrone, N., I. Allegrini, G.J. Keeler, J.O. Nriagu, R. Rossmann, and J.A. Robbins. 1998. Historical atmospheric mercury emissions and depositions in North America compared to mercury accumulations in sedimentary records. *Atmospheric Environment* 32: 929–940.

Pitt, R. 2005. Module 4c introduction: Infiltration as a stormwater control. University of Alabama (Tuscaloosa, AL).

Pitt, R., S.-E. Chen, S.E. Clark, J. Swenson, and C.K. Ong. 2008. Compaction's impacts on urban storm-water infiltration. *Journal of Irrigation and Drainage Engineering* 134: 652–658.

Pitt, R., R. Field, M. Lalor, and M. Brown. 1995. Urban stormwater toxic pollutants: Assessment, sources, and treatability. *Water Environment Research* 67: 260–275.

Pitt, R., J. Lantrip, R. Harrison, C. Henry, and D. Hue. 1999. Infiltration through disturbed urban soils and compost-amended soil effects on runoff quality and quantity. USEPA 600/R-00/016 (Washington, DC).

Pitt, R., A. Maestre, and R. Morquecho. 2004. The national stormwater quality database (NSQD, version 1.1). Department of Civil and Environmental Engineering, University of Alabama (Tuscalousa, AL).

Plummer, A., Woodward, D.E. 1998. The origin and derivation of Ia/S in the runoff curve number system. In: *Water Resources Engineering 1998. Proceedings of the International Water Resources Engineering Conference, August 3–7.* American Society of Civil Engineers (Memphis, TN), pp. 1260–1265.

Poole, G.C., and C.H. Berman. 2001. An ecological perspective on in-stream temperature: Natural heat dynamics and mechanisms of human-caused thermal degradation. *Environmental Management* 27: 787–802.

Potter, K.M., F.W. Cubbage, G.B. Blank, and R.H. Schaberg. 2004. A watershed-scale model for predicting nonpoint pollution risk in North Carolina. *Environmental Management* 34: 62–74.

Powers, C.M., E.D. Levin,, F.J. Seidler, and T.A. Slotkin. 2011. Silver exposure in developing zebrafish produces persistent synaptic and behavioral changes. *Neurotoxicology and Teratology* 33: 329–332.

Pozo, K., T. Harner, S.C. Lee, F. Wania, D.C. Muir, and K.C. Jones. 2009. Seasonally resolved concentrations of persistent organic pollutants in the global atmosphere from the first year of the GAPS study. *Environmental Science and Technology* 43: 796–803.

Prestbo, E.M., and D.A. Gay. 2009. Wet deposition of mercury in the U.S. and Canada, 1996–2005: Results and analysis of the NADP mercury deposition network (MDN). *Atmospheric Environment* 43: 4223–4233.

Purcell, A.H., D.W. Bressler, M.J. Paul, M.T. Barbour, E.T. Rankin, J.L. Carter, and V.H. Resh. 2009. Assessment tools for urban catchments: Developing biological indicators based on benthic macroinvertebrates. *Journal of the American Water Resources Association* 45: 306–319.

Qi, Z., M.J. Helmers, R.D. Christianson, and C.H. Pederson. 2005. Nitrate-nitrogen losses through subsurface drainage under various agricultural land covers. *Journal of Environmental Quality* 40: 1578–1585.

Ragab, R., J. Bromley, P. Rosier, J.D. Cooper, and J.H.C. Gash. 2003. Experimental study of water fluxes in a residential area: 1. Rainfall, roof runoff and evaporation: The effect of slope and aspect. *Hydrological Processes* 17: 2409–2422.

Randall, G.W., and D.J. Mulla. 2001. Nitrate nitrogen in surface waters as influenced by climatic conditions and agricultural practices. *Journal of Environmental Quality* 30: 337–344.

Ratola, N., J.M. Amigo, and A. Alves. 2010. Levels and sources of PAHs in selected sites from Portugal: Biomonitoring with *Pinus pinea* and *Pinus pinaster* needles. *Archives of Environmental Contamination and Toxicology* 58: 631–647.

Reisenauer, A.E. 1963. Methods for solving problems of multidimensional, partially saturated steady flow in soils. *Journal of Geophysical Research* 68: 5725–5733.

Rice, P.J., and B.P. Horgan. 2011. Nutrient loss with runoff from fairway turf: An evaluation of core cultivation practices and their environmental impact. *Environmental Toxicology and Chemistry* 30: 2473–2480.

Richards, L. 2006. Protecting water resources with higher density development. EPA-231-R-06-001 (Washington, DC).

Richardson, S.D., and T.A. Ternes. 2011. Water analysis: Emerging contaminants and current issues. *Analytical Chemistry* 83: 4614–4648.

Riley, S.P.D., G.T. Busteed, L.B. Kats, T.L. Vandergon, L.F.S. Lee, R.G. Dagit, J.L. Kerby, R.N. Fisher, and R.M. Sauvajot. 2005. Effects of urbanization on the distribution and abundance of amphibians and invasive species in southern California streams. *Conservation Biology* 19: 1894–1907.

Roberts, B.J., P.J. Mulholland, and W.R. Hill. 2007. Multiple scales of temporal variability in ecosystem metabolism rates: Results from 2 years of continuous monitoring in a forested headwater stream. *Ecosystems* 10: 588–606.

Rocher, V., S. Azimi, J. Gasperi, L. Beuvin, M. Muller, R. Moilleron, and G. Chebbo. 2004. Hydrocarbons and metals in atmospheric deposition and roof runoff in central Paris. *Water, Air, and Soil Pollution* 159: 67–86.

Rogers, C.E., D.J. Brabander, M.T. Barbour, and H.F. Hemond. 2002. Use of physical, chemical, and biological indices to assess impacts of contaminants and physical habitat alteration in urban streams. *Environmental Toxicology and Chemistry* 21: 1156–1167.

Rose, S., and N.E. Peters. 2001. Effects of urbanization on streamflow in the Atlanta area (Georgia, USA): A comparative hydrological approach. *Hydrological Processes* 15: 1441–1457.

Rosell, M., S. Lacorte, and D. Barceló. 2006. Analysis, occurrence and fate of MTBE in the aquatic environment over the past decade. *Trends in Analytical Chemistry* 25: 1016–1029.

Rosselot, K.S. 2006. Copper released from brake lining wear in the San Francisco Bay area. Final Report. Brake Pad Partnership (Calabasa, CA).

Rossi, L., L. de Alencastro, T. Kupper, and J. Tarradellas. 2004. Urban stormwater contamination by polychlorinated biphenyls (PCBs) and its importance for urban water systems in Switzerland. *Science of the Total Environment* 322: 179–189.

Roy, A.H., A.L. Dybas, K.M. Fritz, and H.R. Lubbers. 2009. Urbanization affects the extent and hydrologic permanence of headwater streams in a midwestern US metropolitan area. *Journal of the North American Benthological Society* 28: 911–928.

Roy, A.H., S.J. Wenger, T.D. Fletcher, C.J. Walsh, A.R. Ladson, W.D. Shuster, H.W. Thurston, and R.R. Brown. 2008. Impediments and solutions to sustainable, watershed-scale urban stormwater management: Lessons from Australia and the United States. *Environmental Management* 42: 344–359.

Rupprecht, R., C. Kilgore, and R. Gunther. 2009. Riparian and wetland buffers for water-quality protection. *Stormwater* 10: 46–51.

Sabin, L.D., J.H. Lim, K.D. Stolzenbach, and K.C. Schiff. 2005. Contribution of trace metals from atmospheric deposition to stormwater runoff in a small impervious urban catchment. *Water Research* 39: 3929–3937.

Sabin, L.D., J.H. Lim, M.T. Venezia, A.M. Winer, K.C. Schiff, and K.D. Stolzenbach. 2006. Dry deposition and resuspension of particle-associated metals near a freeway in Los Angeles. *Atmospheric Environment* 40: 7528–7538.

Sabin, L.D., and K.C. Schiff. 2008. Dry atmospheric deposition rates of metals along a coastal transect in southern California. *Atmospheric Environment* 42: 6606–6613.

Sample, D.J., and J.P. Heaney. 2006. Integrated management of irrigation and urban storm-water infiltration. *Journal of Water Resources Planning and Management* 132: 362–373.

Sample, D.J., and J. Liu. 2014. Optimizing rainwater harvesting systems for the dual purposes of water supply and runoff capture. *Journal of Cleaner Production* 75: 174–194.

Sansalone, J.J., and S.G. Buchberger. 1997. Characterization of solid and metal element distribution in urban highway stormwater. *Water Science and Technology* 36: 155–160.

Sansalone, J.J., and D.W. Glenn. 2002. Accretion of pollutants in snow exposed to urban traffic and winter storm maintenance activities. *Journal of Environmental Engineering* 128: 151–166.

Sansalone, J.J., and J.-Y. Kim. 2008. Transport of particulate matter fractions in urban source area pavement surface runoff. *Journal of Environmental Quality* 37: 1883–1893.

Scanlon, B.R., and R.S. Goldsmith. 1997. Field study of spatial variability in unsaturated flow beneath and adjacent to playas. *Water Resources Research* 33: 2239–2252.

Scanlon, B.R., R.W. Healy, and P.G. Cook. 2002. Choosing appropriate techniques for quantifying groundwater recharge. *Hydrogeology Journal* 10: 18–39.

Schaap, B.D. 1999. Concentrations and possible sources of nitrate in water from the Silurian–Devonian aquifer, Cedar Falls, Iowa. USGS Water-Resources Investigations Report 99–4106 (Iowa City, IA).

Scharenbroch, B.C., J.E. Lloyd, and J.L. Johnson-Maynard. 2005. Distinguishing urban soils with physical, chemical, and biological properties. *Pedobiologia* 49: 283–296.

Scheuhammer, A.M., M.W. Meyer, M.B. Sandheinrich, and M.W. Murray. 2007. Effects of environmental methylmercury on the health of wild birds, mammals, and fish. *Ambio* 36: 12–18.

Schindler, D.W. 1978. Factors regulating phytoplankton production and standing crop in the world's freshwaters. *Limnology and Oceanography* 23: 478–486.

Schlesinger, W.H. 1997. *Biogeochemistry: An analysis of global change.* 2nd ed. Academic Press (San Diego, CA).

Schlesinger, W.H. 2009. On the fate of anthropogenic nitrogen. *Proceedings of the National Academy of Sciences of the USA* 106: 203–208.

Schmidt, C.M., A.T. Fisher, A.J. Racz, B.S. Lockwood, and M. Los Huertos. 2011. Linking denitrification and infiltration rates during managed groundwater recharge. *Environmental Science and Technology* 45: 9634–9640.

Schoonover, J.E., B.G. Lockaby, and B.S. Helms. 2006. Impacts of land cover on stream hydrology in the west Georgia Piedmont, USA. *Journal of Environmental Quality* 35: 2123–2131.

Schueler, T.R. 1994. The importance of imperviousness. *Watershed Protection Techniques* 1: 100–111.

Schueler, T.R., L. Fraley-McNeal, and K. Cappiella. 2009. Is impervious cover still important? Review of recent research. *Journal of Hydrologic Engineering* 14: 309–315.

Schwob, C.E. 1953. Federal water pollution control act: Objectives and policies. *Industrial and Engineering Chemistry* 45: 2648–2652.

Scott, M.C., G.S. Helfman, M.E. McTammany, E.F. Benfield, and P.V. Bolstad. 2002. Multiscale influences on physical and chemical stream conditions across Blue Ridge landscapes. *Journal of the American Water Resources Association* 38: 1379–1392.

Selbig, W.R. 2009. Concentrations of polycyclic aromatic hydrocarbons (PAHs) in urban stormwater, Madison, Wisconsin, 2005-08. USGS Open-File Report 2009–1077.

Selong, J.H., T.E. McMahon, A.V. Zale, and F.T. Barrows. 2001. Effect of temperature on growth and survival of bull trout, with application of an improved method for determining thermal tolerance in fishes. *Transactions of the American Fisheries Society* 130: 1026–1037.

Sepúlveda, M.S., P.C. Frederick, M.G. Spalding, and G.E. Williams Jr. 1999. Mercury contamination in free-ranging great egret nestlings (*Ardea albus*) from southern Florida, USA. *Environmental Toxicology and Chemistry* 18: 985–992.

Shields, C.A., L.E. Band, N. Law, P.M. Groffman, S.S. Kaushal, K. Savvas, G.T. Fisher, and K.T. Belt. 2008. Streamflow distribution of non-point source nitrogen export from urban–rural catchments in the Chesapeake Bay watershed. *Water Resources Research* 44: W09416.

Simcik, M.F., and K.J. Dorweiler. 2005. Ratio of perfluorochemical concentrations as a tracer of atmospheric deposition to surface waters. *Environmental Science and Technology* 39: 8678–8683.

Simmons, M.T., B. Gardiner, S. Windhager, and J. Tinsley. 2008. Green roofs are not created equal: The hydrologic and thermal performance of six different extensive green roofs and reflective and non-reflective roofs in a sub-tropical climate. *Urban Ecosystems* 11: 339–348.

Sklar, L.S., and W.E. Dietrich. 2004. A mechanistic model for river incision into bedrock by saltating bed load. *Water Resources Research* 40: W06301.

Smith, R.A., R.B. Alexander, and G.E. Schwarz. 2003. Natural background concentrations of nutrients in streams and rivers of the conterminous United States. *Environmental Science and Technology* 37: 3039–3047.

Smith, V.H. 2003. Eutrophication of freshwater and coastal marine ecosystems: A global problem. *Environmental Science and Pollution Research* 10: 126–139.

Søballe, D.M., and B.L. Kimmel. 1987. A large-scale comparison of factors influencing phytoplankton abundance in rivers, lakes, and impoundments. *Ecology* 68: 1943–1954.

Soldat, D.J., and A.M. Petrovic. 2008. The fate and transport of phosphorus in turfgrass ecosystems. *Crop Science* 48: 2051–2065.

Soliman, Y.S., and T.L. Wade. 2008. Estimates of PAHs burdens in a population of ampeliscid amphipods at the head of the Mississippi Canyon (N. Gulf of Mexico). *Deep-Sea Research II* 55: 2577–2584.

Sophocleous, M. 2002. Interactions between groundwater and surface water: The state of the science. *Hydrogeology Journal* 10: 52–67.

Sousa, W. P. 1979. Disturbance in marine intertidal boulder fields: The nonequilibrium maintenance of species diversity. *Ecology* 60: 1225–1239.

Squillace, P.J., J.S. Zogorski, W.G. Wilber, and C.V. Price. 1996. Preliminary assessment of the occurrence and possible sources of MTBE in groundwater in the United States, 1993–1994. *Environmental Science and Technology* 30: 1721–1730.

Stapleton, H.M., S. Sharma, G. Getzinger, P.L. Ferguson, M. Gabriel, T.F. Webster, and A. Blum. 2012. Novel and high volume use flame retardants in US couches reflective of the 2005 pentaBDE phase out. *Environmental Science and Technology* 46: 13432–13439.

Starzec, P., B.B. Lind, A. Lanngren, Å. Lindgren, and T. Svenson. 2005. Technical and environmental functioning of detention ponds for the treatment of highway and road runoff. *Water, Air, and Soil Pollution* 163: 153–167.

Steedman, R.J. 1988. Modification and assessment of an index of biotic integrity to quantify stream quality in southern Ontario. *Canadian Journal of Fisheries and Aquatic Sciences* 45: 492–501.

Steuer, J., W. Selbig, N. Hornewer, and J. Prey. 1997. Sources of contamination in an urban basin in Marquette, Michigan and an analysis of concentrations, loads, and data quality. USGS WRI Report 97–4242 (Middleton, WI).

Stock, N.A., V. Furdui, D.C.G. Muir, and S.A. Mabury. 2007. Perfluoroalkyl contaminants in the Canadian Arctic: Evidence of atmospheric transport and local contamination. *Environmental Science and Technology* 41: 3529–3536.

Stoddard, J.L. 1994. Long-term changes in watershed retention of nitrogen: Its causes and consequences. In: *Environmental chemistry of lakes and reservoirs*, L.A. Baker, ed. Advances in Chemistry Series 237. American Chemical Society (Washington, DC), 223–284.

Stout, B., and B. Wallace. 2005. A survey of eight major aquatic insect orders associated with small headwater streams subject to valley fills from mountaintop mining. USEPA (Washington, DC).

Strahler, A.N. 1957. Quantitative analysis of watershed geomorphology. *Transactions of the American Geophysical Union* 8: 913–920.

Struck, S.D., A. Selvakumar, and M. Borst. 2006. Performance of stormwater retention ponds and constructed wetlands in reducing microbial concentrations. USEPA 600/R-06/102 (Washington, DC).

Suarez, S., J.M. Lema, and F. Omil. 2010. Removal of pharmaceutical and personal care products (PPCPs) under nitrifying and denitrifying conditions. *Water Research* 44: 3214–3224.

Sullivan, T.J., C.T. Driscoll, B.J. Cosby, I.J. Fernandez, A.T. Herlihy, J. Zhai, R. Stemberger, K.U. Snyder, J.W. Sutherland, S.A. Nierzwicki-Bauer, C.W. Boylen, T.C. McDonnell, and N.A. Nowicki. 2006. Assessment of the extent to which intensively-studied lakes are representative of the Adirondack Mountain region. Final report 06-17. E&S Environmental Chemistry, Inc. (Corvallis, OR).

Sutton, A.J., T.R. Fisher, and A.B. Gustafson. 2010. Effects of restored stream buffers on water quality in non-tidal streams in the Choptank River basin. *Water, Air and Soil Pollution* 208: 101–118.

Swain, E.B., P.M. Jakus, G. Rice, F. Lupi, P.A. Maxson, J.M. Pacyna, A. Penn, S.J. Spiegel, and M.M. Veiga. 2007. Socioeconomic consequences of mercury use and pollution. *Ambio* 36: 45–61.

Sweeney, B.W., T.L. Bott, J.K. Jackson, L.A. Kaplan, J.D. Newbold, L.J. Standley, W.C. Hession, and R.J. Horwitz. 2004. Riparian deforestation, stream narrowing, and loss of stream ecosystem services. *Proceedings of the National Academy of Sciences of the USA* 101: 14132–14137

Swinbank, W.C. 1951. The measurement of vertical transfer of heat and water vapor by eddies in the lower atmosphere. *Journal of Meteorology* 8: 135–145.

Tan, S.W., J.C. Meiller, and K.R. Mahaffey. 2009. The endocrine effects of mercury in humans and wildlife. *Critical Reviews in Toxicology* 39: 228–269.

Terrell, M.L., A.K. Manatunga, C.M. Small, L.L. Cameron, J. Wirth, H.M. Blanck, R.H. Lyles, and M. Marcus. 2008. A decay model for assessing polybrominated biphenyl exposure among women in the Michigan Long-Term PBB Study. *Journal of Exposure Science and Environmental Epidemiology* 18: 410–420.

Thompson, I.P., I.L. Blackwood, and T.D. Davies. 1987a. The effect of polluted and leached snow melt waters on the soil bacterial community–quantitative response. *Environmental Pollution* 43: 143–154.

Thompson, I.P., I.L. Blackwood, and T.D. Davies. 1987b. Soil bacterial changes upon snowmelt: Laboratory studies of the effects of early and late meltwater fractions. *FEMS Microbiology Ecology* 45: 269–274.

Torgersen, C.E., D.M. Price, H.W. Li, and B.A. McIntosh. 1999. Multiscale thermal refugia and stream habitat associations of chinook salmon in northeastern Oregon. *Ecological Applications* 9: 301–319.

Trenberth, K.E. 2011. Changes in precipitation with climate change. *Climate Research* 47: 123–138.

UNEP (United Nations Environment Programme). 2013. Global mercury assessment 2013: Sources, emissions, releases and environmental transport. UNEP Chemicals Branch (Geneva, Switzerland).

USDA (U.S. Department of Agriculture). 1986. Urban hydrology for small watersheds. 2nd ed., Technical Release 55. Natural Resources Conservation Service, Conservation Engineering Division (Washington, DC).

USEPA. 1997. Mercury study report to congress. EPA-452/R-97-003. USEPA (Washington, DC).

USEPA. 1999. Preliminary data summary of urban storm water best management practices. EPA-821-R-99-012. USEPA (Washington, DC).

USEPA. 2005a. Control of mercury emissions from coal fired electric utility boilers: An update. OAR-2002-0056-6141. USEPA, Office of Research and Development (Research Triangle Park, NC).

USEPA. 2005b. Furniture flame retardancy partnership: Environmental profiles of chemical flame-retardant alternatives for low-density polyurethane foam. EPA-742-R-05-002A. USEPA (Washington, DC).

USEPA. 2005c. National management measures to control nonpoint source pollution from urban areas. EPA-841-B-05-004. USEPA (Washington, DC).

USEPA. 2006. Characterization of mercury-enriched coal combustion residues from electric utilities using enhanced sorbents for mercury control. EPA-600/R-06/008. USEPA, Office of Research and Development (Research Triangle Park, NC).

USEPA. 2007. State of the Great Lakes 2007. EPA-905-R-07-003. Environment Canada and USEPA (Washington, DC).

USEPA. 2009. The national study of chemical residues in lake fish tissue. EPA-823-R-09-006. USEPA, Office of Water (Washington, DC).

U.S. Fish and Wildlife Service. 2001. Economic effects of trout production by national fish hatcheries in the Southeast. Southeast Region (Atlanta, GA).

Utz, R.M., R.H. Hilderbrand, and D.M. Boward. 2009. Identifying regional differences in threshold responses of aquatic invertebrates to land cover gradients. *Ecological Indicators* 9: 556–567.

Van Appledorn, M. 2009. Watershed-scale analysis of riparian buffer function. All Graduate Theses and Dissertations, Paper 419 (digitalcommons.usu.edu/etd/419).

van de Meene, S.J., R.R. Brown, and M.A. Farrelly. 2011. Towards understanding governance for sustainable urban water management. *Global Environmental Change* 21: 1117–1127.

Vandenberg, L.N., T. Colborn, T.B. Hayes, J.J. Heindel, D.R. Jacobs, Jr., D.-H. Lee, T. Shioda, A.M. Soto, F.S. vom Saal, W.V. Welshons, R.T. Zoeller, and J. Peterson Myers. 2012. Hormones and endocrine-disrupting chemicals: Low-dose effects and nonmonotonic dose responses. *Endocrine Reviews* 33: 378–455.

Van Metre, P.C., and B.J. Mahler. 2003. The contribution of particles washed from rooftops to contaminant loading to urban streams. *Chemosphere* 52: 1727–1741.

Van Metre, P.C., and B.J. Mahler. 2010. Contribution of PAHs from coal-tar pavement sealcoat and other sources to 40 U.S. lakes. *Science of the Total Environment* 409: 334–344.

Van Metre, P.C., and B.J. Mahler. 2014a. PAH concentrations in lake sediment decline following ban on coal-tar-based pavement sealants in Austin, Texas. *Environmental Science and Technology* 48: 7222–7228.

Van Metre, P.C., and B.J. Mahler. 2014b. Response to comment on "PAH concentrations in lake sediment decline following ban on coal-tar-based pavement sealants in Austin, Texas." *Environmental Science and Technology* 48: 14063–14064.

Van Metre, P.C., B.J. Mahler, and J.T. Wilson. 2009. PAHs underfoot: Contaminated dust from coal-tar sealcoated pavement is widespread in the United States. *Environmental Science and Technology* 43: 20–25.

VanWoert, N.D., D.B. Rowe, J.A. Andresen, C.L. Rugh, R.T. Fernandez, and L. Xiao. 2005. Green roof stormwater retention: Effects of roof surface, slope, and media depth. *Journal of Environmental Quality* 34: 1036–1044.

Veličković, B. 2005. Colmation as one of the processes in interactions between groundwater and surface water. *Facta Universitatis, Series Architecture and Civil Engineering* 3: 165–172.

Vitòria, L., N. Otero, A. Soler, and À. Canals. 2004. Fertilizer characterization: Isotopic data (N, S, O, C, and Sr). *Environmental Science and Technology* 38: 3254–3262.

Vitousek, P.M., J.D. Aber, R.W. Howarth, G.E. Likens, P.A. Matson, D.W. Schindler, W.H. Schlesinger, and D.G. Tilman. 1997. Human alteration of the global nitrogen cycle: Sources and consequences. *Ecological Applications* 7: 737–750.

Volk, C., L. Wood, B. Johnson, J. Robinson, H.W. Zhu, and L. Kaplan. 2002. Monitoring dissolved organic carbon in surface and drinking waters. *Journal of Environmental Monitoring* 4: 43–47.

Vucetich, J.A., P.M. Outridge, R.O. Peterson, R. Eide, and R. Isrenn. 2009. Mercury, lead and lead isotope ratios in the teeth of moose (*Alces alces*) from Isle Royale, U.S. upper Midwest, from 1952 to 2002. *Journal of Environmental Monitoring* 11: 1352–1359.

Walsh, C.J. 2004. Protection of in-stream biota from urban impacts: Minimise catchment imperviousness or improve drainage design? *Marine and Freshwater Research* 55: 317–326.

Walsh, C.J., and J. Kunapo. 2009. The importance of upland flow paths in determining urban effects on stream ecosystems. *Journal of the North American Benthological Society* 28: 977–990.

Walsh, C.J., T.D. Fletcher, and A.R. Ladson. 2005a. Stream restoration in urban catchments through redesigning stormwater systems: Looking to the catchment to save the stream. *Journal of the North American Benthological Society* 24: 690–705.

Walsh, C.J., A.H. Roy, J.W. Feminella, P.D. Cottingham, P.M. Groffman, and R.P. Morgan. 2005b. The urban stream syndrome: Current knowledge and the search for a cure. *Journal of the North American Benthological Society* 24: 706–723.

Wang, L., and P. Kanehl. 2003. Influences of watershed urbanization and instream habitat on macroinvertebrates in cold-water streams. *Journal of the American Water Resources Association* 39: 1181–1196.

Wang, L., J. Lyons, P. Kanehl, and R. Bannerman. 2001. Impacts of urbanization on stream habitat and fish across multiple spatial scales. *Environmental Management* 28: 255–266.

Wang, L., D.M. Robertson, and P.J. Garrison. 2007. Linkages between nutrients and assemblages of macroinvertebrates and fish in wadeable streams: Implication to nutrient criteria development. *Environmental Management* 39: 194–212.

Ward, J.L., and M.J. Blum. 2012. Exposure to an environmental estrogen breaks down sexual isolation between native and invasive species. *Evolutionary Applications* 5: 901–912.

Watras, C.J., R.C. Back, S. Halvorsen, R.J.M. Hudson, K.A. Morrison, and S.P. Wente. 1998. Bioaccumulation of mercury in pelagic freshwater food webs. *Science of the Total Environment* 219: 183–208.

Weathers, K.C., M.L. Cadenasso, and S.T.A. Pickett. 2001. Forest edges as nutrient and pollutant concentrators: Potential synergisms between fragmentation, forest canopies, and the atmosphere. *Conservation Biology* 15: 1506–1514.

Webb, B.W., D.M. Hannah, R.D. Moore, L.E. Brown, and F. Nobilis. 2008. Recent advances in stream and river temperature research. *Hydrological Processes* 22: 902–918.

Webb, B.W., and D.E. Walling. 1996. Long-term variability in the thermal impact of river impoundment and regulation. *Applied Geography* 16: 211–227.

Webb, B.W., and D.E. Walling. 1997. Complex summer water temperature behaviour below a UK regulating reservoir. *Regulated Rivers: Research and Management* 13: 463–477.

Weier, K.L., J.W. Doran, J.F. Power, and D.T. Walters. 1993. Denitrification and the dinitrogen/nitrous oxide ratio as affected by soil water, available carbon, and nitrate. *Soil Science Society of America Journal* 57: 66–72.

Weiss, P.T., J.S. Gulliver, and A.J. Erickson. 2007. Cost and pollutant removal of storm-water treatment practices. *Journal of Water Resources Planning and Management* 133: 218–229.

Wellington, B.I., and C.T. Driscoll. 2004. The episodic acidification of a stream with elevated concentrations of dissolved organic carbon. *Hydrological Processes* 18: 2663–2680.

Welshons, W.V., K.A. Thayer, B.M. Judy, J.A. Taylor, E.M. Curran, and F.S. vom Saal. 2003. Large effects from small exposures. I. Mechanisms for endocrine-disrupting chemicals with estrogenic activity. *Environmental Health Perspectives* 111: 994–1006.

Wenger, S. 1999. A review of the scientific literature on riparian buffer width, extent, and vegetation. Office of Public Service and Outreach, Institute of Ecology, University of Georgia (Athens, GA).

White, K.D. 1970. Fallowing, crop rotation, and crop yields in Roman times. *Agricultural History* 44: 281–290.

Williams, M., C. Hopkinson, E. Rastetter, J. Vallino, and L. Claessens. 2005. Relationships of land use and stream solute concentrations in the Ipswich River basin, northeastern Massachusetts. *Water, Air, and Soil Pollution* 161: 55–74.

Wilson, W.G. 2011. *Constructed climates: A primer on urban environments.* University of Chicago Press (Chicago, IL).

Winter, T.C. 1999. Relation of streams, lakes, and wetlands to groundwater flow systems. *Hydrogeology Journal* 7: 28–45.

Winter, T.C. 2007. The role of ground water in generating streamflow in headwater areas and in maintaining base flow. *Journal of the American Water Resources Association* 43: 15–25.

Wohl, E., D.A. Cenderelli, K.A. Dwire, S.E. Ryan-Burkett, M.K. Young, and K.D. Fausch. 2010. Large in-stream wood studies: A call for common metrics. *Earth Surface Processes and Landforms* 35: 618–625.

Wollheim, W.M., B.A. Pellerin, C.J. Vörösmarty, and C.S. Hopkinson. 2005. N retention in urbanizing headwater catchments. *Ecosystems* 8: 871–884.

Worrall, F., T. Burt, and R. Shedden. 2003. Long term records of riverine dissolved organic matter. *Biogeochemistry* 64: 165–178.

Xiao, Q., E.G. McPherson, S.L. Ustin, M.E. Grismer, and J.R. Simpson. 2000. Winter rainfall interception by two mature open-grown trees in Davis, California. *Hydrological Processes* 14: 763–784.

Yoder, C.O., and E.T. Rankin. 1998. The role of biological indicators in a state water quality management process *Environmental Monitoring and Assessment* 51: 61–88.

Zanders, J.M. 2005. Road sediment: Characterization and implications for the performance of vegetated strips for treating road run-off. *Science of the Total Environment* 339: 41–47.

Zhu, K., P. Blum, G. Ferguson, K.-D. Balke, and P. Bayer. 2010. The geothermal potential of urban heat islands. *Environmental Research Letters* 5: 044002.

Zorita, S., L. Martensson, and L. Matthiasson. 2009. Occurrence and removal of pharmaceuticals in a municipal sewage treatment system in the south of Sweden. *Science of the Total Environment* 407: 2760–2770.

Place Index

Topic Index

Acronym Index